D1714119

## Communications and Control Engineering

Springer
*London*
*Berlin*
*Heidelberg*
*New York*
*Barcelona*
*Hong Kong*
*Milan*
*Paris*
*Singapore*
*Tokyo*

*Published titles include:*

*Optimal Sampled-Data Control Systems*
Tongwen Chen and Bruce Francis

*Nonlinear Control Systems (3rd edition)*
Alberto Isidori

*Theory of Robot Control*
C. Canudas de Wit, B. Siciliano and G. Bastin (Eds)

*Fundamental Limitations in Filtering and Control*
María M. Seron, Julio Braslavsky and Graham C. Goodwin

*Constructive Nonlinear Control*
R. Sepulchre, M. Jankovic and P.V. Kokotović

*A Theory of Learning and Generalization*
M. Vidyasagar

*Adaptive Control*
I.D. Landau, R. Lozano and M.M'Saad

*Stabilization of Nonlinear Uncertain Systems*
Miroslav Krstić and Hua Deng

*Passivity-based Control of Euler-Lagrange Systems*
Romeo Ortega, Antonio Loría, Per Johan Nicklasson and Hebertt Sira-Ramírez

*Stability and Stabilization of Infinite Dimensional Systems with Applications*
Zheng-Hua Luo, Bao-Zhu Guo and Omer Morgul

*Nonsmooth Mechanics (2nd edition)*
Bernard Brogliato

*Nonlinear Control Systems II*
Alberto Isidori

*$L_2$-Gain and Passivity Techniques in Nonlinear Control*
Arjan van der Schaft

*Control of Linear Systems with Regulation and Input Constraints*
Ali Saberi, Anton A. Stoorvogel and Peddapullaiah Sannuti

*Robust and $H\infty$ Control*
Ben M. Chen

*Computer Controlled Systems*
Efim N. Rosenwasser and Bernhard P. Lampe

Rogelio Lozano, Bernard Brogliato,
Olav Egeland and Bernhard Maschke

# Dissipative Systems Analysis and Control

## Theory and Applications

With 93 Figures

Rogelio Lozano, Doctor
Heudiasyc, Université de Technologie de Compiègne, BP 20529, 60205 Compiègne, France

Bernard Brogliato, Doctor
LAG-CNRS, Domaine Universitaire, BP 46, 38402 Saint Martin d'Hères, France

Olav Egeland, Professor
Department of Engineering Cybernetics, Norwegian University of Science and Technology, O.S. Bragstads plass 2D, N-7491 Trondheim, Norway

Bernard Maschke, Doctor
Laboratoire d'Automatisme Industriel, Conservatoire national des arts et métiers, 292 rue Saint Martin, F-75141 Paris Cedex 03, France

*Series Editors*

E.D. Sontag • M. Thoma

ISSN 0178-5354

ISBN 1-85233-285-9 Springer-Verlag London Berlin Heidelberg

British Library Cataloguing in Publication Data
Dissipative systems analysis and control. - (Communications
and control engineering)
1.Systems analysis 2.Automatic control - Mathematics
I.Lozano, Rogelio
621.3'92
ISBN 1852332859

Library of Congress Cataloging-in-Publication Data
Dissipative systems analysis and control / Rogelio Lozano ... [et al.].
p. cm.
ISBN 1-85233-285-9 (alk. paper)
1. Automatic control. 2. Systems analysis I. Lozano, R. (Rogelio), 1954-
TJ213 .D5435 2000
629.8--dc21 00-028464


Printed in Great Britain



Typesetting: Camera ready by authors
Printed and bound at the Athenæum Press Ltd., Gateshead, Tyne & Wear
69/3830-543210 Printed on acid-free paper SPIN 10745157

# Preface

Thank you for opening this book devoted to dissipative systems theory and applications in control of linear and nonlinear systems. The physical concept of dissipativity (or passivity) is ubiquituos in Physics, Applied Mathematics, Mechanics etc. Actually the notion of energy, which is at the core of the developments presented in this monograph, has long been used in these scientific areas. Its study and application in the field of Systems and Control Theory is more recent, and traces back to the fundamental contributions by Kalman, Popov, Yakubovich (hyperstability, characterization of positive real transfer functions, Popov's criterion), and Willems (who settled the fundamental definitions and properties of dissipative systems), in the sixties. The basic and important motivation was (and is still!) the stabilization of controlled dynamical systems and the obtaining of Lyapunov functions. Indeed it is known that one of the cornerstones in modern Systems and Control Theory, is the use of Lyapunov functions, whose construction is generally difficult. Dissipativity theory provides one with a somewhat systematic way to build Lyapunov functions: this direction of research explains its success in the past forty years.

Later on, significant advances were realized by Hill, Moylan and Anderson, especially concerning the links between input-output and state space approaches (Lyapunov stability), and between optimal control and dissipative systems (the so-called inverse optimal control problem). In the eighties, dissipative systems theory was again pushed forward by the application made in the field of Robotics and the control of mechanical systems. In parallel the characterization of a class of nonlinear dynamical systems, using tools form differential geometry and extending the results concerning positive real transfer functions, was made by Byrnes, Isidori and Willems.

This monograph is to be considered as an advanced introduction to the field of Dissipative Systems Theory. It also contains a chapter devoted to the application of so-called Passivity-Based feedback controllers on two mechanical devices. Results of the experiments are reported and commented. Throughout the book, the theoretical developments are accompanied by many physical examples and illustrations. This monograph is therefore expected to be used by graduate students interested in control and stabilization of linear and nonlinear systems, as well as more mature researchers who wish to find in a single volume the main results about dissipative systems. Practitioners will also find some elements that may help them improve their knowledge on advanced control theory. Due to the goal we assigned ourselves, it is clear that not all the works published in the literature on the topic have been exposed nor cited. We are open to any comments or remarks about our work, that may allow us to improve it.

The writing of this monograph emerged initially from a set of hand-written notes by the first author (that were used by several people during several seminars, including a workshop at the IEEE Conference on Decision and Control, and a workshop at the University of Technology of Compiegne). It would not have been possible without the help of the following persons, who are gratefully acknowledged here:

- Springer Verlag London and the series Editor Prof. Sontag, who motivated us to accomplish this project.
- Isabelle Fantoni whose help on developing the control of the inverted pendulum and on handling the figures of the book was precious.
- David Hill for many interesting discussions and providing us with a nice example of an unstable dissipative system.
- The second author is indebted to Prof. Bernat, University Autonoma of Barcelona, Spain, Dept. of Mathematics, who provided him with useful references and comments concerning Kalman conjecture counterexample.
- Ioan-Dore Landau, without whom the first and second authors would perhaps never have been involved and interested in passive systems.
- Arjan van der Schaft, thanks to whom and with whom the fourth author got the opportunity to work intensively on Hamiltonian systems theory.
- Dorothée Normand-Cyrot, GDR-Automatique, for supporting a seminar on Passivity in Compiègne in 1999.

We wish to thank also J. Collado, A. Dzul, R. Galindo, T. Hamel, R. Johansson, R. Mahony, S. Niculescu, V. Rasvan and F. Ruiz for their valuable discussions and suggestions to improve the book.

Rogelio Lozano
Bernard Brogliato
Olav Egeland
Bernhard Maschke

# Contents

# Chapter 1

# Introduction

Dissipativity theory gives a framework for the design and analysis of control systems using an input-output description based on energy-related considerations. Dissipativity is a notion which can be used in many areas of science, and it allows the control engineer to relate a set of efficient mathematical tools to well known physical phenomena. The insight gained in this way is very useful for a wide range of control problems. In particular the input-output description allows for a modular approach to control systems design and analysis. Dissipativity has proven useful or even indispensable for control applications like robotics, active vibration damping, electromechanical systems, combustion engines, circuit theory, and for control techniques like adaptive control, nonlinear $H^\infty$, and inverse optimal control.

The main idea behind this is that many important physical systems have certain input-output properties related to the conservation, dissipation and transport of energy. Before introducing precise mathematical definitions we will somewhat loosely refer to such input-output properties as dissipative properties, and systems with dissipative properties will be termed dissipative systems. When modeling dissipative systems it may be useful to develop the state-space or input-output models so that they reflect the dissipativity of the system, and thereby ensure that the dissipativity of the model is invariant with respect to model parameters, and to the mathematical representation used in the model. The aim of this book is to give a comprehensive presentation of how the energy-based notion of dissipativity can be used to establish the input-output properties of models for dissipative systems. Also it will be shown how these results can be used in controller design. Moreover it will appear clearly how these results can be generalized to a dissipativity theory where conservation of other physical properties, and even abstract quantities can be handled.

Models for use in controller design and analysis are usually derived from the basic laws of physics (electrical systems, dynamics, thermodynamics). Then

a controller can be designed based on this model. An important problem in controller design is the issue of robustness which relates to how the closed loop system will perform when the physical system differs either in structure or in parameters from the design model. For a system where the basic laws of physics imply dissipative properties, it may make sense to define the model so that it possesses the same dissipative properties regardless of the numerical values of the physical parameters. Then if a controller is designed so that stability relies on the dissipative properties only, the closed-loop system will be stable whatever the values of the physical parameters. Even a change of the system order will be tolerated provided it does not destroy the dissipativity.
Parallel interconnections and feedback interconnections of dissipative systems inherit the dissipative properties of the connected subsystems, and this simplifies analysis by allowing for manipulation of block diagrams, and provides guidelines on how to design control systems. A further indication of the usefulness of dissipativity theory is the fact that the PID controller is a dissipative system, and a fundamental result that will be presented is the fact that the stability of a dissipative system with a PID controller can be established using dissipativity arguments. Note that such arguments rely on the structural properties of the physical system, and are not sensitive to the numerical values used in the design model. The technique of controller design using dissipativity theory can therefore be seen as a powerful generalization of PID controller design.
There is another aspect of dissipativity which is very useful in practical applications. It turns out that dissipativity considerations are helpful as a guide for the choice of a suitable variable for output feedback. This is helpful for selecting where to place sensors for feedback control.
Throughout the book we will treat dissipativity for state space and input-output models, but first we will investigate simple examples which illustrate some of the main ideas to be developed more deeply in the sequel.

## 1.1 Example 1: System with mass spring and damper

Consider a one dimensional simple mechanical system with a mass, a spring and a damper. The equation of motion is

$$m\ddot{x} + D\dot{x} + Kx = F$$

where $m$ is the mass, $D$ is the damper constant, $K$ is the spring stiffness, $x$ is the position of the mass and $F$ is the force acting on the mass. The energy of the system is

$$V(x, \dot{x}) = \frac{1}{2}m\dot{x}^2 + \frac{1}{2}Kx^2$$

The time derivative of the energy when the system moves is

$$\frac{d}{dt}V(x,\dot{x}) = m\ddot{x}\dot{x} + Kx\dot{x}$$

Inserting the equation of motion we get

$$\frac{d}{dt}V(x,\dot{x}) = F\dot{x} - D\dot{x}^2$$

Integration of this equation from $t = 0$ to $t = T$ gives

$$V[x(T),\dot{x}(T)] = V[x(0),\dot{x}(0)] + \int_0^T F(t)\dot{x}(t)\,dt - \int_0^T D\dot{x}^2(t)\,dt$$

This means that the energy at time $t = T$ is the initial energy plus the energy supplied to the system by the control force $F$ minus the energy dissipated by the damper. Note that if the input force $F$ is zero, and if there is no damping, then the energy $V$ of the system is constant. Here $D \geq 0$ and $V[x(0),\dot{x}(0)] > 0$, and it follows that the integral of the force $F$ and the velocity $v = \dot{x}$ satisfies

$$\int_0^T F(t)v(t)\,dt \geq -V[x(0),\dot{x}(0)] \tag{1.1}$$

The physical interpretation of this inequality is seen from the equivalent inequality

$$-\int_0^T F(t)v(t)\,dt \leq V[x(0),\dot{x}(0)] \tag{1.2}$$

which shows that the energy $-\int_0^T F(t)\dot{x}(t)\,dt$ that can be extracted from the system is less than or equal to the initial energy stored in the system. We will show later that (1.1) implies that the system with input $F$ and output $v$ is passive. The Laplace transform of the equation of motion is

$$\left(ms^2 + Ds + K\right)x(s) = F(s)$$

which leads to the transfer function

$$\frac{v}{F}(s) = \frac{s}{ms^2 + Ds + K}$$

It is seen that the transfer function is stable, and that for $s = j\omega$ the phase of the transfer function has absolute value less or equal to $90°$, that is,

$$\left|\angle\frac{v}{F}(j\omega)\right| \leq 90° \quad \Rightarrow \quad \mathbf{Re}\left[\frac{v}{F}(j\omega)\right] \geq 0 \tag{1.3}$$

We will see in the following that these properties of the transfer function are consequences of the condition (1.1), and that they are important in controller design.

## 1.2 Example 2: *RLC* circuit

Consider a simple electrical system with a resistor $R$, inductance $L$ and a capacitor $C$ with current $i$ and voltage $u$. The differential equation for the circuit is

$$L\frac{di}{dt} + Ri + Cx = u$$

where

$$x(t) = \int_0^t i(t')\,dt'$$

The energy stored in the system is

$$V(x,i) = \frac{1}{2}Li^2 + \frac{1}{2}Cx^2$$

The time derivative of the energy when the system evolves is

$$\frac{d}{dt}V(x,i) = L\frac{di}{dt}i + Cxi$$

Inserting the differential equation of the circuit we get

$$\frac{d}{dt}V(x,i) = ui - Ri^2$$

Integration of this equation from $t = 0$ to $t = T$ gives

$$V[x(T), i(T)] = V[x(0), i(0)] + \int_0^T u(t)\,i(t)\,dt - \int_0^T Ri^2(t)\,dt$$

Similarly to the previous example this means that the energy at time $t = T$ is the initial energy plus the energy supplied to the system by the voltage $u$ minus the energy dissipated by the resistor. Note that if the input voltage $u$ is zero, and if there is no resistance, then the energy $V$ of the system is constant. Here $R \geq 0$ and $V[x(0), \dot{x}(0)] > 0$, and it follows that the integral of the voltage $u$ and the current $i$ satisfies

$$\int_0^T u(t)\,i(t)\,dt \geq -V[x(0), i(0)] \tag{1.4}$$

The physical interpretation of this inequality is seen from the equivalent inequality

$$-\int_0^T u(t)\,i(t)\,dt \leq V[x(0), i(0)] \tag{1.5}$$

which shows that the energy $-\int_0^T u(t)\,i(t)\,dt$ that can be extracted from the system is less than or equal to the initial energy stored in the system. We will show later that (1.4) implies that the system with input $u$ and output $i$ is passive. The Laplace transform of the differential equation of the circuit is

$$\left(Ls^2 + Rs + C\right) x(s) = u\left(s\right)$$

which leads to the transfer function

$$\frac{i}{u}(s) = \frac{s}{Ls^2 + Rs + C}$$

It is seen that the transfer function is stable, and that for $s = j\omega$ the phase of the transfer function has absolute value less or equal to 90°, that is,

$$\left|\angle\frac{i}{u}(j\omega)\right| \leq 90^\circ \quad \Rightarrow \quad \mathbf{Re}\left[\frac{i}{u}(j\omega)\right] \geq 0 \tag{1.6}$$

We see that in both examples we arrive at transfer functions that are stable, and that have positive real parts on the $j\omega$ axis. This motivates for further investigations on whether there is some fundamental connection between conditions on the energy flow in equations associated with the control equations (1.1,1.4) and the conditions on the transfer functions (1.3,1.6). This will be made clear in chapter 2.

## 1.3 Example 3: A mass with a PD controller

Consider the mass $m$ with the external control force $u$. The equation of motion is

$$m\ddot{x} = u$$

Suppose that a PD controller

$$u = -K_P x - K_D \dot{x}$$

is used. Then the closed loop dynamics are

$$m\ddot{x} + K_D\dot{x} + K_P x = 0$$

A purely mechanical system with the same dynamics as this system is called a mechanical analog. The mechanical analog for this system is a mass $m$ with a spring with stiffness $K_P$ and a damper with damping constant $K_D$. We see that the proportional action corresponds to the spring force, and that the derivative action corresponds to the damper force. Similarly as in example 1 we can define an energy function

$$V\left(x, \dot{x}\right) = \frac{1}{2}m\dot{x}^2 + \frac{1}{2}K_P x^2$$

which is the total energy of the mechanical analog. In the same way as in example 1, the derivative action will dissipate the virtual energy that is initially stored in the system, and intuitively, we may accept that the system will converge to the equilibrium $x = 0$, $\dot{x} = 0$. This can also be seen from the Laplace transform

$$\left(ms^2 + K_D s + K_P\right) x(s) = 0$$

which implies that the poles of the system have negative real parts. The point we are trying to make is that for this system the stability of the closed loop system with a PD controller can be established using energy arguments. Moreover, it is seen that stability is ensured for any positive gains $K_P$ and $K_D$ independently of the physical parameter $m$. There are many important results derived from energy considerations in connection with PID control, and this will be investigated in chapter 2.

## 1.4 Example 4: Adaptive control

We consider a simple first order system given by

$$\dot{x} = a^\star x + u$$

where the parameter $a^\star$ is unknown. An adaptive tracking controller can be designed using the control law

$$u = -Ke - \hat{a}x + \dot{x}_d, \quad e = x - x_d$$

where $x_d$ is the desired trajectory to be tracked, $\hat{a}$ is the estimate of the parameter $a^\star$, and $K$ is the feedback gain. The differential equation for the tracking error $e$ is

$$\begin{aligned}
\frac{de}{dt} &= a^\star x + u - \dot{x}_d \\
&= a^\star x - Ke - \hat{a}x + \dot{x}_d - \dot{x}_d \\
&= -Ke - \tilde{a}x
\end{aligned}$$

where $\tilde{a} = \hat{a} - a^\star$ is the estimation error. We now define

$$\psi = -\tilde{a}x$$

which leads to the following description of the tracking error dynamics

$$\frac{de}{dt} + Ke = \psi$$

We define a function $V_e$ which plays the role of an abstract energy function related to the tracking error $e$:

$$V_e(e) = \frac{1}{2}e^2$$

The time derivative of $V_e$ along the solutions of the system is

$$\dot{V}_e(e) = e\psi - Ke^2$$

Note that this time derivative has a similar structure to that seen in examples 1 and 2. In particular, the $-Ke^2$ term is a dissipation term, and if we think of $\psi$ as the input and $e$ as the output, then the $e\psi$ term is the rate of (abstract) energy supplied from the input. We note that this implies that the following inequality holds for the dynamics of the tracking error:

$$\int_0^T e(t)\psi(t)dt \geq -V_e\left[e\left(0\right)\right]$$

To proceed, we define one more energy-like function. Suppose that we are able to select an adaptation law so that there exists an energy-like function $V_a\left(\tilde{a}\right) \geq 0$ with a time derivative

$$\dot{V}_a\left(\tilde{a}\right) = -e\psi \tag{1.7}$$

We note that this implies that the following inequality holds for the adaptation law

$$\int_0^T \left[-\psi(t)\right] e(t)dt \geq -V_a\left[\tilde{a}(0)\right]$$

Then the sum of the energy functions

$$V\left(e, \tilde{a}\right) = V_e\left(e\right) + V_a\left(\tilde{a}\right)$$

has a time derivative along the solutions of the system given by

$$\dot{V}\left(e, \tilde{a}\right) = -Ke^2$$

This means that the energy function $V\left(e, \tilde{a}\right)$ is decreasing as long as $e$ is nonzero, and by invoking additional arguments from Barbalat's lemma (see 9), we can show that this implies that $e$ tends to zero. The required adaptation law for (1.7) to hold can be selected as the simple gradient update

$$\frac{d\hat{a}}{dt} = xe$$

and the associated energy-like function is

$$V_a\left(\tilde{a}\right) = \frac{1}{2}\tilde{a}^2$$

Note that the convergence of the adaptive tracking controller was established using energy-like arguments, and that other adaptation laws can be used as long as they satisfy the energy-related requirement (1.7).

# Chapter 2

# Positive Real Systems

In the study and control of general dynamical systems the energy stored in the system is an important information. Some autonomous dynamical systems (i.e. with input equal to zero) have the important property that the amount of energy stored in the system remains unchanged or even decreases. This gives rise to the definition of Dissipative systems which is very useful in the synthesis of control systems. Different definitions are introduced to distinguish different classes of systems. Linear systems and nonlinear systems are two important classes of dynamical systems which for simplicity reasons have been usually studied separately in what concerns their dissipative properties. Linear systems whose stored energy remains constant or decreases are called Positive Real (PR) systems or Strictly Positive Real (SPR) respectively while we usually reserve the term dissipative for nonlinear systems in general. We can certainly use the term dissipative for a linear system however the term positive real is only reserved for transfer functions. Furthermore, there exists several important subclasses of linear and nonlinear systems with slightly different properties which are important in the analysis. This will lead us to introduce a series of different definitions of PR and dissipative systems. This chapter deals with Positive Real systems. The next chapter will be devoted to dissipative nonlinear systems.

The definition of Positive Real systems has been motivated by the study of linear electric circuits composed of resistors, inductors and capacitors. The driving point impedance from any point to any other point of such electric circuits is always positive real. The result holds also in the sense that any PR transfer function can be realized with an electric circuit using only resistors, inductors and capacitors. The same result holds for any analogous mechanical or hydraulic systems.This idea can be extended to study analogous electric circuits with nonlinear passive components and even magnetic couplings as done by Arimoto [9] to study dissipative nonlinear systems.

Energy related concepts like positive real systems and passivity (or dissipativity) are widely used in the study and control of physical systems. These control tools have lead to significant progress in the study of dynamic properties and stabilization of nonlinear systems. This chapter reviews the main results available for positive real linear systems and passive nonlinear systems. The applications of these ideas to stability analysis, stabilization and adaptive control of nonlinear systems will also be presented. It is clear that the study of positive real systems and dissipative systems is useful when dealing with systems having non-increasing internal stored energy. Typical dissipative systems include mechanical systems, electric circuits with passive elements, etc. However it should be noted that the properties obtained for dissipative systems can be used to analyze or to control nonlinear systems in general.

**Remark 2.1** The approach presented in this book is devoted to continuous-time systems. However there exists also discrete-time counterparts to most of the results in continuous-time, see for instance an appendix in [96]. The scope of application of passivity concepts in discrete-time seems however more limited than in continuous-time domain. As pointed out in [120], a strictly causal discrete-time system (i.e. having a delay $d \geq 1$) cannot be passive. To illustrate this result consider a unitary delay system described by $y(k) = u(k-1)$. It can be readily seen that there exists input sequences such that $\sum_i y(i)u(i)$ has not a finite lower bound which means that the system is not passive. The result in [120] generalizes this result for any strictly causal system.

**Remark 2.2** Throughout this book we shall assume that the controlled dynamical systems we deal with are well-posed, i.e. the basic requirements for the existence and uniqueness of solutions (i.e. of the state) are satisfied. For the basic definitions of the mathematical objects called "controlled systems", the reader is referred to the book [185] where several and general definitions are provided in great detail. It is clear that from a general point of view, the systems we deal with are dynamical systems whose well-posedness implies some restrictions on the input space. When the analyzed systems depart from the classical systems with continuous or even smooth –i.e. infinitely differentiable– vector fields, we refer the reader to references in which he (or she) will be able to find suitable mathematical developments. In this book we shall encounter linear, nonlinear, smooth, discontinuous, finite and infinite dimensional systems. Providing a general definition of these objects would take us too far from the rest of the presentation, and we prefer to keep it as simple as possible.

# 2.1 Definitions

**Definition 2.1** A system with input $u$ and output $y$ where $u(t), y(t) \in I\!R^n$ is passive if there is a constant $\beta$ such that

$$\int_0^T y^T(t)u(t)dt \geq \beta \tag{2.1}$$

for all functions $u$ and all $T \geq 0$. If, in addition, there are constants $\delta \geq 0$ and $\epsilon \geq 0$ such that

$$\int_0^T y^T(t)u(t)dt \geq \beta + \delta \int_0^T u^T(t)u(t)dt + \epsilon \int_0^T y^T(t)y(t)dt \tag{2.2}$$

for all functions $u$, and all $T \geq 0$, then the system is input strictly passive if $\delta > 0$, output strictly passive if $\epsilon > 0$, and very strictly passive if $\delta > 0$ and $\epsilon > 0$ ♠♠

Obviously $\beta \leq 0$ as the inequality is to be valid for all functions $u$ and in particular the control $u(t) = 0$ for all $t \geq 0$, which gives $0 = \int_0^T y^T(t)u(t)dt \geq \beta$. The importance of the form of $\beta$ in (2.1) will be illustrated in example 4.3.

**Theorem 2.1** Assume that there is a continuous function $V(t) \geq 0$ such that

$$V(T) - V(0) \leq \int_0^T y(t)^T u(t)dt \tag{2.3}$$

for all functions $u$, for all $T \geq 0$ and all $V(0)$. Then the system with input $u(t)$ and output $y(t)$ is passive. Assume, in addition, that there are constants $\delta \geq 0$ and $\epsilon \geq 0$ such that

$$V(T) - V(0) \leq \int_0^T y^T(t)u(t)dt - \delta \int_0^T u^T(t)u(t)dt - \epsilon \int_0^T y^T(t)y(t)dt \tag{2.4}$$

for all functions $u$, for all $T \geq 0$ and all $V(0)$. Then the system is input strictly passive if there is a $\delta > 0$, it is output strictly passive if there is an $\epsilon > 0$, and very strictly passive is there is a $\delta > 0$ and an $\epsilon > 0$ such that the inequality holds. ♠♠

**Proof**
It follows from the assumption $V(T) \geq 0$ that

$$\int_0^T y^T(t)u(t)dt \geq -V(0)$$

for all functions $u$ and all $T \geq 0$, so that (2.1) is satisfied with $\beta := -V(0) \leq 0$. Input strict passivity, output strict passivity and very strict passivity are shown in the same way. 

This indicates that the constant $\beta$ is related to the initial conditions of the system, see also example 4.3 for more informations on the role played by $\beta$.

**Corollary 2.1** Assume that there is a continuously differentiable function $V(t) \geq 0$ and a $d(t)$ such that $\int_0^T d(t)dt \geq 0$ for all $T \geq 0$. Then

1. if
$$\dot{V}(t) \leq y^T(t)u(t) - d(t) \tag{2.5}$$
for all $t \geq 0$ and all functions $u$, then the system is passive.

2. if there is a $\delta > 0$ such that
$$\dot{V}(t) \leq y^T(t)u(t) - \delta u^T(t)u(t) - d(t) \tag{2.6}$$
for all $t \geq 0$ and all functions $u$, then the system is input strictly passive (ISP).

3. if there is a $\epsilon > 0$ such that
$$\dot{V}(t) \leq y^T(t)u(t) - \epsilon y^T(t)y(t) - d(t) \tag{2.7}$$
for all $t \geq 0$ and all functions $u$, then the system is output strictly passive (OSP).

4. if there is a $\delta > 0$ and a $\epsilon > 0$ such that
$$\dot{V}(t) \leq y^T(t)u(t) - \delta u^T(t)u(t) - \epsilon y^T(t)y(t) - d(t) \tag{2.8}$$
for all $t \geq 0$ and all functions $u$, then the system is very strictly passive (VSP). ♠♠

If $V$ is the total energy of the system, then $\langle u, y \rangle = \int_0^T y^T(t)u(t)$ can be seen as the power supplied to the system from the control, while $d(t)$ can be seen as the power dissipated by the system. This means that the condition $\int_0^T d(t)dt \geq 0$ for all $T \geq 0$ means that the system is dissipating energy. The term $w(u, y) = u^T y$ is called the *supply rate* of the system.

## 2.2 Interconnections of passive systems

A useful result for passive systems is that parallel and feedback interconnections of passive systems are passive, and that certain strict passivity properties are inherited.

To explore this we consider two passive systems with scalar inputs and outputs. Similar results are found for multivariable systems. System 1 has input $u_1$ and output $y_1$, and system 2 has input $u_2$ and output $y_2$. We make the following assumptions:

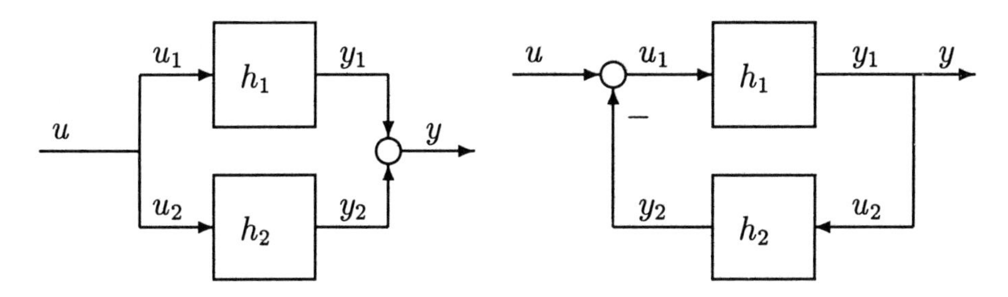


Figure 2.1: Parallel and feeback interconnection

1. There are continuous differentiable functions $V_1(t) \geq 0$ and $V_2(t) \geq 0$.
2. There are functions $d_1$ and $d_2$ such that $\int_0^T d_1(t)dt \geq 0$ and $\int_0^T d_2(t)dt \geq 0$ for all $T \geq 0$.
3. There are constants $\delta_1 \geq 0$, $\delta_2 \geq 0$, $\epsilon_1 \geq 0$ and $\epsilon_2 \geq 0$ such that

$$\dot{V}_1(t) = y_1(t)u_1(t) - \delta_1 u_1^2(t) - \epsilon_1 y_1^2 - d_1(t) \tag{2.9}$$

$$\dot{V}_2(t) = y_2(t)u_2(t) - \delta_2 u_2^2(t) - \epsilon_2 y_2^2 - d_2(t) \tag{2.10}$$

Assumption 3 implies that both systems are passive, and that system $i$ is strictly passive in some sense if any of the constants $\delta_i$ or $\epsilon_i$ are greater than zero. For the parallel interconnection we have $u_1 = u_2 = u$, $y = y_1 + y_2$,and

$$yu = (y_1 + y_2)u = y_1 u + y_2 u = y_1 u_1 + y_2 u_2 \tag{2.11}$$

By adding (2.9) (2.10) and (2.11), there exists a $V = V_1 + V_2 \geq 0$ and a $d_p = d_1 + d_2 + \epsilon_1 y_1^2 + \epsilon_2 y_2^2$ such that $\int_0^T d_p(t)dt \geq 0$ for all $T \geq 0$, and

$$\dot{V}(t) = y(t)u(t) - \delta u^2(t) - d_p(t) \tag{2.12}$$

where $\delta = \delta_1 + \delta_2 \geq 0$. This means that the parallel interconnection system having input $u$ and output $y$ is passive and strictly passive if $\delta_1 > 0$ or $\delta_2 > 0$. For the feedback interconnection we have $y_1 = u_2 = y$, $u_1 = u - y_2$, and

$$yu = y_1(u_1 + y_2) = y_1 u_1 + y_1 y_2 = y_1 u_1 + u_2 y_2 \tag{2.13}$$

Again by adding (2.9) (2.10) and (2.11) we find that there is a $V = V_1 + V_2 \geq 0$ and a $d_{fb} = d_1 + d_2 + \delta_1 u_1^2$ such that $\int_0^T d_{fb}(t)dt \geq 0$ for all $T \geq 0$ and

$$\dot{V}(t) = y(t)u(t) - \epsilon y^2(t) - d_{fb}(t) \tag{2.14}$$

where $\epsilon = \epsilon_1 + \epsilon_2 + \delta_2$. This means that the feedback interconnection is passive, and in addition output strictly passive if $\epsilon_1 > 0$, $\epsilon_2 > 0$, or $\delta_2 > 0$.
By induction it can be shown that any combination of passive systems in parallel or feedback interconnection is passive.

## 2.3 Linear systems

Parseval's theorem is very useful in the study of passive linear systems, as shown next. It is now recalled for the sake of completeness.

**Theorem 2.2 (Parseval's theorem)** Provided that the integrals exist, the following relation holds

$$\int_{-\infty}^{\infty} x(t)y^*(t)dt = \frac{1}{2\pi}\int_{-\infty}^{\infty} x(j\omega)y^*(j\omega)d\omega \tag{2.15}$$

where $y^*$ denotes the complex conjugate of $y$ and $x(j\omega)$ is the Fourier transform of $x(t)$. $x(t)$ is a complex function of $t$, Lebesgue integrable. ♠♠

**Proof** The result is established as follows: the Fourier transform of the time function $x(t)$ is

$$x(j\omega) = \int_{-\infty}^{\infty} x(t)e^{-j\omega t}dt \tag{2.16}$$

while the inverse Fourier transform is

$$x(t) = \frac{1}{2\pi}\int_{-\infty}^{\infty} x(j\omega)e^{j\omega t}d\omega \tag{2.17}$$

Insertion of (2.17) in (2.15) gives

$$\int_{-\infty}^{\infty} x(t)y^*(t)dt = \int_{-\infty}^{\infty}\left[\frac{1}{2\pi}\int_{-\infty}^{\infty} x(j\omega)e^{j\omega t}d\omega\right]y^*(t)dt \tag{2.18}$$

By changing the order of integration this becomes

$$\int_{-\infty}^{\infty} x(t)y^*(t)dt = \frac{1}{2\pi}\int_{-\infty}^{\infty} x(j\omega)\left[\int_{-\infty}^{\infty} y^*(t)e^{j\omega t}dt\right]d\omega \tag{2.19}$$

Here

$$\int_{-\infty}^{\infty} y^*(t)e^{j\omega t}dt = \left[\int_{-\infty}^{\infty} y(t)e^{-j\omega t}dt\right]^* = y^*(j\omega) \tag{2.20}$$

and the result follows. ♠♠

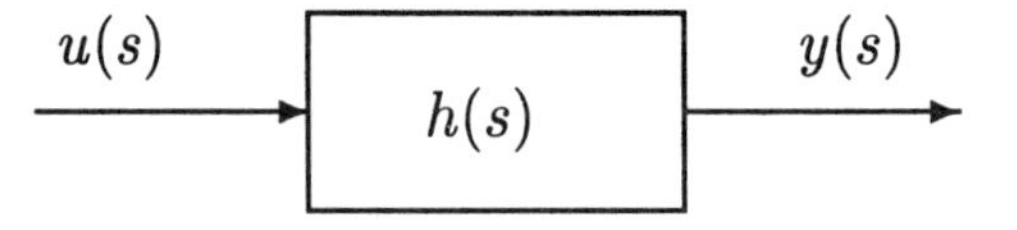


Figure 2.2: Linear time-invariant system

We will now present an important property of linear time-invariant passive systems. **Re**[·] denotes the real part and **Im**[·] denotes the imaginary part

**Theorem 2.3** Given a linear time-invariant linear system

$$y(s) = h(s)u(s) \tag{2.21}$$

with a rational transfer function $h(s)$. Assume that all the poles of $h(s)$ have real parts less than zero. Then the following assertions hold:

1. The system is passive $\Leftrightarrow$ $\mathbf{Re}[h(j\omega)] \geq 0$ for all $\omega$.

2. The system is input strictly passive $\Leftrightarrow$ There is a $\delta > 0$ such that $\mathbf{Re}[h(j\omega)] \geq \delta > 0$ for all $\omega$.

3. The system is output strictly passive $\Leftrightarrow$ There is an $\epsilon > 0$ such that

$$\begin{array}{rcl} \mathbf{Re}[h(j\omega)] & \geq & \epsilon \left|h(j\omega)\right|^2 \\ & \Updownarrow & \\ \left(\mathbf{Re}[h(j\omega)] - \frac{1}{2\epsilon}\right)^2 + \left(\mathbf{Im}[h(j\omega)]\right)^2 & \leq & \left(\frac{1}{2\epsilon}\right)^2 \end{array}$$

♠♠

**Remark 2.3** The notation *for all* $\omega$ means *for all* $\omega \in I\!R \cup \{-\infty, \infty\}$, which is the extended real line.

**Proof**
The proof is based on the use of Parseval's theorem. In this theorem the time integration is over $t \in [0, \infty)$. In the definition of passivity there is an integration over $t \in [0, T]$. To be able to use Parseval's theorem in this proof we introduce the truncated function

$$u_T(t) = \begin{cases} u(t) & \text{when} \quad t \leq T \\ 0 & \text{when} \quad t > T \end{cases} \tag{2.22}$$

which is equal to $u(t)$ for all $t$ less than or equal to $T$, and zero for all $t$ greater than $T$. The Fourier transform of $u_T(t)$, which is denoted $u_T(j\omega)$, will be used in Parseval's theorem.
Without loss of generality we will assume that $y(t)$ and $u(t)$ are equal to zero for all $t \leq 0$. Then according to Parseval's theorem

$$\int_0^T y(t)u(t)dt = \int_{-\infty}^{\infty} y(t)u_T(t)dt = \frac{1}{2\pi} \int_{-\infty}^{\infty} y(j\omega)u_T^*(j\omega)d\omega \tag{2.23}$$

Insertion of $y(j\omega) = h(j\omega)u_T(j\omega)$ gives

$$\int_0^T y(t)u(t)dt = \frac{1}{2\pi} \int_{-\infty}^{\infty} h(j\omega)u_T(j\omega)u_T^*(j\omega)d\omega \tag{2.24}$$

where

$$h(j\omega)u_T(j\omega)u_T^*(j\omega) = \{\mathbf{Re}[h(j\omega)] + j\mathbf{Im}[h(j\omega)]\}|u_T(j\omega)|^2 \tag{2.25}$$

The left side of (2.24) is real, and it follows that the imaginary part on the right hand side is zero. This implies that

$$\int_0^T u(t)y(t)dt = \frac{1}{2\pi}\int_{-\infty}^{\infty} \mathbf{Re}[h(j\omega)]|u_T(j\omega)|^2 d\omega \tag{2.26}$$

First, assume that $\mathbf{Re}[h(j\omega)] \geq \delta \geq 0$ for all $\omega$. Then

$$\int_0^T u(t)y(t)dt \geq \frac{\delta}{2\pi}\int_{-\infty}^{\infty} |u_T(j\omega)|^2 d\omega = \delta\int_0^T u^2(t)dt \tag{2.27}$$

the equality is implied by Parseval's theorem. It follows that the system is passive, and in addition input strictly passive if $\delta > 0$.
Then, assume that the system is passive. Thus there exists a $\delta \geq 0$ so that

$$\int_0^T y(t)u(t)dt \geq \delta\int_0^T u^2(t)dt = \frac{\delta}{2\pi}\int_{-\infty}^{\infty} |u_T(j\omega)|^2 d\omega \tag{2.28}$$

for all $u$, where the initial conditions have been selected so that $\beta = 0$. Here $\delta = 0$ for a passive system, while $\delta > 0$ for a strictly passive system. Then

$$\frac{1}{2\pi}\int_{-\infty}^{\infty} \mathbf{Re}[h(j\omega)]|u_T(j\omega)|^2 d\omega \geq \frac{\delta}{2\pi}\int_{-\infty}^{\infty} |u_T(j\omega)|^2 d\omega \tag{2.29}$$

and

$$\frac{1}{2\pi}\int_{-\infty}^{\infty} (\mathbf{Re}[h(j\omega)] - \delta)|u_T(j\omega)|^2 d\omega \geq 0 \tag{2.30}$$

If there exists a $\omega_0$ so that $\mathbf{Re}[h(j\omega_0)] < \delta$, then inequality will not hold for all $u$ because the integral on the left hand side can be made arbitrarily small if the control signal is selected to be $u(t) = U\cos\omega_0 t$. The results 1 and 2 follow.
To show result 3 we first assume that the system is output strictly passive, that is, there is an $\epsilon > 0$ such that

$$\int_0^T y(t)u(t)dt \geq \epsilon\int_0^T y^2(t)dt = \frac{\epsilon}{2\pi}\int_{-\infty}^{\infty} |h(j\omega)|^2|u_T(j\omega)|^2 d\omega \tag{2.31}$$

This gives the inequality (see (2.26))

$$\mathbf{Re}[h(j\omega)] \geq \epsilon\,|h(j\omega)|^2 \tag{2.32}$$

which is equivalent to

$$\epsilon\left[(\mathbf{Re}[h(j\omega)])^2 + (\mathbf{Im}[h(j\omega)])^2\right] - \mathbf{Re}[h(j\omega)] \leq 0 \tag{2.33}$$

and the second inequality follows by straighforward algebra. The converse result is shown similarly as the result for input strict passivity. ♠♠
Note that according to the theorem a passive system will have a transfer function which satisfies

$$|\angle h(j\omega)| \leq 90^\circ \quad \text{for all} \quad \omega \tag{2.34}$$

In a Nyquist diagram the theorem states that $h(j\omega)$ is in the closed half plane $\mathbf{Re}\,[s] \geq 0$ for passive systems, $h(j\omega)$ is in $\mathbf{Re}\,[s] \geq \delta > 0$ for input strictly passive systems, and for output strictly passive systems $h(j\omega)$ is inside the circle with center in $s = 1/\,(2\epsilon)$ and radius $1/\,(2\epsilon)$. This is a circle that crosses the real axis in $s = 0$ and $s = 1/\epsilon$.

**Remark 2.4** A transfer function $h(s)$ is rational if it is the fraction of two polynomials in the complex variable $s$, that is if it can be written in the form

$$h(s) = \frac{Q(s)}{R(s)} \tag{2.35}$$

where $Q(s)$ and $R(s)$ are polynomials in $s$. An example of a transfer function that is not rational is $h(s) = \tanh s$ which appears in connection with systems described by partial differential equations.

**Example 2.1** Note the difference between the condition $\mathbf{Re}[h(j\omega)] > 0$ and the condition for input strict passivity that there exists a $\delta > 0$ so that $\mathbf{Re}[h(j\omega_0)] \geq \delta > 0$ for all $\omega$. An example of this is

$$h_1(s) = \frac{1}{1 + Ts} \tag{2.36}$$

We find that $\mathbf{Re}[h_1(j\omega)] > 0$ for all $\omega$ because

$$h_1(j\omega) = \frac{1}{1 + j\omega T} = \frac{1}{1 + (\omega T)^2} - j\frac{\omega T}{1 + (\omega T)^2} \tag{2.37}$$

However there is no $\delta > 0$ that ensures $\mathbf{Re}[h(j\omega_0)] \geq \delta > 0$ for all $\omega$. This is seen from the fact that for any $\delta > 0$ we have

$$\mathbf{Re}[h_1(j\omega)] = \frac{1}{1 + (\omega T)^2} < \delta \quad \text{for all} \quad \omega > \sqrt{\frac{1-\delta}{\delta}} \tag{2.38}$$

This implies that $h_1(s)$ is not input strictly passive.
We note that for this system

$$|h_1(j\omega)|^2 = \frac{1}{1 + (\omega T)^2} = \mathbf{Re}[h_1(j\omega)] \tag{2.39}$$

which means that the system is output strictly passive with $\epsilon = 1$.

**Example 2.2** Consider a system with the transfer function

$$h_2(s) = \frac{s+c}{(s+a)(s+b)} \tag{2.40}$$

where $a$, $b$ and $c$ are positive constants. We find that

$$\begin{aligned} h_2(j\omega) &= \frac{j\omega+c}{(j\omega+a)(j\omega+b)} \\ &= \frac{(c+j\omega)(a-j\omega)(b-j\omega)}{(a^2+\omega^2)(b^2+\omega^2)} \\ &= \frac{abc+\omega^2(a+b-c)+j[\omega(ab-ac-bc)-\omega^3]}{(a^2+\omega^2)(b^2+\omega^2)} \end{aligned}$$

From this it is clear that

1. If $c \leq a+b$, then $\mathbf{Re}[h_2(j\omega)] > 0$ for all $\omega$. As $\mathbf{Re}[h_2(j\omega)] \to 0$ when $\omega \to \infty$, *the* system is not input strictly passive.
2. If $c > a+b$, then $h_2(s)$ is not passive because $\mathbf{Re}[h_2(j\omega)] < 0$ for $\omega > \sqrt{abc/(c-a-b)}$.

**Example 2.3** The systems with the transfer functions

$$h_3(s) = 1 + Ts \tag{2.41}$$

$$h_4(s) = \frac{1+T_1 s}{1+T_2 s}, \quad T_1 < T_2 \tag{2.42}$$

are input strictly passive because

$$\mathbf{Re}[h_3(j\omega)] = 1 \tag{2.43}$$

and

$$\mathbf{Re}[h_4(j\omega)] = \frac{1+\omega^2 T_1 T_2}{1+(\omega T_2)^2} \in \left(\frac{T_1}{T_2}, 1\right] \tag{2.44}$$

Moreover $|h_4(j\omega)|^2 \leq 1$, so that

$$\mathbf{Re}[h_4(j\omega)] \geq \frac{T_1}{T_2} \geq \frac{T_1}{T_2} |h_4(j\omega)|^2 \tag{2.45}$$

which shows that the system is output strictly passive with $\epsilon = T_1/T_2$. The reader may verify from a direct calculation of $|h_4(j\omega)|^2$ and some algebra that it is possible to have $\mathbf{Re}[h_4(j\omega)] \geq |h_4(j\omega)|^2$, that is, $\epsilon = 1$. This agrees with the Nyquist plot of $h_4(j\omega)$.

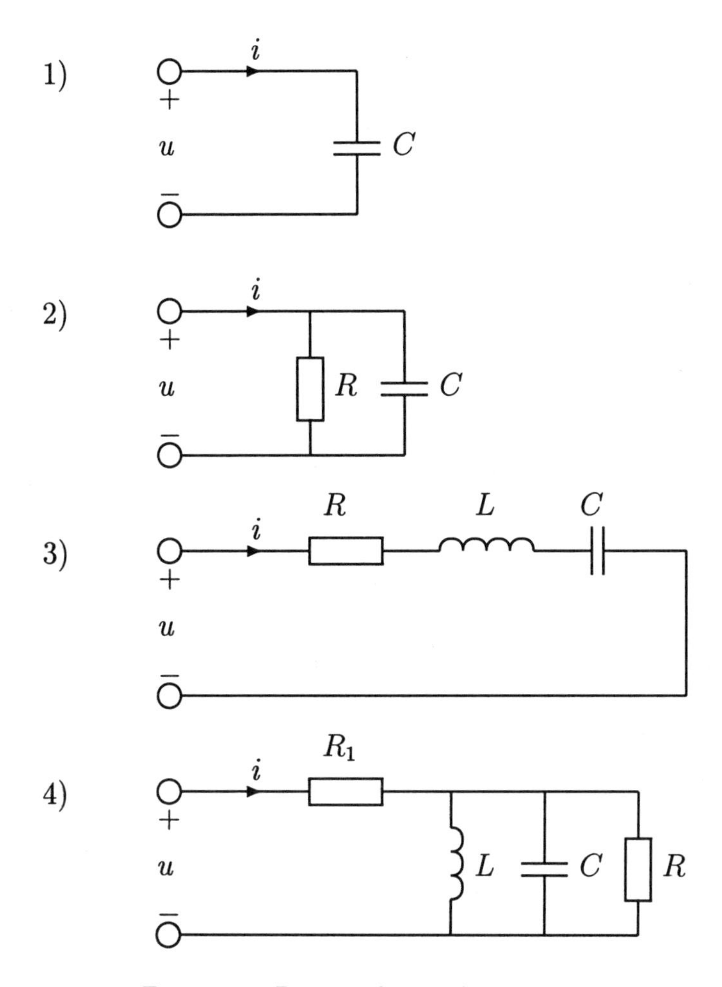


Figure 2.3: Passive electrical one-ports

**Example 2.4** A dynamic system describing an electrical one-port with resistors, inductors and capacitors is passive if the voltage over the port is input and the current into the port is output, or *vice versa.* In figure 2.3 different passive one-port are shown. We consider the voltage over the port to be the input and the current into the port as the output. The resulting transfer functions are admittances, which are the inverses of the impedances. Circuit 1 is a capacitor, circuit 2 is a resistor in parallel with a capacitor, circuit 3 is a resistor in series with a inductor and a capacitor, while circuit 4 is a resistor in series with a parallel connection of an inductor, a capacitor and a resistor. The transfer functions are:

$$h_1(s) = Cs \tag{2.46}$$

$$h_2(s) = \frac{1}{R}(1 + RCs) \tag{2.47}$$

$$h_3(s) = \frac{Cs}{1 + RCs + LCs^2} \tag{2.48}$$

$$h_4(s) = \frac{1}{R_1} \frac{1 + \frac{L}{R}s + LCs^2}{1 + (\frac{L}{R_1} + \frac{L}{R})s + LCs^2} \tag{2.49}$$

Systems 1, 2, 3 and 4 are all passive as the poles have real parts that are less than zero, and in addition $\mathbf{Re}[h_i(j\omega)] \geq 0$ for all $\omega$ and $i \in \{1, 2, 3, 4\}$. It follows that the transfer functions have phases that satisfy $|\angle h_i(j\omega)| \leq 90°$. In addition system 2 is input strictly passive as $\mathbf{Re}[h_2(j\omega)] = 1/R > 0$ for all $\omega$. For system 4 we find that

$$\mathbf{Re}[h_4(j\omega)] = \frac{1}{R_1} \frac{(1 - \omega^2 LC)^2 + \omega^2 \frac{L^2}{R_1(R_1+R)}}{(1 - \omega^2 LC)^2 + \omega^2 \frac{L^2}{(R_1+R)^2}} \geq \frac{1}{R_1 + R} \tag{2.50}$$

which means that system 4 is input strictly passive. ♠♠

So far we have only considered systems where the transfer functions $h(s)$ have poles with negative real parts. There are however passive systems that have transfer functions with poles on the imaginary axis. This is demonstrated in the following example:

**Example 2.5** Consider the system $\dot{y}(t) = u(t)$ which is represented in transfer function description by $y(s) = h(s)u(s)$ where $h(s) = \frac{1}{s}$. This means that the transfer function has a pole at the origin, which is on the imaginary axis. For this system $\mathbf{Re}[h(j\omega)] = 0$ for all $\omega$. However, we cannot establish passivity using theorem 2.3 as this theorem only applies to systems where all the poles have negative real parts. Instead, consider

$$\int_0^T y(t)u(t)dt = \int_0^T y(t)\dot{y}(t)dt \tag{2.51}$$

A change of variables $\dot{y}dt = dy$ gives

$$\int_0^T y(t)u(t)dt = \int_{y(0)}^{y(T)} y(t)dy = \frac{1}{2}[y(T)^2 - y(0)^2] \geq -\frac{1}{2}y(0)^2 \tag{2.52}$$

and passivity is shown with $\beta = -\frac{1}{2}y(0)^2$. ♠♠

It turns out to be relatively involved to find necessary and sufficient conditions on $h\,(j\omega)$ for the system to be passive when we allow for poles on the imaginary axis. The conditions are relatively simple and are given in the following theorem.

**Theorem 2.4** Consider a linear time-invariant system with a rational transfer function $h(s)$. The system is passive if and only if

1. $h(s)$ has no poles in $\mathbf{Re}[s] > 0$.
2. $\mathbf{Re}[h(j\omega)] \geq 0$ for all $\omega$ such that $j\omega$ is not a pole of $h(s)$.
3. If $j\omega_0$ is a pole of $h(s)$, then it is a simple pole, and the residual in $s = j\omega_0$ is real and greater than zero, that is, $Res_{s=j\omega_0} h(s) = \lim_{s \to j\omega_0}(s - j\omega_0)h(j\omega) > 0$.

♠♠

The above result is established in section 2.11.

**Corollary 2.2** If a system with transfer function $h(s)$ is passive, then $h(s)$ has no poles in $\mathbf{Re}[s] > 0$. ♠♠

**Proposition 2.1** Consider a rational transfer function

$$h(s) = \frac{(s + z_1)(s + z_2)\ldots}{s(s + p_1)(s + p_2)\ldots} \tag{2.53}$$

where $\mathbf{Re}[p_i] > 0$ and $\mathbf{Re}[z_i] > 0$ which means that $h(s)$ has one pole at the origin and the remaining poles in $\mathbf{Re}[s] < 0$, while all the zeros are in $\mathbf{Re}[s] < 0$. Then the system with transfer function $h(s)$ is passive if and only if $\mathbf{Re}[h(j\omega)] \geq 0$ for all $\omega$. ♠♠

**Proof**

The residual of the pole on the imaginary axis is

$$\mathrm{Res}_{s=0} h(s) = \frac{z_1 z_2 \ldots}{p_1 p_2 \ldots} \tag{2.54}$$

Here the constants $z_i$ and $p_i$ are either real and positive, or they appear in complex conjugated pairs where the products $z_i z_i^* = |z_i|^2$ and $p_i p_i^* = |p_i|^2$ are real and positive. It is seen that the residual at the imaginary axis is real and positive. As $h(s)$ has no poles in $\mathbf{Re}[s] < 0$ by assumption, it follows that the system is passive if and only if $\mathbf{Re}[h(j\omega)] \geq 0$ for all $\omega$. ♠♠

**Example 2.6** Consider two systems with transfer functions

$$h_1(s) = \frac{s^2 + a^2}{s(s^2 + \omega_0^2)}, \quad a \neq 0, \omega_0 \neq 0 \tag{2.55}$$

$$h_2(s) = \frac{s}{s^2 + \omega_0^2}, \quad \omega_0 \neq 0 \tag{2.56}$$

where all the poles are on the imaginary axis. Thus condition 1 in theorem 2.4 is satisfied. Moreover,

$$h_1(j\omega) = -j\frac{a^2 - \omega^2}{\omega(\omega_0^2 - \omega^2)} \tag{2.57}$$

$$h_2(j\omega) = j\frac{\omega}{\omega_0^2 - \omega^2} \tag{2.58}$$

so that also condition 2 holds in view of $\mathbf{Re}[h_1(j\omega)] = \mathbf{Re}[h_2(j\omega)] = 0$ for all $\omega$ so that $j\omega$ is not a pole in $h(s)$. We now calculate the residual, and find that

$$\text{Res}_{s=0}h_1(s) = \frac{a^2}{\omega_0^2} \tag{2.59}$$

$$\text{Res}_{s=\pm j\omega_0}h_1(s) = \frac{\omega_0^2 - a^2}{2\omega_0^2} \tag{2.60}$$

$$\text{Res}_{s=\pm j\omega_0}h_2(s) = \frac{1}{2} \tag{2.61}$$

We see that according to theorem 2.4 the system with transfer function $h_2(s)$ is passive, while $h_1(s)$ is passive whenever $a < \omega_0$.

**Example 2.7** Consider a system with transfer function

$$h(s) = -\frac{1}{s} \tag{2.62}$$

The transfer function has no poles in $\mathbf{Re}[s] > 0$, and $\mathbf{Re}[h(j\omega)] \geq 0$ for all $\omega \neq 0$. However, $Res_{s=0}h(s) = -1$, and theorem 2.4 shows that the system is not passive This result agrees with the observation

$$\int_0^T y(t)u(t)dt = -\int_{y(0)}^{y(T)} y(t)dy = \frac{1}{2}[y(0)^2 - y(T)^2] \tag{2.63}$$

where the right hand side has no lower bound as $y(T)$ can be arbitrarily large.

## 2.4 Passivity of the PID controllers

**Proposition 2.2** Assume that $0 \leq T_d < T_i$ and $0 \leq \alpha \leq 1$. Then the PID controller

$$h_r(s) = K_p\frac{1 + T_i s}{T_i s}\frac{1 + T_d s}{1 + \alpha T_d s} \tag{2.64}$$

is passive. This follows from proposition 2.1.

♠♠

**Proposition 2.3** Consider a PID controller with transfer function

$$h_r(s) = K_p\beta \frac{1+T_i s}{1+\beta T_i s} \frac{1+T_d s}{1+\alpha T_d s} \tag{2.65}$$

where $0 \le T_d < T_i$, $1 \le \beta < \infty$ and $0 < \alpha \le 1$. Then the controller is passive, and in addition, the transfer function gain has an upper bound $|h_r(j\omega)| \le \frac{K_p\beta}{\alpha}$ and the real part of the transfer function is bounded away from zero according to $\mathbf{Re}\,[h_r(j\omega)] \ge K_p$ for all $\omega$.

♠♠

It follows from Bode diagram techniques that

$$|h_r(j\omega)| \le K_p\beta \cdot 1 \cdot \frac{1}{\alpha} = \frac{K_p\beta}{\alpha} \tag{2.66}$$

The result on the $\mathbf{Re}\,[h_r(j\omega)]$ can be established using Nyquist diagram, or by direct calculation of $\mathbf{Re}\,[h_r(j\omega)]$. ♠♠

## 2.5 Stability of a passive feedback interconnection

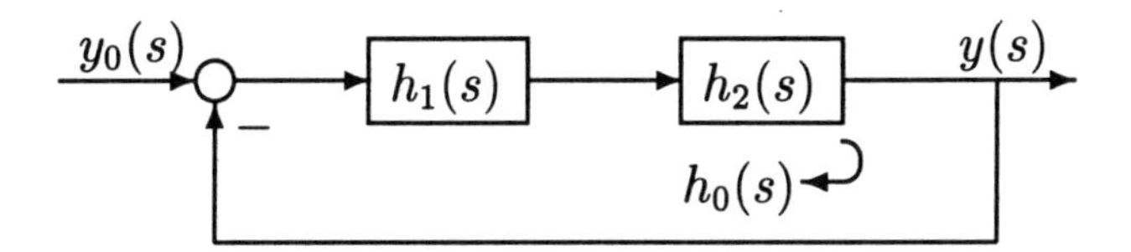


Figure 2.4: Interconnection of a passive system $h_1(s)$ and a strictly system $h_2(s)$

Consider a feedback loop with loop transfer function $h_0(s) = h_1(s)h_2(s)$ as shown in figure 2.4. If $h_1$ is passive and $h_2$ is strictly passive, then the phases of the transfer functions satisfy

$$|\angle h_1(j\omega)| \le 90^\circ \quad \text{and} \quad |\angle h_2(j\omega)| < 90^\circ \tag{2.67}$$

It follows that the phase of the loop transfer function $h_0(s)$ is bounded by

$$|\angle h_0(j\omega)| < 180^\circ \tag{2.68}$$

As $h_1$ and $h_2$ are passive, it is clear that $h_0(s)$ has no poles in $\mathbf{Re}\,[s] > 0$. Then according to standard Bode-Nyquist stability theory the system is asymptotically stable and BIBO stable. The same result is obtained if instead $h_1$ is strictly passive and $h_2$ is passive.
We note that in view of proposition 2.3 a PID controller with limited integral action is strictly stable. This implies that

- A passive linear system with a PID controller with limited integral action is BIBO stable.

For an important class of systems passivity or strict passivity is a structural property which is not dependent on the numerical values of the parameters of the system. Then passivity considerations may be used to establish stability even if there are large uncertainties or large variations in the system parameters. This is often referred to as robust stability. When it comes to performance it is possible to use any linear design technique to obtain high performance for the nominal parameters of the system. The resulting system will have high performance under nominal conditions, and in addition robust stability under large parameter variations.

## 2.6 Mechanical analogs for PD controllers

In this section we will study how PD controllers for position control can be represented by mechanical analogs when the input to the system is force and the output is position. Note that when force is input and position is output, then the physical system is not passive. We have a passive physical system if force is input and velocity is the output, and then a PD controller from position corresponds to PI controller from velocity. For this reason we might have referred to the controllers in this section as PI controllers for velocity control.
We consider a mass $m$ with position $x$ and velocity $v = \dot{x}$. The dynamics is given by $m\ddot{x} = u$ where the force $u$ is the input. The desired position is $x_d$, while the desired velocity is $v_d = \dot{x}_d$. A PD controller $u = K_p(1 + T_d s)\,[x_d(s) - x(s)]$ is used. The control law can be written

$$u = K_p(x_d - x) + D(v_d - v) \tag{2.69}$$

where $D = K_p T_d$. The mechanical analog appears from the observation that this control force is the force that results if the mass $m$ with position $x$ is connected to the position $x_d$ with a parallel interconnection of a spring with stiffness $K_p$ and a damper with coefficient $D$ as shown in figure 2.5.

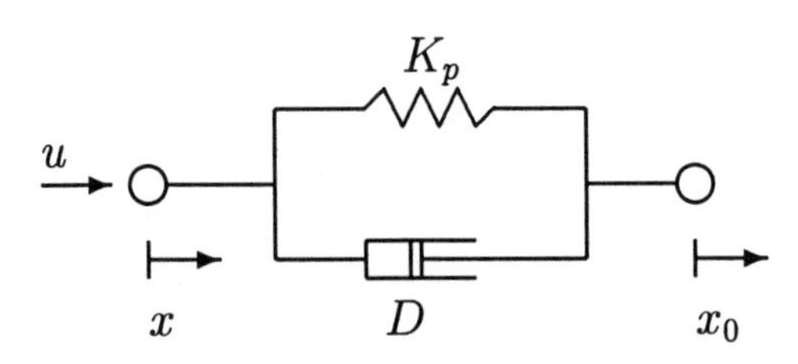


Figure 2.5: Mechanical analog of PD controller with feedback from position

If the desired velocity is not available, and the desired position is not smooth a PD controller of the type

$$u = K_p x_d - K_p(1 + T_d s)x$$

can be used. Then the control law is

$$u = K_p(x_d - x) - Dv \tag{2.70}$$

This is the force that results if the mass $m$ is connected to the position $x_d$ with a spring of stiffness $K_p$ and a damper with coefficient $D$ as shown in figure 2.6.

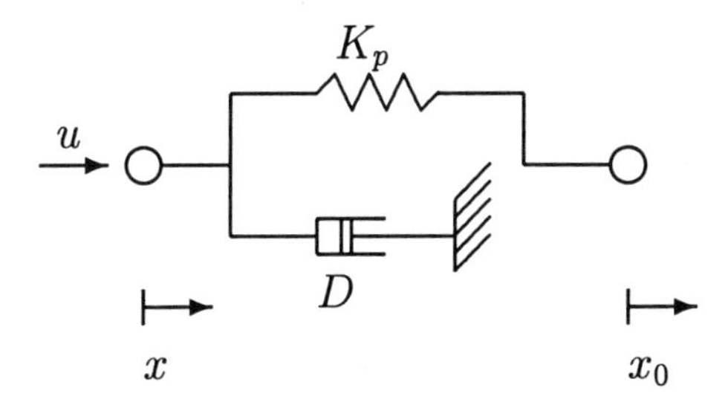


Figure 2.6: Mechanical analog of a PD controller without desired velocity input.

If the velocity is not measured the following PD controller can be used

$$u(s) = K_p \frac{1 + T_d s}{1 + \alpha T_d s}[x_d(s) - x(s)] \tag{2.71}$$

where $0 \leq \alpha \leq 1$ is the filter parameter. We will now demonstrate that this transfer function appears by connecting the mass $m$ with position $x$ to a spring with stiffness $K_1$ in series with a parallel interconnection of a spring with stiffness $K$ and a damper with coefficient $D$ as shown in figure 2.7.
To find the expression for $K_1$ and $K$ we let $x_1$ be the position of the connection point between the spring $K_1$ and the parallel interconnection. Then the force is $u = K_1(x_1 - x)$, which implies that $x_1(s) = x(s) + u(s)/K_1$. As there is no mass in the point $x_1$ there must be a force of equal magnitude in the opposite direction from the parallel interconnection, so that

$$u(s) = K[x_d(s) - x_1(s)] + D[v_d(s) - v_1(s)] = (K + Ds)[x_d(s) - x_1(s)] \tag{2.72}$$

Insertion of $x_1(s)$ gives

$$u(s) = (K + Ds)[x_d(s) - x(s) - \frac{1}{K_1}u(s)] \tag{2.73}$$

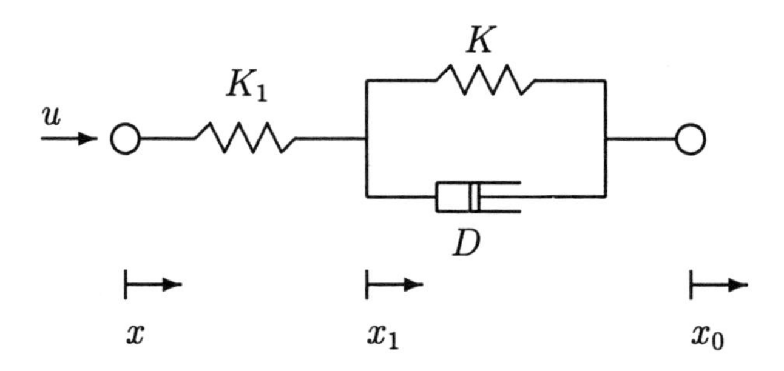


Figure 2.7: Mechanical analog of a PD controller without velocity measurement

We solve for $u(s)$ and the result is

$$\begin{aligned} u(s) &= K_1 \frac{K+Ds}{K_1+K+Ds}[x_d(s) - x(s)] \\ &= \frac{K_1 K}{K_1+K} \frac{1+\frac{D}{K}s}{1+\frac{K}{K_1+K}\frac{D}{K}s}[x_d(s) - x(s)] \end{aligned}$$

We see that this is a PD controller without velocity measurement where

$$\begin{aligned} K_p &= \frac{K_1 K}{K_1+K} \\ T_d &= \frac{D}{K} \\ \alpha &= \frac{K}{K_1+K} \in [0,1) \end{aligned}$$

## 2.7 Multivariable linear systems

**Theorem 2.5** Consider a linear time-invariant system

$$y(s) = H(s)u(s) \tag{2.74}$$

with a rational transfer function matrix $H(s)$, input $u(t) \in \mathbb{R}^m$ and input $y(t) \in \mathbb{R}^m$. Assume that all the poles of $H(s)$ are in $\mathbf{Re}[s] < 0$. Let $\lambda_{min}[H(j\omega) + H^*(j\omega)]$ the smallest eigenvalue of the Hermitian matrix $H(j\omega) + H^*(j\omega)$. Then,

1. The system is passive $\Leftrightarrow$ $\lambda_{min}[H(j\omega) + H^*(j\omega)] \geq 0$ for all $\omega$.

2. The system is input strictly passive $\Leftrightarrow$ There is a $\delta > 0$ so that $\lambda_{min}[H(j\omega) + H^*(j\omega)] \geq \delta > 0$ for all $\omega$.

♠♠

**Proof**

Let $A \in \mathbb{C}^{m\times m}$ be some Hermitian matrix with eigenvalues $\lambda_i(A)$. Let $x \in \mathbb{C}^m$ be an arbitrary vector with complex entries. It is well-known from linear algebra that $x^*Ax$ is real, and that $x^*Ax \geq \lambda_{min}(A)|x|^2$. From Parseval's theorem we have

$$\begin{aligned}
\int_0^\infty y^T(t)u_T(t)dt &= \sum_{i=1}^m \int_0^\infty y_i(t)(u_i)_T(t)dt \\
&= \sum_{i=1}^m \frac{1}{2\pi}\int_{-\infty}^\infty y_i^*(j\omega)(u_i)_T(j\omega)d\omega \\
&= \frac{1}{2\pi}\int_{-\infty}^\infty y^*(j\omega)u_T(j\omega)d\omega
\end{aligned}$$

This leads to

$$\begin{aligned}
\int_0^T y(t)^T u(t)dt &= \int_0^\infty y^T(t)u_T(t)dt \\
&= \frac{1}{2\pi}\int_{-\infty}^\infty y^*(j\omega)u_T(j\omega)d\omega \\
&= \frac{1}{4\pi}\int_{-\infty}^\infty [u_T^*(j\omega)y(j\omega) + y^*(j\omega)u_T(j\omega)]d\omega \\
&= \frac{1}{4\pi}\int_{-\infty}^\infty u_T^*(j\omega)[H(j\omega) + H^*(j\omega)]u_T(j\omega)d\omega
\end{aligned}$$

Because $H(j\omega) + H^*(j\omega)$ is Hermitian we find that

$$\int_0^T y(t)^T u(t)dt \geq \frac{1}{4\pi}\int_{-\infty}^\infty \lambda_{\min}[H(j\omega) + H^*(j\omega)]|u_T(j\omega)|^2 d\omega \tag{2.75}$$

The result can be established along the lines of theorem 2.3. ♠♠

# 2.8 The scattering formulation

By a change of variables an alternative description can be established where passivity corresponds to small gain. We will introduce this idea with an example from linear circuit theory. Consider a linear time-invariant system describing an electrical one-port with voltage $e$, current $i$ and impedance $z(s)$ so that

$$e(s) = z(s)i(s) \tag{2.76}$$

Define the wave variables

$$a = e + z_0 i \quad \text{and} \quad b = e - z_0 i \tag{2.77}$$

where $z_0$ is a positive constant. The Laplace transform is

$$\begin{aligned}
a(s) &= [z(s) + z_0]i(s) \\
b(s) &= [z(s) - z_0]i(s)
\end{aligned}$$

Combining the two equations we get

$$b(s) = g(s)a(s) \tag{2.78}$$

where

$$g(s) = \frac{z(s) - z_0}{z_0 + z(s)} = \frac{\frac{z(s)}{z_0} - 1}{1 + \frac{z(s)}{z_0}} \tag{2.79}$$

is the *scattering function* of the system. The terms wave variable and scattering function originate from the description of transmission lines where $a$ can be seen as the incident wave and $b$ can be seen as the reflected wave.
If the electrical circuit has only passive elements, that is, if the circuit is an interconnection of resistors, capacitors and inductors, the passivity inequality satisfies

$$\int_0^T e(t)i(t)dt \geq 0 \tag{2.80}$$

where it is assumed that the initial energy stored in the circuit is zero. We note that

$$a^2 - b^2 = (e + z_0 i)^2 - (e - z_0 i)^2 = 4z_0 ei \tag{2.81}$$

which implies

$$\int_0^T b^2(t)dt = \int_0^T a^2(t)dt - 4z_0 \int_0^T e(t)i(t)dt \tag{2.82}$$

From this it is seen that passivity of the system with input $i$ and output $e$ corresponds to small gain for the system with input $a$ and output $b$ in the sense that

$$\int_0^T b^2(t)dt \leq \int_0^T a^2(t)dt \tag{2.83}$$

This small gain condition can be interpreted loosely in the way that the energy content $b^2$ of the reflected wave is smaller than the energy $a^2$ of the incident wave.
For the general linear time-invariant system

$$y(s) = h(s)u(s) \tag{2.84}$$

introduce the wave variables

$$a = y + u \quad \text{and} \quad b = y - u \tag{2.85}$$

where, as above, $a$ is the incident wave and $b$ is the reflected wave. As for electrical circuits it will usually be necessary to include a constant $z_0$ so that $a = y + z_0 u$ $b = y - z_0 u$ so that the physical units agree. We tacitly suppose that this is done by letting $z_0 = 1$ with the appropriate physical unit. The scattering function is defined by

$$g(s) := \frac{b}{a}(s) = \frac{y - u}{y + u}(s) = \frac{h(s) - 1}{1 + h(s)} \tag{2.86}$$

**Theorem 2.6** Consider a system with rational transfer function $h(s)$ with no poles in $\mathbf{Re}[s] \geq 0$, and scattering function $g(s)$ given by (2.86). Then

1. The system is passive if and only if $|g(j\omega)| \leq 1$ for all $\omega$.
2. The system is strictly passive, and there is a $\gamma$ so that $|h(j\omega)| \leq \gamma$ for all $\omega$ if and only if there is a $\gamma' \in (0,1)$ so that $|g(j\omega)|^2 \leq 1-\gamma'$

♠♠

**Proof**
Consider the following computation

$$\begin{aligned} |g(j\omega)|^2 &= \frac{|h(j\omega)-1|^2}{|h(j\omega)+1|^2} \\ &= \frac{|h(j\omega)|^2-2\mathbf{Re}[h(j\omega)]+1}{|h(j\omega)|^2+2\mathbf{Re}[h(j\omega)]+1} \\ &= 1-\frac{4\mathbf{Re}[h(j\omega)]}{|h(j\omega)+1|^2} \end{aligned} \tag{2.87}$$

It is seen that $|g(j\omega)| \leq 1$ if and only if $\mathbf{Re}[h(j\omega)] \geq 0$. Result 1 then follows as the necessary and sufficient condition for the system to be passive is that $\mathbf{Re}[h(j\omega)] \geq 0$ for all $\omega$.
Concerning the second result we show the "if" part. Assume that there is a $\delta$ so that $\mathbf{Re}[h(j\omega)] \geq \delta > 0$ and a $\gamma$ so that $|h(j\omega)| \leq \gamma$ for all $\omega$. Then

$$|g(j\omega)|^2 \geq 1-\frac{4\delta}{(\gamma+1)^2} \tag{2.88}$$

and the result follows with $0<\gamma'<\min\left(1, \frac{4\delta}{(\gamma+1)^2}\right)$.
Next assume that $g(j\omega)|^2 \leq 1-\gamma'$ for all $\omega$. Then

$$4\mathbf{Re}\left[h(j\omega)\right] \geq \gamma'\left(|h(j\omega)|^2+2\mathbf{Re}[h(j\omega)]+1\right) \tag{2.89}$$

and strict passivity follows from

$$\mathbf{Re}\left[h(j\omega)\right] \geq \frac{\gamma'}{4-2\gamma'} > 0 \tag{2.90}$$

Finite gain of $h(j\omega)$ follows from

$$\gamma'|h(j\omega)|^2-(4-2\gamma')\,\mathbf{Re}[h(j\omega)]+\gamma' \leq 0 \tag{2.91}$$

which in view of the general result $|h(j\omega)| > \mathbf{Re}[h(j\omega)]$ gives the inequality

$$|h(j\omega)|^2-\frac{(4-2\gamma')}{\gamma'}|h(j\omega)|+1 \leq 0 \tag{2.92}$$

This implies that

$$|h(j\omega)| \leq \frac{(4-2\gamma')}{\gamma'} \tag{2.93}$$

## 2.9 Impedance matching

In this section we will briefly review the concept of impedance matching. Again an electrical one-port is studied. The one-port has a voltage source $e$, serial impedance $z_0$, output voltage $v$ and current $i$. The circuit is coupled to the load which is a passive one-port with driving point impedance $z_L(s)$ as shown in figure 2.8.

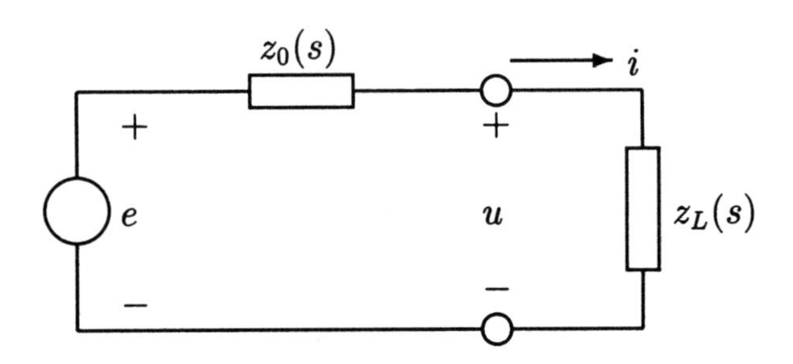


Figure 2.8: Impedance matching

The following problem will be addressed: suppose $z_0(s)$ is given and that $e(t) = E \sin \omega_e t$. Select $z_L(s)$ so that the power dissipated in $z_L$ is maximized. The current is given by

$$i(s) = \frac{e(s)}{z_0(s) + z_L(s)} \tag{2.94}$$

while the voltage over $z_L$ is

$$u(s) = z_L(s)i(s) \tag{2.95}$$

The power dissipated in $z_L$ is therefore

$$\begin{aligned} P(\omega_e) &= \tfrac{1}{2}\mathbf{Re}[u_L(j\omega_e)i^*(j\omega_e)] \\ &= \tfrac{1}{2}\mathbf{Re}[z_L(j\omega_e)]i(j\omega_e)i^*(j\omega_e) \\ &= \frac{1}{2}\frac{\mathbf{Re}[z_L(j\omega_e)]}{[z_0(j\omega_e)+z_L(j\omega_e)]^*[z_0(j\omega_e)+z_L(j\omega_e)]}E^2 \end{aligned}$$

where $(\cdot)^*$ denotes the complex conjugate. Denote

$$z_0(j\omega_e) = \alpha_0 + j\beta_0 \quad \text{or} \quad z_L(j\omega_e) = \alpha_L + j\beta_L \tag{2.96}$$

This gives

$$P = \frac{1}{2}\frac{\alpha_L E^2}{(\alpha_0 + \alpha_L)^2 + (\beta_0 + \beta_L)^2} \tag{2.97}$$

We see that if $\alpha_L = 0$, then $P = 0$, whereas for nonzero $\alpha_L$ then $|\beta_L| \to \infty$, gives $P \to 0$. A maximum for $P$ would be expected somewhere between these extremes. Differentiation with respect to $\beta_L$ gives

$$\frac{\partial P}{\partial \beta_L} = \frac{E^2}{2} \frac{-2\alpha_L(\beta_0 + \beta_L)}{[(\alpha_0 + \alpha_L)^2 + (\beta_0 + \beta_L)^2]^2} \tag{2.98}$$

which implies that the maximum *of* $P$ appears for $\beta_L = -\beta_0$. Differentiation with respect to $\alpha_L$ with $\beta_L = -\beta_0$ gives

$$\frac{\partial P}{\partial a} = \frac{E^2}{2} \frac{\alpha_0^2 - {\alpha_L}^2}{[(\alpha_0 + \alpha_L)^2 + (\beta_0 + \beta_L)^2]^2} \tag{2.99}$$

and it is seen that the maximum is found for $\alpha_L = \alpha_0$. This means that the maximum power dissipation in $z_L$ is achieved with

$$z_L(j\omega_e) = z_0^*(j\omega_e) \tag{2.100}$$

This particular selection of $z_L(j\omega_e)$ is called impedance matching. If the voltage source $e(t)$ is not simply a sinusoid but a signal with a arbitrary spectrum, then it is not possible to find a passive impedance $z_L(s)$ which satisfies the impedance matching condition or a general series impedance $z_0(j\omega)$. This is because the two impedances are required to have the same absolute values, while the phase have opposite signs. This cannot be achieved for one particular $z_L(s)$ for all frequencies.
However, if $z_0(j\omega) = z_0$ is a real constant, then impedance matching at all frequencies is achieved with $z_L = z_0$. We now assume that $z_0$ is a real constant, and define the wave variables to be

$$a = u + z_0 i \quad \text{or} \quad b = u - z_0 i \tag{2.101}$$

Then it follows that

$$a = e \tag{2.102}$$

for the system in figure 2.8.
A physical interpretation of the incident wave $a$ is as follows: let $u$ be the input voltage to the one-port and let $i$ be the current into the port. Consider the extended one-port where a serial impedance $z_0$ is connected in to the one-port as shown in figure 2.9. Then $a$ is the input voltage of the extended one-port. The physical interpretation of the reflected wave $b$ is shown in figure 2.10. We clearly see that if $z_L = z_0$, then

$$u = z_0 i \quad \Rightarrow \quad b = 0 \tag{2.103}$$

This shows that if impedance matching is used with $z_0$ being constant, then the scattering function is

$$g(s) = \frac{b(s)}{a(s)} = 0 \tag{2.104}$$

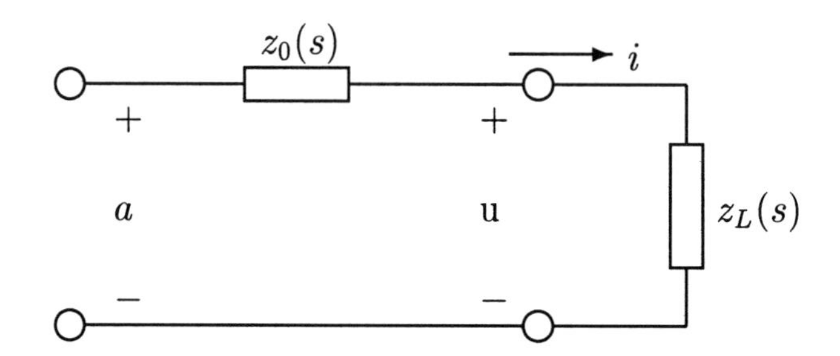


Figure 2.9: Extended one-port with a serial impedance $z_0$

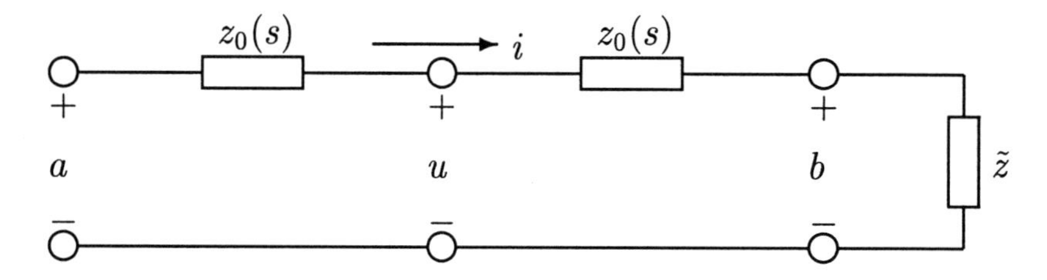


Figure 2.10: Physical interpretation of the reflected wave $b$ where $\tilde{z} = z_L(s) - z_0(s)$.

## 2.10 Feedback loop

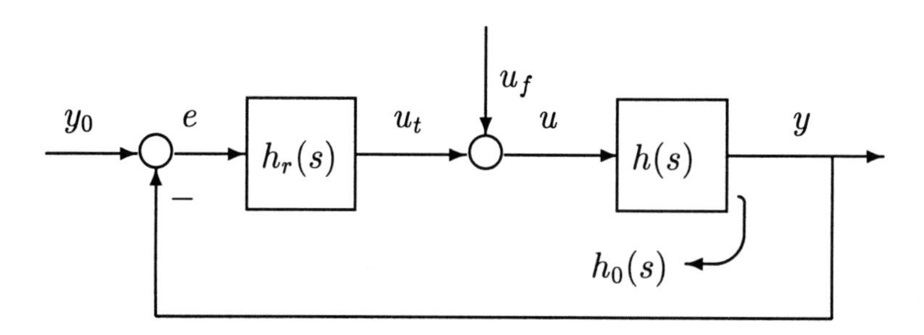


Figure 2.11: Feedback interconnection of two passive systems

A feedback interconnection of two passive linear time-invariant systems is shown in figure 2.11 where signals are given by

$$y(s) = h(s)u(s), \qquad u_t(s) = h_r(s)e(s) \tag{2.105}$$

$$u(t) = u_f(t) + u_t(t), \qquad e(t) = y_0(t) - y(t) \tag{2.106}$$

We can think of $h(s)$ as describing the plant to be controlled, and $h_r(s)$ as describing the feedback controller. Here $u_t$ is the feedback control and $u_f$ is

the feedforward control. We assume that the plant $h(s)$ and that the feedback controller $h_r(s)$ are strictly passive with finite gain. Then, as shown in section 2.5 we have $\angle|h_0(j\omega)| < 180°$ where $h_0(s) := h(s)h_r(s)$ is the loop transfer function, and the system is BIBO stable.
A change of variables in now introduced to bring the system into a scattering formulation. The new variables are

$$a := y + u \quad \text{and} \quad b := y - u$$

for the plant and

$$a_r := u_t + e \quad \text{and} \quad b_r := u_t - e$$

for the feedback controller. In addition input variables

$$a_0 := y_0 + u_f \quad \text{and} \quad b_0 := y_0 - u_f$$

are defined. We find that

$$a_r = u_t + y_0 - y = u - u_f + y_0 - y = b_0 - b \tag{2.107}$$

and

$$b_r = u_t - y_0 + y = u - u_f - y_0 + y = a - a_0 \tag{2.108}$$

The associated scattering functions are

$$g(s) := \frac{h(s) - 1}{1 + h(s)} \quad \text{and} \quad g_r(s) := \frac{h_r(s) - 1}{1 + h_r(s)}$$

Now, $h(s)$ and $h_r(s)$ are passive by assumption, and as a consequence they cannot have poles in $\mathbf{Re}[s] > 0$. Then it follows that $g(s)$ and $g_r(s)$ cannot have poles in $\mathbf{Re}[s] > 0$ because $1+h(s)$ is the characteristic equations for $h(s)$ with a unity negative feedback, which obviously is a stable system. Similar arguments apply for $1 + h(s)$.
The system can then be represented as in figure 2.12 where

$$b(s) = g(s)a(s), \qquad b_r(s) = g_r(s)a_r(s) \tag{2.109}$$

$$a(t) = b_r(t) + a_0(t), \qquad a_r(t) = b_0(t) - b(t) \tag{2.110}$$

In the passivity setting, stability was ensured when two passive systems were interconnected in a feedback structure because the loop transfer function $h_0(j\omega)$ had a phase limitation so that $\angle h_0(j\omega) > -180°$. We would now like to check if there is an interpretation for the scattering formulation that is equally simple. This indeed turns out to be the case.
We introduce the loop transfer function

$$g_0(s) := g(s)g_r(s) \tag{2.111}$$

of the scattering formulation. The function $g_0(s)$ cannot have poles in $\mathbf{Re}[s] > 0$ as $g(s)$ and $g_r(s)$ have no poles in $\mathbf{Re}[s] > 0$ by assumption. Then we have from theorem 2.6

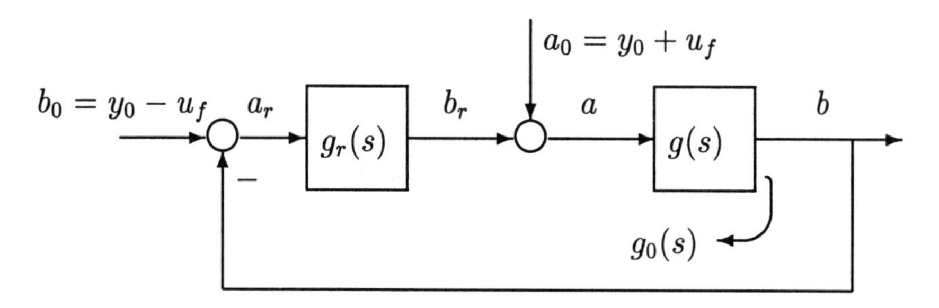


Figure 2.12: Equivalent system

1. $|g(j\omega)| \leq 1$ for all $\omega$ because $h(s)$ is passive.
2. $|g_r(j\omega)| < 1$ for all $\omega$ because $h_r(s)$ is strictly passive with finite gain.

As a consequence of this

$$|g_0(j\omega)| < 1 \tag{2.112}$$

for all $\omega$, and according to the Nyquist stability criterion the system is BIBO stable.

## 2.11 Bounded real and positive real

Bounded real and positive real are two important properties of transfer functions related to passive systems that are linear and time-invariant. We will in this section show that a linear time-invariant system is passive if and only if the transfer function of the system is positive real. To do this we first show that a linear time-invariant system is passive if and only if the scattering function, which is the transfer function of the wave variables, is bounded real. Then we show that the scattering function is bounded real if and only if the transfer function of the system is positive real. We will also discuss different aspects of these results for rational and irrational transfer functions.

We consider a linear time-invariant system $y(s) = h(s)u(s)$ with input $u$ and output $y$. The incident wave is denoted $a := y + u$, and the reflected wave is denoted $b := y - u$. The scattering function $g(s)$ is given by

$$g(s) = \frac{h(s) - 1}{1 + h(s)} \tag{2.113}$$

and satisfies $b(s) = g(s)a(s)$. We note that

$$u(t)y(t) = \frac{1}{4}[a^2(t) - b^2(t)] \tag{2.114}$$

For linear time-invariant systems the properties of the system do not depend on the initial conditions, as opposed to nonlinear systems. We therefore assume for simplicity that the initial conditions are selected so that the energy function $V(t)$ is zero for initial time, that is $V(0) = 0$. The passivity inequality is then

$$0 \le V(T) = \int_0^T u(t)y(t)dt = \frac{1}{4}\int_0^T [a^2(t) - b^2(t)]dt \tag{2.115}$$

The properties bounded real and positive real will be defined for functions that are analytic in the open right half plane $\mathbf{Re}[s] > 0$. We recall that a function $f(s)$ is *analytic* in a domain only if it is defined and infinitely differentiable for all points in the domain. A point where $f(s)$ ceases to be analytic is called a *singular point*, and we say that $f(s)$ has a *singularity* at this point. If $f(s)$ is rational, then $f(s)$ has a finite number of singularities, and the singularities are called *poles*. The poles are the roots of the denominator polynomial $R(s)$ if $f(s) = Q(s)/R(s)$, and a pole is said to be simple pole if it is not a multiple root in $R(s)$.

**Definition 2.2** A function $g(s)$ is said to be bounded real if

1. $g(s)$ is analytic in $\mathbf{Re}[s] > 0$.
2. $g(s)$ is real for real and positive $s$.
3. $|g(s)| \le 1$ for all $\mathbf{Re}[s] > 0$.

♠♠

**Theorem 2.7** Consider a linear time-invariant system described by $y(s) = h(s)u(s)$, and the associated scattering function $a = y + u$, $b = y - u$ and $b(s) = g(s)a(s)$ where

$$g(s) = \frac{h(s) - 1}{1 + h(s)} \tag{2.116}$$

which satisfies $b(s) = g(s)a(s)$ $a = y + u$ and $b = y - u$. Then the system $y(s) = h(s)u(s)$ is passive if and only if $g(s)$ is bounded real.

♠♠

**Proof**
Assume that $y(s) = h(s)u(s)$ is passive. Then (2.115) implies that

$$\int_0^T a^2(t)dt \ge \int_0^T b^2(t)dt \tag{2.117}$$

for all $T \ge 0$. It follows that $g(s)$ cannot have any singularities in $\mathbf{Re}[s] > 0$ as this would result in exponential growth in $b(t)$ for any small input $a(t)$. Thus, $g(s)$ must satisfy condition 1 in the definition of bounded real.
Let $\sigma_0$ be an arbitrary real and positive constant, and let $a(t) = e^{\sigma_0 t}\mathbf{1}(t)$ where $\mathbf{1}(t)$ is the unit step function. Then the Laplace transform of $a(t)$ is

$a(s) = \frac{1}{s-\sigma_0}$, while $b(s) = \frac{g(s)}{s-\sigma_0}$. Suppose that the system is not initially excited so that the inverse Laplace transform for rational $g(s)$ gives

$$b(t) = \sum_{i=1}^{n} \left( \text{Res}_{s=s_i} \frac{g(s)}{s-\sigma_0} \right) e^{s_i t} + \left( \text{Res}_{s=\sigma_0} \frac{g(s)}{s-\sigma_0} \right) e^{\sigma_0 t}$$

where $s_i$ are the poles of $g(s)$, that satisfy $\mathbf{Re}\,[s_i] < 0$, and $\text{Res}_{s=\sigma_0} \frac{g(s)}{s-\sigma_0} = g(\sigma_0)$. When $t \to \infty$ the term including $e^{\sigma_0 t}$ will dominate the terms including $e^{s_i t}$, and $b(t)$ will tend to $g(\sigma_0)e^{\sigma_0 t}$. The same limit for $b(t)$ will also be found for irrational $g(s)$. As $a(t)$ is real, it follows that $g(\sigma_0)$ is real, and it follows that $g(s)$ must satisfy condition 2 in the definition of bounded real.
Let $s_0 = \sigma_0 + j\omega_0$ be an arbitrary point in $\mathbf{Re}[s] > 0$, and let the input be $a(t) = \mathbf{Re}[e^{s_0 t}\mathbf{1}(t)]$. Then $b(t) \to \mathbf{Re}[g(s_0)e^{s_0 t}]$ as $t \to \infty$ and the power

$$P(t) := \frac{1}{4}[a^2(t) - b^2(t)] \tag{2.118}$$

will tend to

$$P(t) = \frac{1}{4}[e^{2\sigma_0 t}\cos^2 \omega_0 t - |g(s_0)|^2 e^{2\sigma_0 t}\cos^2(\omega_0 t + \phi)]$$

where $\phi = \arg[g(s_0)]$. This can be rewritten using $\cos^2\alpha = \frac{1}{2}(1 + \cos 2\alpha)$, and the result is

$$\begin{aligned} 8P(t) &= (1 + \cos 2\omega_0 t)e^{2\sigma_0 t} - |g(s_0)|^2[1 + \cos(2\omega_0 t + 2\phi)]e^{2\sigma_0 t} \\ &= [1 - |g(s_0)|^2]e^{2\sigma_0 t} + \mathbf{Re}[\left(1 - g(s_0)^2\right) e^{2s_0 t}] \end{aligned}$$

In this expression $s_0$ and $\sigma_0$ are constants, and we can integrated $P(t)$ to get the energy function $V(T)$:

$$\begin{aligned} V(T) &= \int_{-\infty}^{T} P(t)dt \\ &= \tfrac{1}{16\sigma_0}[1 - |g(s_0)|^2]e^{2\sigma_0 T} + \tfrac{1}{16}\mathbf{Re}\{\tfrac{1}{s_0}[1 - g(s_0)^2]e^{2s_0 T}\} \end{aligned}$$

First it is assumed that $\omega_0 \neq 0$. Then $\mathbf{Re}\{\frac{1}{s_0}[1 - g(s_0)^2]e^{2s_0 T}\}$ will be a sinusoidal function which becomes zero for certain values of $T$. For such values of $T$ the condition $V(T) \geq 0$ implies that

$$\frac{1}{16\sigma_0}[1 - |g(s_0)|^2]e^{2\sigma_0 T} \geq 0$$

which implies that

$$1 - |g(s_0)|^2 \geq 0$$

Next it is assumed that $\omega_0 = 0$ such that $s_0 = \sigma_0$ is real. Then $g(s_0)$ will be real, and the two terms in $V(T)$ become equal. This gives

$$0 \leq V(T) = \frac{1}{8\sigma_0}[1 - g^2(s_0)]e^{2\sigma_0 T}$$

and with this it is established that for all $s_0$ in $\mathbf{Re}[s] > 0$ we have

$$1 - |g(s_0)|^2 \geq 0 \Rightarrow |g(s_0)| \leq 1$$

To show the converse we assume that $g(s)$ is bounded real and consider

$$g(j\omega) = \lim_{\substack{\sigma \to 0 \\ \sigma > 0}} g(\sigma + j\omega) \tag{2.119}$$

Because $g(s)$ is bounded and analytic for all $\mathbf{Re}\,[s] > 0$ it follows that this limit exists for all $\omega$, and moreover

$$|g(j\omega)| \leq 1$$

Then it follows from Parseval's theorem that with $a_T$ being the truncated version of $a$ we have

$$\begin{aligned} 0 &\leq \tfrac{1}{8\pi} \textstyle\int_{-\infty}^{\infty} |a_T(j\omega)|^2 \left(1 - |g(j\omega)|^2\right) d\omega \\ &= \tfrac{1}{4} \textstyle\int_0^T [a^2(t) - b^2(t)] dt \\ &= \textstyle\int_0^T u(t) y(t) dt \end{aligned}$$

which shows that the system must be passive. ♠♠

Define the contour $C$ which encloses the right half plane as shown in figure 2.13. The maximum modulus theorem is as follows: let $f(s)$ be a function that is analytic inside the contour $C$. Let $M$ be the upper bound on $|f(s)|$ on $C$. Then $|f(s)| \leq M$ inside the contour, and equality is achieved at some point inside $C$ if and only if $f(s)$ is a constant. This means that if $g(s)$ is bounded real, and $|g(s)| = 1$ for some point in $\mathbf{Re}[s] > 0$, then $|g(s)|$ achieves its maximum inside the contour $C$, and it follows that $g(s)$ is a constant in $\mathbf{Re}[s] \geq 0$. Because $g(s)$ is real for real $s > 0$, this means that $g(s) = 1$ for all $s$ in $\mathbf{Re}[s] \geq 0$. In view of this $[1 - g(s)]^{-1}$ has singularities in $\mathbf{Re}[s] > 0$ if and only if $g(s) = 1$ for all $s$ in $\mathbf{Re}[s] \geq 0$.

If $g(s)$ is assumed to be a rational function the maximum modulus theorem can be used to reformulate the condition on $|g\,(s)|$ to be a condition on $|g\,(j\omega)|$. The reason for this is that for a rational transfer function satisfying $|g(j\omega)| \leq 1$ for all $\omega$ will also satisfy

$$\lim_{\omega \to \infty} |g(j\omega)| = \lim_{|s| \to \infty} |g(s)| \tag{2.120}$$

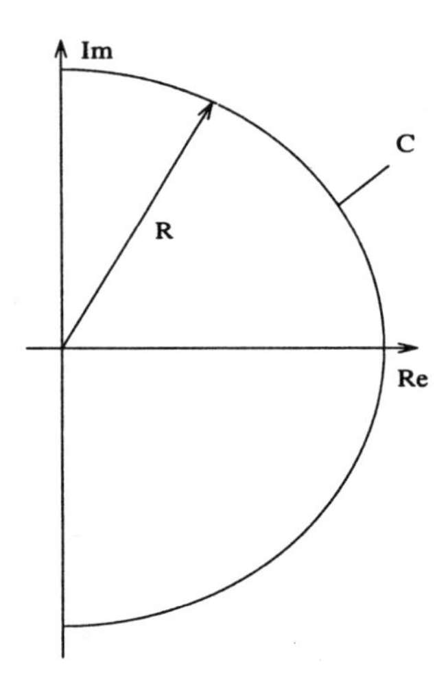


Figure 2.13: Contour in the right half plane.

Therefore, for a sufficiently large contour $C$ we have that $|g(j\omega)| \leq 1$ implies $|g(s)| \leq 1$ for all $\mathbf{Re}[s] > 0$ whenever $g(s)$ is rational. This leads to the following result:

**Theorem 2.8** A rational function $g(s)$ is bounded real if and only if

1. $g(s)$ has no poles in $\mathbf{Re}[s] \geq 0$.
2. $|g(j\omega)| \leq 1$ for all $\omega$.

♠♠

**Definition 2.3** A transfer function $h(s)$ is said to be positive real (PR) if

1. $h(s)$ is analytic in $\mathbf{Re}[s] > 0$.
2. $h(s)$ is real for positive real $s$.
3. $\mathbf{Re}[h(s)] \geq 0$ for all $\mathbf{Re}[s] > 0$.

♠♠

The last condition above is illustrated in figure 2.14 where the Nyquist plot of a PR transfer function $H(s)$ is shown.

**Theorem 2.9** Consider the linear time-invariant system $y(s) = h(s)u(s)$, and the scattering formulation $a = y + u$, $b = y - u$ and $b(s) = g(s)a(s)$ where

$$g(s) = \frac{h(s) - 1}{1 + h(s)} \tag{2.121}$$

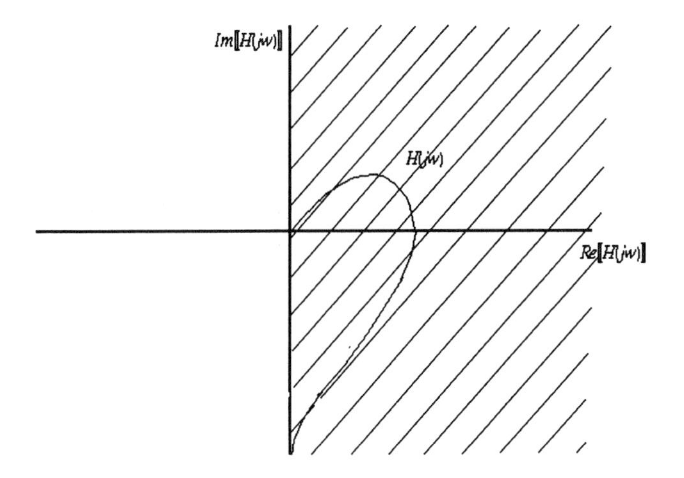


Figure 2.14: Positive real transfer function.

Assume that $g(s) \neq 1$ for all $\mathbf{Re}[s] > 0$. Then $h(s)$ is positive real if and only if $g(s)$ is bounded real. ♠♠

**Proof**
Assume that $g(s)$ is bounded real and that $g(s) \neq 1$ for all $\mathbf{Re}[s] > 0$. Then $[1 - g(s)]^{-1}$ exists for all $s$ in $\mathbf{Re}[s] > 0$. From (2.121) we find that

$$h(s) = \frac{1 + g(s)}{1 - g(s)} \tag{2.122}$$

where $h(s)$ is analytic in $\mathbf{Re}[s] > 0$ as $g(s)$ is analytic in $\mathbf{Re}[s] > 0$, and $[1 - g(s)]^{-1}$ is nonsingular by assumption in $\mathbf{Re}[s] > 0$. To show that $\mathbf{Re}[h(s)] \geq 0$ for all $\mathbf{Re}[s] > 0$ the following computation is used:

$$\begin{aligned} 2\mathbf{Re}[h(s)] &= h^*(s) + h(s) \\ &= \tfrac{1+g^*(s)}{1-g^*(s)} + \tfrac{1+g(s)}{1-g(s)} \\ &= 2\tfrac{1-g^*(s)g(s)}{[1-g^*(s)][1-g(s)]} \end{aligned} \tag{2.123}$$

We see that $\mathbf{Re}[h(s)] \geq 0$ for all $\mathbf{Re}[s] > 0$ whenever $g(s)$ is bounded real.
Next assume that $h(s)$ is positive real. Then $h(s)$ is analytic in $\mathbf{Re}[s] > 0$, and $[1 + h(s)]$ is nonsingular in $\mathbf{Re}[s] > 0$ as $\mathbf{Re}[h(s)] \geq 0$ in $\mathbf{Re}[s] > 0$. It follows that $g(s)$ is analytic in $\mathbf{Re}[s] > 0$. From (2.123) it is seen that $|g(s)| \leq 1$ in $\mathbf{Re}[s] > 0$, it follows that $g(s)$ is bounded real. ♠♠
From theorem 2.7 and theorem 2.9 it follows that:

**Corollary 2.3** A system with transfer function $h(s)$ is passive if and only if the transfer function $h(s)$ is positive real. ♠♠

**Example 2.8** A fundamental result in electrical circuit theory is that if the transfer function $h(s)$ is rational and positive real, then there exists an electrical one-port built from resistors, capacitors and inductors so that $h(s)$ is the impedance of the one-port [46, p. 815]. If $e$ is the voltage over the one-port and $i$ is the current into the one-port, then $e(s) = h(s)i(s)$. The system with input $i$ and output $e$ must be passive because the total stored energy of the circuit must satisfy

$$\dot{V}(t) = e(t)i(t) - g(t) \tag{2.124}$$

where $g(t)$ is the dissipated energy.

**Example 2.9** The transfer function

$$h(s) = \frac{1}{\tanh s} \tag{2.125}$$

is irrational, and positive realness of this transfer function cannot be established from conditions on the frequency response $h(j\omega)$.
We note that $\tanh s = \sinh s / \cosh s$, where $\sinh s = \frac{1}{2}(e^s - e^{-s})$ and $\cosh s = \frac{1}{2}(e^s + e^{-s})$. First we investigate if $h(s)$ is analytic in the right half plane. The singularities are given by

$$\sinh s = 0 \Rightarrow e^s - e^{-s} = 0 \Rightarrow e^s(1 - e^{-2s}) = 0$$

Here $|e^s| \geq 1$ for $\mathbf{Re}[s] > 0$, while

$$e^s(1 - e^{-2s}) = 0 \Rightarrow e^{-2s} = 1$$

Therefore the singularities are found to be

$$s_k = jk\pi, \quad k \in \{0, \pm 1, \pm 2 \ldots\} \tag{2.126}$$

which are on the imaginary axis. This means that $h(s)$ is analytic in $\mathbf{Re}[s] > 0$. Obviously, $h(s)$ is real for real $s > 0$.
Finally we check if $\mathbf{Re}\,[h(s)]$ is positive in $\mathbf{Re}[s] > 0$. Let $s = \sigma + j\omega$. Then

$$\begin{aligned} \cosh s &= \tfrac{1}{2}[e^{\sigma}(\cos\omega + j\sin\omega) + e^{-\sigma}(\cos\omega - j\sin\omega)] \\ &= \cosh\sigma\cos\omega + j\sinh\sigma\sin\omega \end{aligned}$$

while

$$\sinh s = \sinh\sigma\cos\omega + j\cosh\sigma\sin\omega \tag{2.127}$$

This gives

$$\mathbf{Re}[h(s)] = \frac{\cosh\sigma\sinh\sigma}{|\sinh s|^2} > 0, \quad \mathbf{Re}\,[s] > 0 \tag{2.128}$$

where it is used that $\sigma = \mathbf{Re}\,[s]$, and the positive realness of $h(s)$ has been established.

♠♠

Consider a linear system represented by a rational function $H(s)$ of the complex variable $s = \sigma + j\omega$

$$H(s) = \frac{b_m s^m + \ldots + b_0}{s^n + a_{n-1}s^{n-1} + \ldots + a_0} \tag{2.129}$$

where $a_i, b_i \in I\!R$ are the system parameters $n$ is the order of the system and $n^* = n - m$ is the relative degree. For rational transfer functions it is possible to find conditions on the frequency response $h(j\omega)$ for the transfer function to be positive real. The result is presented in the following theorem:

**Theorem 2.10** A rational function $h(s)$ is positive real if and only if

1. $h(s)$ has no poles in $\mathbf{Re}[s] > 0$.
2. $\mathbf{Re}[h(j\omega)] \geq 0$ for all $\omega$ such that $j\omega$ is not a pole in $h(s)$.
3. If $s = j\omega_0$ is a pole in $h(s)$, then it is a simple pole, and $\omega_0$ is finite, then the residual

$$\mathrm{Res}_{s=j\omega_0} h(s) = \lim_{s \to j\omega_0} (s - j\omega_0)h(s)$$

is real and positive. If $\omega_0$ is infinite, then the limit

$$R_\infty := \lim_{\omega \to \infty} \frac{h(j\omega)}{j\omega}$$

is real and positive.

♠♠

**Proof**

The proof can be established by showing that conditions 2 and 3 in this theorem are equivalent to the condition

$$\mathbf{Re}[h(s)] \geq 0 \tag{2.130}$$

for all $\mathbf{Re}[s] > 0$ for $h(s)$ with no poles in $\mathbf{Re}[s] > 0$.
First assume that conditions 2 and 3 hold. We use a contour $C$ as shown in figure 2.15 which goes from $-j\Omega$ to $j\Omega$ along the $j\omega$ axis with small semicircular indentations into the right half plane around points $j\omega_0$ that are poles of $h(s)$. The contour $C$ is closed with a semicircle into the right half plane. On the part of $C$ that is on the imaginary axis $\mathbf{Re}[h(s)] \geq 0$ by assumption. On the small indentations

$$h(s) \approx \frac{\mathrm{Res}_{s=j\omega_0} h(s)}{s - j\omega_0} \tag{2.131}$$

As $\mathbf{Re}[s] \geq 0$ on the small semi-circles and $\mathrm{Res}_{s=j\omega_0} h(s)$ is real and positive according to condition 3, it follows that $\mathbf{Re}[h(s)] \geq 0$ on these semi-circles.

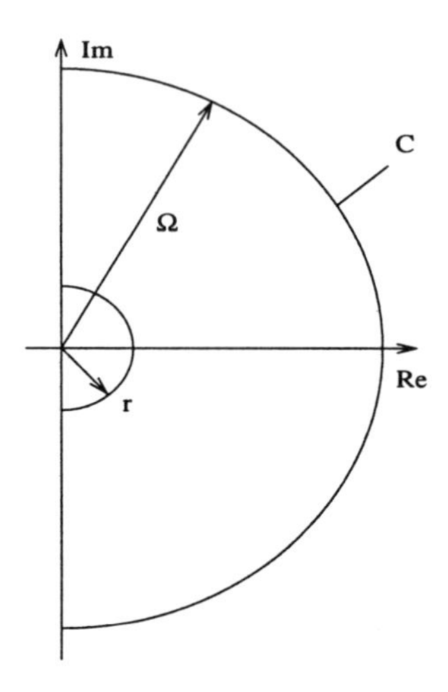


Figure 2.15: Contour $C$ of $h(s)$ in the right half plane.

On the large semi-circle into the right half plane with radius $\Omega$ we also have $\mathbf{Re}[h(s)] \geq 0$ and the value is a constant equal to $\lim_{\omega \to \infty} \mathbf{Re}[h(j\omega)]$, unless $h(s)$ has a pole at infinity at the $j\omega$ axis, in which case $h(s) \approx sR_\infty$ on the large semi-circle. Thus we may conclude that $\mathbf{Re}[h(s)] \geq 0$ on $C$. Define the function

$$f(s) = e^{-\mathbf{Re}[h(s)]}$$

Then $|f(s)| \leq 1$ on $C$, and in view of the maximum modulus theorem, $|f(s)| \leq 1$ for all $s \in \mathbf{Re}[s] > 0$. It follows that $\mathbf{Re}[h(s)] \geq 0$ in $\mathbf{Re}[s] > 0$, and the result is shown.

Next assume that $\mathbf{Re}[h(s)] \geq 0$ for all $\mathbf{Re}[s] > 0$. Then condition 2 follows because

$$h(j\omega) = \lim_{\substack{\sigma \to 0 \\ \sigma > 0}} h(\sigma + j\omega)$$

exists for all $\omega$ such that $j\omega$ is not a pole in $h(s)$. To show condition 3 we assume that $\omega_0$ is a pole of multiplicity $m$ for $h(s)$. On the small indentation with radius $r$ into the right half plane we have $s - j\omega_0 = re^{j\theta}$ where $-\pi/2 \leq \theta \leq \pi/2$. Then

$$h(s) \approx \frac{\mathrm{Res}_{s=j\omega_0} h(s)}{r^m e^{jm\theta}} = \frac{\mathrm{Res}_{s=j\omega_0} h(s)}{r^m} e^{-jm\theta} \tag{2.132}$$

Clearly, here it is necessary that $m = 1$ to achieve $\mathbf{Re}[h(s)] \geq 0$ because the term $e^{-jm\theta}$ gives an angle from $-m\pi/2$ to $m\pi/2$ in the complex plane. Moreover, it is necessary that $\mathrm{Res}_{s=j\omega_0} h(s)$ positive and real because $e^{-jm\theta}$ gives and angle from $-\pi/2$ to $\pi/2$ when $m = 1$. The result follows. ♠♠

## 2.12 Lossless transmission lines

The dynamics of a lossless transmission line is given, as we will see in the next section, by the wave equation

$$\frac{\partial^2 v}{\partial t^2}(x,t) = c^2 \frac{\partial^2 v}{\partial x^2}(x,t) \tag{2.133}$$

where $v(x,t)$ is the voltage along the line. The wave equations appear in many applications, and the results for lossless transmission lines that are presented in the following will have relevance to other interesting control problems. In particular the wave equation leads to irrational transfer functions, and we will show how passivity can be used for such systems.

### 2.12.1 Dynamic model

A dynamic model for a transmission line is obtained from Kirchhoff's voltage and current law for a length element $dx$ of the transmission line. The transmission line is modelled by a distributed serial inductance $Ldx$ and a parallel capacitance $Cdx$. The voltage along the line is denoted $v(x,t)$ while the current is denoted $i(x,t)$ where $x$ is the coordinate along the line.
The dynamics is given by

$$\begin{aligned} \tfrac{\partial v}{\partial x}(x,t) &= -L\tfrac{\partial i}{\partial t}(x,t) \\ \tfrac{\partial i}{\partial x}(x,t) &= -C\tfrac{\partial v}{\partial t}(x,t) \end{aligned}$$

or, in term of the Laplace transform

$$\begin{aligned} \tfrac{\partial v}{\partial x}(x,s) &= -Lsi(x,t) \\ \tfrac{\partial i}{\partial x}(x,s) &= -Csv(x,t) \end{aligned}$$

We introduce the wave variables $a$ and $b$ defined by

$$\begin{aligned} a(x,s) &= v(x,s) + z_0 i(x,s) \\ b(x,s) &= v(x,s) - z_0 i(x,s) \end{aligned}$$

where

$$z_0 = \sqrt{\frac{L}{C}} \tag{2.134}$$

is the characteristic impedance of the transmission line. We also define the wave propagation speed

$$c = \frac{1}{\sqrt{LC}} \tag{2.135}$$

We then find that the wave variables satisfy

$$\frac{\partial a}{\partial x}(x,s) = -\frac{s}{c}a(x,s)$$

$$\frac{\partial b}{\partial x}(x,s) = \frac{s}{c}b(x,s)$$

which has solutions

$$a(x,s) = a(0,s)\exp\left[-\frac{x}{c}s\right]$$

$$b(x,s) = b(\ell,s)\exp\left[-\frac{\ell - x}{c}s\right]$$

Inverse Laplace transformation then gives

$$a(x,t) = a(0,t - \frac{x}{c})$$

$$b(x,t) = b(\ell,t - \frac{\ell - x}{c})$$

The physical interpretation of this is that $a$ is a wave moving in the positive $x$ direction with velocity $c$, and $b$ is a wave moving in the negative $x$ direction with velocity $c$.
The voltage $v$ and current $i$ are found from

$$v(x,s) = \frac{1}{2}\left(a(x,s) + b(x,s)\right)$$

$$i(x,s) = \frac{1}{2z_0}\left(a(x,s) - b(x,s)\right)$$

With the boundary conditions $v_2 = v(\ell,s)$ and $i_2 = i(0,s)$, and with $\sinh x = \frac{1}{2}(e^x - e^{-x})$, $\cosh x = \frac{1}{2}(e^x + e^{-x})$ and $\tanh x = \sinh x/\cosh x$.This gives the following two-port description of the lossless transmission line

$$\begin{pmatrix} v_2(s) \\ i_2(s) \end{pmatrix} = \begin{pmatrix} \cosh[Ts] & -z_0\sinh[Ts] \\ -\frac{1}{z_0}\sinh[Ts] & \cosh[Ts] \end{pmatrix} \begin{pmatrix} v_1(s) \\ i_1(s) \end{pmatrix} \tag{2.136}$$

where $T := \ell/c$ is the wave propagation time along the line from $x = 0$ to $x = \ell$.
If a passive load impedance $z_L(s)$ is connected at $x = \ell$ so that $v_2(s) = z_L(s)i_2(s)$, then the transfer function from $i_1(s)$ to $v_1(s)$ is found to be

$$\frac{v_1}{i_1}(s) = z_0(s)\frac{z_L(s)\cosh[Ts] + z_0(s)\sinh[Ts]}{z_0(s)\sinh[Ts] + z_L(s)\cosh[Ts]} \tag{2.137}$$

In terms of wave variables the load impedance is described by $b_2 = g_L(s)a_2$ where $a_2 = a(\ell, s)$ and $b_2 = b(\ell, s)$ and

$$g_L(s) = \frac{\frac{z_L(s)}{z_0} - 1}{1 + \frac{z_L(s)}{z_0}} \tag{2.138}$$

is the scattering function corresponding to the transfer function $z_L(s)/z_0$. Because $z_L(s)$ is positive real, it follows that $g_L(s)$ is bounded real. We find that

$$\begin{aligned} b_1(s) &= \exp[-Ts]\, b_2(s) = \exp[-Ts]\, g_L(s) a_2(s) \\ &= \exp[-Ts]\, g_L(s) \exp[-Ts]\, a_1(s) \end{aligned}$$

which leads to

$$\frac{b_1}{a_1}(s) = \exp[-2Ts]\, g_L(s) \tag{2.139}$$

which is a time delay of $2T$ which is the wave propagation time from $x = 0$ to $x = \ell$ and back, in series with the load scattering function. It is seen that $\frac{b_1}{a_1}(s)$ is bounded real as it is the product of two bounded real functions. This means that $\frac{v_1}{i_1}(s)$ is positive real, which is not immediately obvious from (2.137).
We consider the following three important cases:

1. First consider impedance matching for the transmission line, which is achieved with $z_L(s) = z_0$. Then

$$\frac{v_1}{i_1}(s) = z_0 \tag{2.140}$$

whereas $g_L(s) = 0$, and accordingly

$$\frac{b_1}{a_1}(s) = 0 \tag{2.141}$$

This shows that all the wave energy is absorbed in the load impedance.

2. Next assume that the load is a short circuit, which means that $v_2 = 0$ and $z_L(s) = 0$. Then

$$\frac{v_1}{i_1}(s) = z_0 \tanh[Ts] \tag{2.142}$$

while $g_L(s) = -1$ and

$$\frac{b_1}{a_1}(s) = -\exp[-2Ts] \tag{2.143}$$

In this case the wave is reflected with opposite sign, and the wave transfer function is a sign reversal and a time delay of $2T$.

3. Finally, assume that the load is an open circuit, so that $i_2 = 0$ and $z_L(s) = \infty$. Then

$$\frac{v_1}{i_1}(s) = z_0 \frac{1}{\tanh[Ts]} \tag{2.144}$$

while $g_L(s) = 1$ and

$$\frac{b_1}{a_1}(s) = \exp[-2Ts] \tag{2.145}$$

It is seen that the wave is reflected with, and the wave transfer function is a time delay of $2T$.

### 2.12.2 Energy considerations

We have established that the transfer function from $i_1(s)$ to $v_1(s)$ is positive real whenever the load impedance $z_L(s)$ is positive real. This was done using wave variables. The same result follows easily from an energy consideration where $V$ denotes the energy stored in the transmission line and the load impedance. The time derivative of the energy function as the system evolves is

$$\dot{V}(t) = i_1(t)v_1(t) - d(t) \tag{2.146}$$

where $d(t)$ is the power dissipated in the load. This shows that the system with input $i_1$ and output $v_1$ is passive, which is equivalent to the transfer function $v_1(s)/i_1(s)$ being positive real.

## 2.13 Pipeline with compressible fluid

A pipeline of length $\ell$ and cross section $A$ is filled with a fluid of bulk modulus $\beta$. The pressure at the inlet is $p_1$, and the volumetric flow is $q_1$. At the outlet the pressure is $p_2$ and the volumetric flow is $q_2$. It is assumed that the pipeline is connected to a passive load in the sense that $p_2(s) = z_L(s)q_2(s)$ where $z_L(s)$ is positive real.

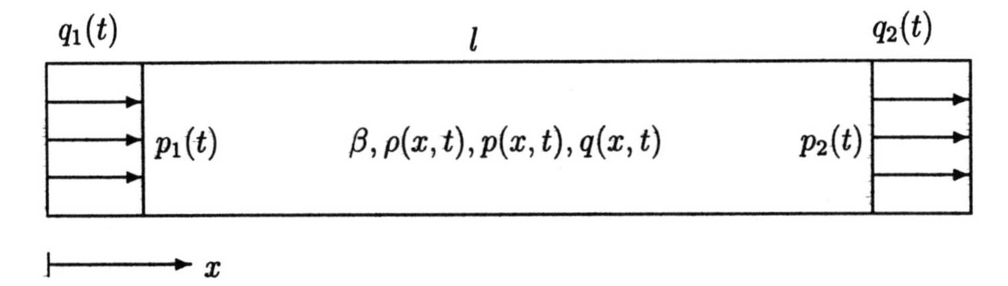


Figure 2.16: Pipeline connected to a passive load

### 2.13.1 Energy equation

The total energy in the pipeline and load is denoted $V$. This is the sum of kinetic and potential energy in the system. The power supplied from the inlet is $q_1 p_1$. The energy dynamics can then be written

$$\dot{V} = q_1 p_1 - d \tag{2.147}$$

where $d$ is the power dissipated in the system. Because it is assumed that there is no energy-generating component in the system it follows that

$$\int_0^T d(t)dt \geq 0 \tag{2.148}$$

for all $T \geq 0$. This means that the system with input $q_1$ and output $p_1$ is passive.

### 2.13.2 Dynamics

The continuity equation for a volume element $Adx$ is

$$\frac{\partial \rho}{\partial t} + \frac{q}{A}\frac{\partial \rho}{\partial x} = -\frac{\rho}{A}\frac{\partial q}{\partial x} \tag{2.149}$$

Insertion of $d\rho = (\rho/\beta)dp$ gives

$$\frac{\partial p}{\partial t} + \frac{q}{A}\frac{\partial p}{\partial x} = -cz_0\frac{\partial q}{\partial x} \tag{2.150}$$

where $c = \sqrt{\beta/\rho}$ is the speed of sound and $z_0 = \rho c/A$ is the characteristic impedance of the pipeline. The momentum equation is

$$\frac{\partial q}{\partial t} + \frac{q}{A}\frac{\partial q}{\partial x} = -\frac{c}{z_0}\frac{\partial p}{\partial x} \tag{2.151}$$

Around zero volumetric flow, that is around $q = 0$ this can be represented by

$$\frac{\partial p}{\partial t} = -cz_0\frac{\partial q}{\partial x}$$

$$\frac{\partial q}{\partial t} = -\frac{c}{z_0}\frac{\partial p}{\partial x}$$

We note in passing that this leads to the wave equation

$$\frac{\partial^2 p}{\partial t^2} = c^2\frac{\partial^2 p}{\partial x^2} \tag{2.152}$$

As for the electrical transmission lines we find that

$$\frac{p_1}{q_1}(s) = z_0\frac{z_L(s)\cosh(Ts) + z_0\sinh(Ts)}{z_0\cosh(Ts) + z_L(s)\sinh(Ts)} \tag{2.153}$$

where

$$T = \frac{\ell}{c} = \ell\sqrt{\frac{\rho}{\beta}} \tag{2.154}$$

is the wave propagation time along the pipeline. A scattering formulation is obtained by introducing $a_i = p_i + z_0 q_i$ and $b = p_i - z_0 q_i$ for $i \in \{1, 2\}$. With a positive real load impedance $z_L$ the scattering function from $a_1$ to $b_1$ becomes as for lossless electrical transmission lines

$$g_1(s) = \frac{b_1}{a_1}(s) = \exp(-2Ts) g_L(s) \tag{2.155}$$

where

$$g_L(s) = \frac{\frac{z_L(s)}{z_0} - 1}{\frac{z_L(s)}{z_0} + 1} \tag{2.156}$$

is the scattering function for the impedance $z_L(s)/z_0$. It is seen that $g_1(s)$ is bounded real for all positive real $z_L(s)$. It follows that the system with input $q_1$ and output $p_1$ is passive whenever $z_L(s)$ is positive real.
Consider the following special cases:

1. Impedance matching can be achieved with a restriction at the outlet satisfying

$$\frac{p_2}{q_2}(s) = z_0 \tag{2.157}$$

which gives $z_L = z_0$. This gives

$$g_1(s) = 0 \tag{2.158}$$

which means that all wave energy is dissipated in the restriction.

2. If the pipeline is open so that $p_2 = 0$ and $z_L = 0$ the transfer function becomes

$$\frac{p_1}{q_1}(s) = z_0 \tanh(Ts) \tag{2.159}$$

and the scattering function is

$$g_1(s) = -\exp(-2Ts) \tag{2.160}$$

In this case the wave is reflected from the load with a change of sign.

3. If the pipeline is closed at the outlet we have $q_2 = 0$ and $z_L(s) = \infty$, and

$$\frac{p_1}{q_1}(s) = \frac{z_0}{\tanh(Ts)} \tag{2.161}$$

while the scattering function is

$$g_1(s) = \exp(-2Ts) \tag{2.162}$$

This shows that the wave is reflected from the load.

## 2.14 Mechanical resonances

### 2.14.1 Motor and load with elastic transmission

An interesting and important type of system is a motor that is connected to a load with an elastic transmission. The motor has moment of inertia $J_m$, the load has moment of inertia $J_L$, while the transmission has spring constant $K$ and damper coefficient $D$.
The dynamics of the motor is given by

$$J_m\ddot{\theta}_m = T_m - T_L \tag{2.163}$$

where $\theta_m$ is the motor angle, $T_m$ is the motor torque, which is considered to be the control variable, and $T_L$ is the torque from the transmission. The dynamics of the load is

$$J_L\ddot{\theta}_L = T_L \tag{2.164}$$

The transmission torque is given by

$$T_L = -D\left(\dot{\theta}_L - \dot{\theta}_m\right) - K\left(\theta_L - \theta_m\right) \tag{2.165}$$

The load dynamics can then be written in Laplace transform form as

$$\left(J_L s^2 + Ds + K\right)\theta_L(s) = (Ds + K)\,\theta_m(s) \tag{2.166}$$

which gives

$$\frac{\theta_L}{\theta_m}(s) = \frac{1 + 2Z\frac{s}{\Omega_1}}{1 + 2Z\frac{s}{\Omega_1} + \frac{s^2}{\Omega_1^2}} \tag{2.167}$$

where

$$\Omega_1^2 = \frac{K}{J_L} \tag{2.168}$$

and

$$\frac{2Z}{\Omega_1} = \frac{D}{K} \tag{2.169}$$

By adding the dynamics of the motor and the load we get

$$J_m\ddot{\theta}_m + J_L\ddot{\theta}_L = T_m \tag{2.170}$$

which leads to

$$J_m s^2\theta_m(s) + J_L s^2 \frac{1 + 2Z\frac{s}{\Omega_1}}{1 + 2Z\frac{s}{\Omega_1} + \frac{s^2}{\Omega_1^2}}\theta_m(s) = T_m(s) \tag{2.171}$$

and from this

$$\frac{\theta_m}{T_m}(s) = \frac{1 + 2Z\frac{s}{\Omega_1} + \frac{s^2}{\Omega_1^2}}{Js^2(1 + 2\zeta\frac{s}{\omega_1} + \frac{s^2}{\omega_1^2})} \tag{2.172}$$

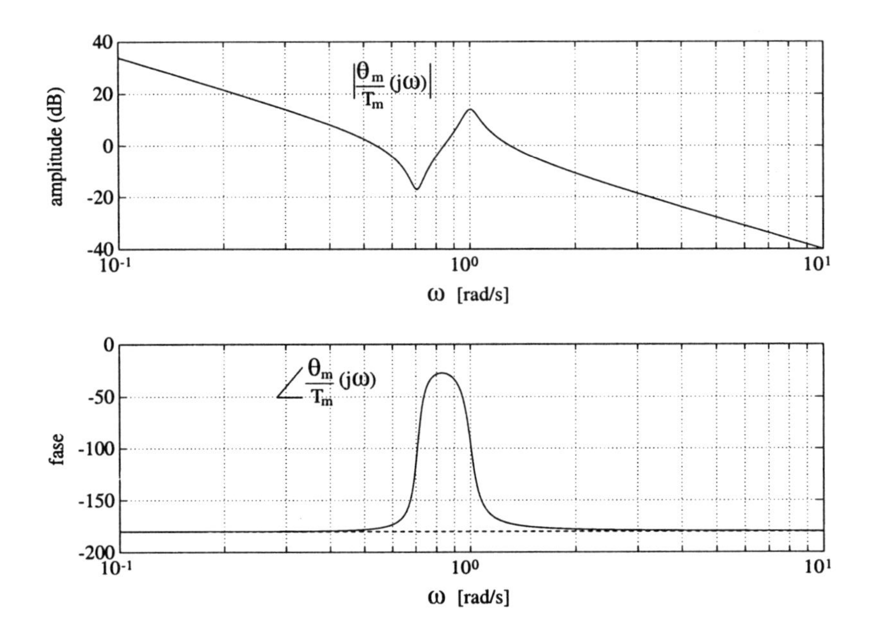


Figure 2.17: Frequency response of $\theta_m(s)/T_m(s)$.

where

$$J = J_m + J_L \tag{2.173}$$

is the total inertia of motor and load, and the resonant frequency $\omega_1$ is given by

$$\omega_1^2 = \frac{1}{1-\frac{J_L}{J}}\Omega_1^2 = \frac{J}{J_m}\Omega_1^2 \tag{2.174}$$

while the relative damping is given by

$$\zeta = \sqrt{\frac{J}{J_m}}Z \tag{2.175}$$

We note that the parameters $\omega_1$ and $\zeta$ depend on both motor and load parameters, while the parameters $\Omega_1$ and $Z$ depend only on the load.
The main observation in this development is the fact that $\Omega_1 < \omega_1$. This means that in the transfer function $\theta_m(s)/T_m(s)$ has a complex conjugated pair of zeros with resonant frequency $\Omega_1$, and a pair of poles at the somewhat higher resonant frequency $\omega_1$. The frequency response is shown in figure 2.17 when $K = 20$, $J_m = 20$, $J_L = 15$ and $D = 0.5$. Note that the elasticity does not give any negative phase contribution.

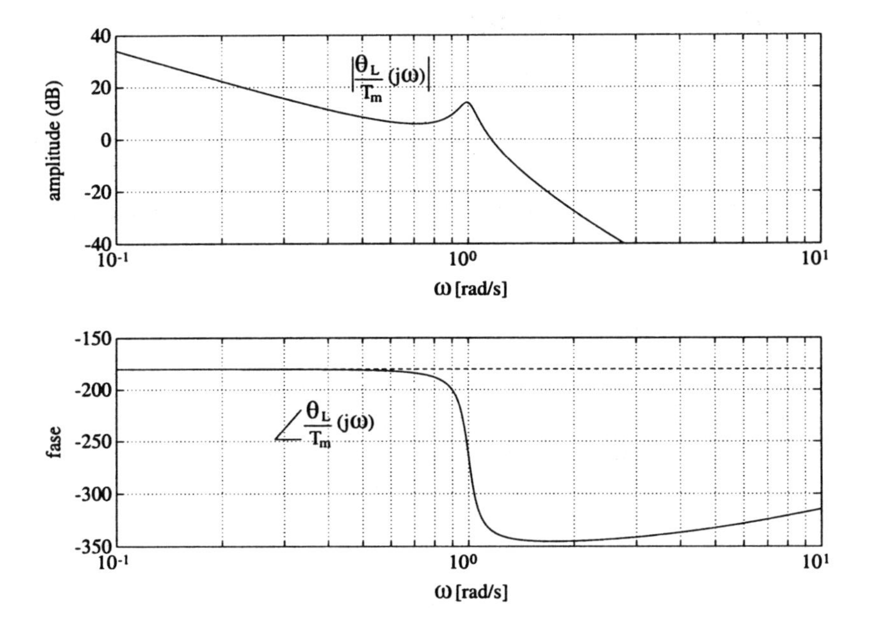


Figure 2.18: Frequency response of $\theta_L(s)/\theta_m(s)$ .

By multiplying the transfer functions $\theta_L(s)/\theta_m(s)$ and $\theta_m(s)/T_m(s)$ the transfer function

$$\frac{\theta_L}{T_m}(s) = \frac{1 + 2Z\frac{s}{\Omega_1}}{Js^2(1 + 2\zeta\frac{s}{\omega_1} + \frac{s^2}{\omega_1^2})} \tag{2.176}$$

is found from the motor torque to the load angle.
The resulting frequency response is shown in figure 2.18. In this case the elasticity results in a negative phase contribution for frequencies above $\omega_1$.

**Example 2.10** Typically the gear is selected so that $J_m = J_L$. This gives

$$\Omega_1 = \frac{1}{\sqrt{2}}\omega_1 = 0.707\omega_1 \tag{2.177}$$

**Example 2.11** Let $Z = 0.1$ and $J_m = J_L$. In this case

$$\frac{\theta_L}{T_m}(s) = \frac{1 + \frac{s}{3.535\omega_1}}{Js^2(1 + 2\zeta\frac{s}{\omega_1} + \frac{s^2}{\omega_1^2})} \tag{2.178}$$

### 2.14.2 Passivity inequality

The total energy of motor and load is given by

$$V(t) = \frac{1}{2}J_m\omega_m^2(t) + \frac{1}{2}J_L\omega_L^2(t) + \frac{1}{2}K[\theta_L(t) - \theta_m(t)]^2 \tag{2.179}$$

where $\omega_m = \dot{\theta}_m$ and $\omega_L = \dot{\theta}_L$. The rate of change of the total energy is equal to the power supplied from the control torque $T_m$ minus the power dissipated in the system. This is written

$$\dot{V}(t) = \omega_m(t)T_m(t) - D[\omega_L(t) - \omega_m(t)]^2 \tag{2.180}$$

We see that the power dissipated in the system is $D[\omega_L(t) - \omega_m(t)]^2$ which is the power loss in the damper. Clearly the energy function $V(t) \geq 0$ and the power loss satisfies $D[\Delta\omega(t)]^2 \geq 0$,. It follows that

$$\int_0^T \omega_m(t)T_m(t)dt = V(T) - V(0) + \int_0^T D[\Delta\omega(t)]^2dt \geq -V(0) \tag{2.181}$$

which implies that the system with input $T_m$ and output $\omega_m$ is passive. It follows that

$$\mathbf{Re}[h_m(j\omega)] \geq 0 \tag{2.182}$$

for all $\omega$. From energy arguments we have been able to show that

$$-180^\circ \leq \angle\frac{\theta_m}{T_m}(j\omega) \leq 0^\circ \tag{2.183}$$

which corresponds to the result derived in section 2.14.1.

## 2.15 Systems with several resonances

### 2.15.1 Passivity

Consider a motor driving $n$ inertias in a serial connection with springs and dampers. Denote the motor torque by $T_m$ and the angular velocity of the motor shaft by $\omega_m$. The energy in the system is

$$\begin{aligned} V \;=\; & \tfrac{1}{2}J_m\omega_m^2 + \tfrac{1}{2}K_{01}(\theta_m - \theta_{L1})^2 \\ & + \tfrac{1}{2}J_{L1}\omega_{L1}^2 + \tfrac{1}{2}K_{12}(\theta_{L1} - \theta_{L2})^2 + \ldots \\ & + \tfrac{1}{2}J_{L,n-1}\omega_{L,n-1}^2 + \tfrac{1}{2}K_{n-1,n}(\theta_{L,n-1} - \theta_{Ln})^2 \\ & + \tfrac{1}{2}J_{Ln}\omega_{Ln}^2 \end{aligned}$$

Clearly, $V \geq 0$. Here $J_m$ is the motor inertia, $\omega_{Li}$ is the velocity of inertia $J_{Li}$, while $K_{i-1,i}$ is the spring connecting inertia $i-1$ and $i$ and $D_{i-1,i}$ is the coefficient of the damper in parallel with $K_{i-1,i}$. The index runs over $i = 1, 2, \ldots, n$.
The system therefore satisfies the equation

$$\dot{V} = T_m \omega_m - d \tag{2.184}$$

where

$$d(t) = D_{12}(\omega_{L1} - \omega_{L2})^2 + \ldots + D_{n-1,n}(\omega_{L,n-1} - \omega_{Ln})^2 \geq 0 \tag{2.185}$$

represents the power that is dissipated in the dampers, and it follows that the system with input $T_m$ and output $\omega_m$ is passive.
If the system is linear, then the passivity implies that the transfer function

$$h_m(s) = \frac{\omega_m}{T_m}(s) \tag{2.186}$$

has the phase constraint

$$|\angle h_m(j\omega)| \leq 90^\circ \tag{2.187}$$

for all $\omega$.
It is quite interesting to note that the only information that is used to find this phase constraint on the transfer function is that the system is linear, and that the load is made up from passive mechanical components. It is not even necessary to know the order of the system dynamics, as the result holds for an arbitrary $n$.

## 2.16 Two motors driving an elastic load

In this section we will see how passivity considerations can be used as a guideline for how to control two motors that actuate on the same load through elastic interconnections consisting of inertias, springs and dampers as shown in figure 2.19.
The motors have inertias $J_{mi}$, angle $q_{mi}$ and motor torque $T_{mi}$ where $i \in \{1, 2\}$. Motor 1 is connected to the inertia $J_{L1}$ with a spring with stiffness $K_{11}$ and a damper $D_{11}$. Motor 2 is connected to the inertia $J_{L2}$ with a spring with stiffness $K_{22}$ and a damper $D_{22}$. Inertia $J_{Li}$ has angle $q_{Li}$. The two inertias are connected with a spring with stiffness $K_{12}$ and a damper $D_{12}$.
The total energy of the system is

$$\begin{aligned} V &= \tfrac{1}{2}[J_{m1}q_{m1}^2 + J_{m2}q_{m2}^2 + J_{L1}q_{L1}^2 + J_{L2}q_{L2}^2 \\ &\quad + K_{11}(q_{m1} - q_{L1})^2 + K_{22}(q_{m2} - q_{L2})^2 + K_{12}(q_{L1} - q_{L2})^2] \end{aligned}$$

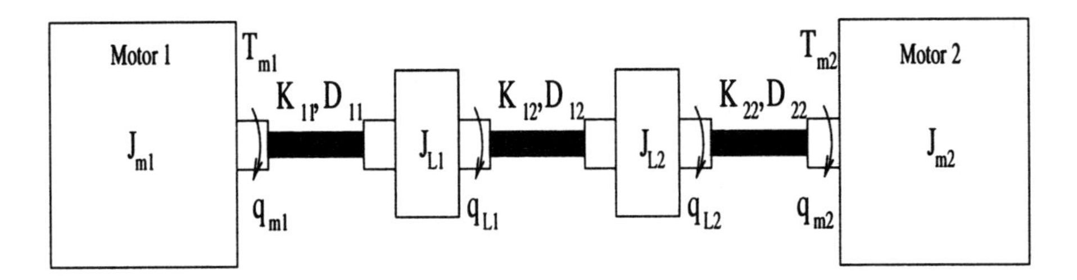


Figure 2.19: Two motors actuating on one load.

and the time derivative of the energy when the system evolves is

$$\begin{aligned}\dot{V} &= T_{m1}\dot{q}_{m1} + T_{m2}\dot{q}_{m2} - D_{11}(\dot{q}_{m1} - \dot{q}_{L1})^2 + D_{22}(\dot{q}_{m2} - \dot{q}_{L2})^2 \\ &\quad + D_{12}(\dot{q}_{L1} - \dot{q}_{L2})^2\end{aligned}$$

It is seen that the system is passive from $(T_{m1}, T_{m2})^T$ to $(\dot{q}_{m1}, \dot{q}_{m2})^T$.
The system is multivariable, with controls $T_{m1}$ and $T_{m2}$ and outputs $q_{m1}$ and $q_{m2}$. A controller can be designed using multivariable control theory, and passivity might be a useful tool in this connection. However, here we will close one control loop at a time to demonstrate that independent control loops can be constructed using passivity arguments.
The desired outputs are assumed to be $q_{m1} = q_{m2} = 0$. Consider the PD controller

$$T_{m2} = -K_{p2}q_{m2} - K_{v2}\dot{q}_{m2} \tag{2.188}$$

for motor 2 which is passive from $\dot{q}_{m2}$ to $-T_{m2}$. The mechanical analog of this controller is a spring with stiffness $K_{p2}$ and a damper $K_{v2}$ which is connected between the inertia $J_{m2}$ and a fixed point. The total energy of the system with this mechanical analog is

$$\begin{aligned}V &= \tfrac{1}{2}[J_{m1}q_{m1}^2 + J_{m2}q_{m2}^2 + J_{L1}q_{L1}^2 + J_{L2}q_{L2}^2 \\ &\quad + K_{11}(q_{m1} - q_{L1})^2 + K_{22}(q_{m2} - q_{L2})^2 \\ &\quad + K_{12}(q_{L1} - q_{L2})^2 + K_{p2}q_2^2]\end{aligned}$$

and the time derivative is

$$\begin{aligned}\dot{V} &= vT_{m1}\dot{q}_{m1} - D_{11}(\dot{q}_{m1} - \dot{q}_{L1})^2 + D_{22}(\dot{q}_{m2} - \dot{q}_{L2})^2 \\ &\quad + D_{12}(\dot{q}_{L1} - \dot{q}_{L2})^2 - K_{v2}\dot{q}_2^2\end{aligned}$$

It follows that the system with input $T_{m1}$ and output $\dot{q}_{m1}$ is passive when the PD controller is used to generate the control $T_{m2}$.

The following controller can then be used

$$T_1(s) = K_{v1}\beta\frac{1+T_is}{1+\beta T_is}\dot{q}_1(s) = K_{v1}[1+(\beta-1)\frac{1}{1+\beta T_is}]sq_1(s) \tag{2.189}$$

This is a PI controller with limited integral action if $\dot{q}_1$ is considered as the output of the system.
The resulting closed loop system will be BIBO stable independently from system and controller parameters, although in practice, unmodelled dynamics and motor torque saturation dictate some limitations on the controller parameters. As the system is linear, stability is still ensured even if the phase of the loop transfer function becomes less that $-180°$ for certain frequency ranges. Integral effect from the position can therefore be included for one of the motors, say motor 1. The resulting controller is

$$T_1(s) = K_{p1}\frac{1+T_is}{T_is}q_1(s) + K_{v1}sq_1 \tag{2.190}$$

In this case the integral time constant $T_i$ must be selected e.g. by bode diagram techniques so that stability is ensured.

## 2.17 Distributed elasticity

Systems with distributed elasticity are described by partial differential equations, as opposed to systems with lumped elasticity which are described by ordinary differential equations. Distributed elasticity leads to irrational transfer functions. These irrational transfer functions are often approximated by rational transfer function where the numerator and denominator polynomials are of infinite order, leading to the term infinite dimensional systems for systems described by irrational transfer functions.
We will show here that passivity may be a very useful tool in the control of systems with distributed elasticity. In particular we will see that passivity analysis demonstrates that collocation of sensor and actuators may simplify the control problem in many cases, as this may lead to passivity of the infinite dimensional system to be controlled.

### 2.17.1 Euler-Bernoulli beam

We consider the lateral elastic deformations of a beam using the Euler-Bernoulli beam model [162]. The beam is shown in figure 2.20.
The deformations satisfy the partial differential equation

$$c^2\frac{\partial^4 w}{\partial x^4}(x,t) + \frac{\partial^2 w}{\partial t^2}(x,t) = 0 \tag{2.191}$$

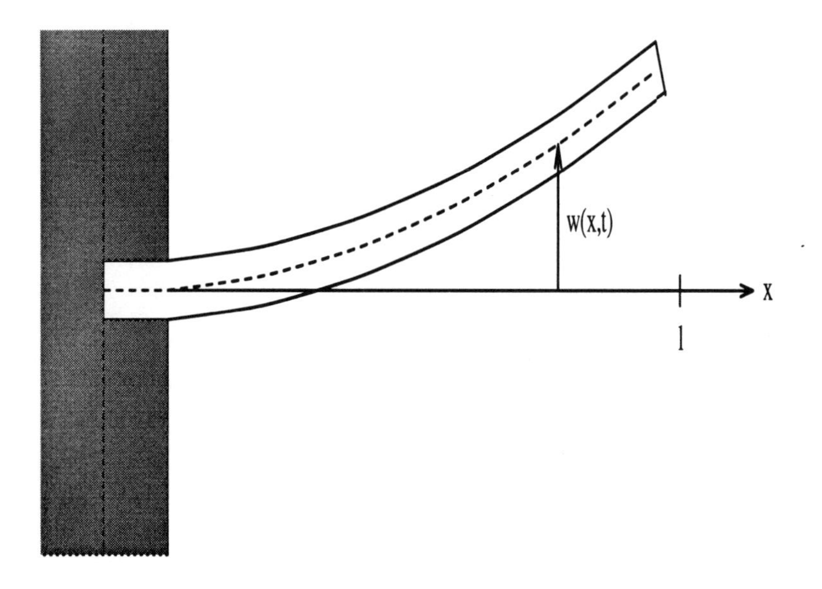


Figure 2.20: Elastic beam.

where

$$c^2 = \frac{EI}{\rho} \tag{2.192}$$

$E$ is Young's modulus, $I$ is the moment of inertia of the beam cross section, and $\rho$ is mass per length unit of the beam. This partial differential equation can be solved by separation of variables using

$$w(x,t) = \phi(x)q(t) \tag{2.193}$$

This gives

$$\frac{c^2}{\phi(x)}\frac{d^4\phi(x)}{dx^4} = -\frac{1}{q(t)}\frac{d^2q(t)}{dt^2} = \omega^2 \tag{2.194}$$

where $\omega^2$ is a constant. This implies that the spatial and time descriptions can be separated into two ordinary differential equations

$$\frac{d^4\phi(x)}{dx^4} - \beta^4\phi(x) = 0 \tag{2.195}$$

$$\frac{d^2q(t)}{dt^2} + \omega^2 q(t) = 0$$

where

$$\beta^4 = \frac{\omega^2}{c^2} = \frac{\rho\omega^2}{EI} \tag{2.196}$$

Equation (2.195) has a discrete set of eigenvalues $\beta_i$, $i \in \{1, 2, \ldots\}$ for which there exists orthogonal eigenfunctions $\phi_i(x)$ so that the eigenvalues $\beta_i$ and the corresponding eigenfunctions $\phi_i(x)$ depend on the boundary conditions, and for each $\beta_i$ there is a resonant frequency $\omega_i = \beta_i^4 c^2$. The solution can therefore be written

$$w(x,t) = \sum_{i=1}^{\infty} w_i(x,t) \tag{2.197}$$

where

$$w_i(x,t) = \phi_i(x) q_i(t) \tag{2.198}$$

and $q_i(t)$ is a solution of

$$\frac{d^2 q_i(t)}{dt^2} + \omega_i^2 q_i(t) = 0 \tag{2.199}$$

The function $w_i(x,t)$ is called mode $i$.
We now include a control force $u$ which is assumed to be perpendicular to the beam at the point $x_u$. This is modelled by a Dirac measure $\delta(x)$ in the partial differential equation:

$$\rho c^2 \frac{\partial^4 w}{\partial x^4}(x,t) + \rho \frac{\partial^2 w}{\partial t^2}(x,t) = \delta(x - x_u) u \tag{2.200}$$

To account for the control force we use the fact that the eigenfunctions are orthogonal, which in this connection means that

$$\int_0^{\ell} \rho \phi_i(x) \phi_j(x) dx = \delta_{ij} \tag{2.201}$$

where

$$\delta_{ij} = \begin{cases} 1, & i = j \\ 0, & i \neq j \end{cases} \tag{2.202}$$

is the Kronecker delta.
Using $w(x,t) = \sum_{i=1}^{\infty} q_i(t)\phi_i(x)$ the partial differential equation can be written

$$\sum_{i=1}^{\infty} [\rho c^2 q_i(t) \frac{\partial^4 \phi_i(x)}{\partial x^4}(x,t) + \rho \frac{\partial^2 q_i(t)}{\partial t^2}(x,t)\phi_i(x)] = \delta(x - x_u) u(t) \tag{2.203}$$

Insertion of (2.195) and $\omega_i^2 = c^2 \beta_i^4$ gives

$$\sum_{i=1}^{\infty} \rho \phi_i(x) [\omega_i^2 q_i(t) + \ddot{q}_i(t)] = \delta(x - x_u) u(t) \tag{2.204}$$

Multiplication by the function $\phi_j(x)$ and integration of the product over $x \in [0,\ell]$ gives

$$\int_0^\ell \phi_j(x)\sum_{i=1}^{\infty}\rho\phi_i(x)dx[\omega_i^2 q_i(t)+\ddot{q}_i(t)] = \int_0^\ell \phi_j(x)\delta(x-x_u)dxu(t) \quad (2.205)$$

Using the orthogonality of the $\phi_i(x)$ as given by (2.201) and the shifting property of the Dirac measure we get

$$\omega_i^2 q_i(t)+\ddot{q}_i(t) = \phi_i(x_u)u(t) \quad (2.206)$$

with Laplace transform

$$q_i(s) = \frac{\phi_i(x_u)}{\omega_i^2+s^2}u(s) \quad (2.207)$$

The measurement is assumed to be the velocity

$$y(t) = \dot{w}(x_y,t) \quad (2.208)$$

of the deformation at position $x_y$, which can be written

$$y(t) = \sum_{i=1}^{\infty}\dot{q}_i(t)\phi_i(x_y) \quad (2.209)$$

The resulting transfer function from $u$ to $y$ is found to be

$$\frac{y}{u}(s) = \sum_{i=1}^{\infty}\frac{s\phi_i(x_y)\phi_i(x_u)}{\omega_i^2+s^2} \quad (2.210)$$

The following observations are done:

1. If the control and the measurement are collocated, that is if $x_u = x_y = x_0$, then the transfer function $y(s)/u(s)$ is positive real as it can be written as

$$\frac{y}{u}(s) = \sum_{i=1}^{\infty}\frac{s\phi_i^2(x_0)}{\omega_i^2+s^2} \quad (2.211)$$

   which is the sum of passive transfer functions. This result agrees with an energy argument where $V$ is the sum of kinetic and potential energy. Then

$$\dot{V}(t) = y(t)u(t) - d(t) \quad (2.212)$$

   where $\int_0^T d(t)dt \geq 0$ is the dissipated energy in the system. It follows that the system with input $u$ and output $y$ is passive.

2. If measurement and control are not collocated, then $x_u \neq x_y$, and it may be that $\phi_i(x_y)$ and $\phi_i(x_u)$ have opposite signs for certain $i$. In this case the transfer function $y(s)/u(s)$ will not be positive real. This may cause difficulties in designing a controller to damp out vibrations.

3. If $\phi_i(x_u) = 0$, the control $u$ will have no influence on mode $i$. In the state-space terminology this means that mode $i$ is not controllable with the control $u$.

4. If $\phi_i(x_y) = 0$, then mode $i$ will not be noticeable in the measurement $y$. This means that mode $i$ is not observable from the measurement $y$.

**Example 2.12** Consider a homogeneous beam of aluminium with a rectangular cross section, length $\ell = 2$ m, width $b = 0.05$ m, height $h = 0.01$ m, density $\rho = b \cdot h \cdot 2700$ kg/m$^3$ $= 1.35$ kg/m, Young's modulus $E = 70 \cdot 10^9$ N/m$^2$ and moment of inertia $I = bh^3/12 = 4.167 \cdot 10^{-9}$ m$^4$. The beam is clamped at $x = 0$ and free at $x = \ell$.
The shape functions can be found to be equal to:

$$\begin{aligned}\phi_1(x) &= -0.6086 \cdot \{\cos(\beta_1 x) - \cosh(\beta_1 x) \\ &\quad -0.7341 \cdot [\sin(\beta_1 x) - \sinh(\beta_1 x)]\}\end{aligned}$$

and

$$\begin{aligned}\phi_2(x) &= -0.6086 \cdot \{\cos(\beta_2 x) - \cosh(\beta_2 x) \\ &\quad -1.0185 \cdot [\sin(\beta_2 x) - \sinh(\beta_2 x)\}\end{aligned}$$

First collocation is tried with $x_u = x_y = 2$ m. Then

$$\phi_1(x_u) = \phi_1(x_y) = 1.22, \quad \phi_2(x_u) = \phi_2(x_y) = -1.22$$

and the transfer function

$$\frac{y}{u}(s) = \frac{1.5s}{12.8^2 + s^2} + \frac{1.5s}{80.1^2 + s^2} = 3.0 \frac{s(57.4^2 + s^2)}{(12.8^2 + s^2)(80.1^2 + s^2)} \tag{2.213}$$

results, which is passive. Note that the complex conjugated zeros at $s = \pm j57.4$ are located between the poles in $s = \pm j12.8$ and $s = \pm j80.1$. A simple P controller

$$u = -K_p y \tag{2.214}$$

will give stability, with a power dissipation of $u(t)y(t) = -K_p y(t)^2$. The gain $K_p$ is only limited by noise, quantization and discretization effects.
Next noncollocation is tried with $x_u = 0.5$ m and $x_y = 2$ m. Then

$$\phi_1(x_u) = 0.12 \quad \phi_1(x_y) = 1.22$$

$$\phi_2(x_u) = 0.51 \quad \phi_2(x_y) = -1.22$$

and the transfer function is

$$\frac{y}{u}(s) = \frac{0.15s}{12.8^2 + s^2} - \frac{0.62s}{80.1^2 + s^2} = -0.47 \frac{s(47.6^2 - s^2)}{(12.8^2 + s^2)(80.1^2 + s^2)} \tag{2.215}$$

This transfer function has a zero in the right half plane at $s = 47.6$. This limits the bandwidth of the system. Alternatively, we see that the transfer function is the sum of two transfer functions that are not very different, except that they have opposite signs. Thus if a P controller is used in a negative feedback, this will give stabilization and power dissipation for mode 1, while it will destabilize and add power to mode 2. In fact the only possibility for stabilization is that the mode with positive feedback has gain less that unity which ensures stability according to Bode-Nyquist stability theory.

## 2.18 Strictly positive real systems

Consider again the definition of Positive Real transfer function in definition (2.3). The following is the standard definition of Strictly Positive Real (SPR) transfer functions.

**Definition 2.4 (Strictly Positive Real)** A rational transfer function $H(s)$ is strictly positive real (SPR) if $H(s-\epsilon)$ is PR for some $\epsilon > 0$. ♠♠
Let us now consider two simple examples:

**Example 2.13** The transfer function of an asymptotically stable first order system is given by

$$H(s) = \frac{1}{s+\lambda} \tag{2.216}$$

where $\lambda > 0$. Replacing $s$ by $\sigma + j\omega$ we get

$$H(s) = \frac{1}{(\sigma+\lambda)+j\omega} = \frac{\sigma+\lambda-j\omega}{(\sigma+\lambda)^2+\omega^2} \tag{2.217}$$

Note that $\forall \mathbf{Re}[s] = \sigma > 0$ we have $\mathbf{Re}[H(s)] \geq 0$. Therefore $H(s)$ is PR. Furthermore $H(s-\epsilon)$ for $\epsilon = \frac{\lambda}{2}$ is also PR and thus $H(s)$ is also SPR.

**Example 2.14** Consider now a simple integrator (*i.e.* take $\lambda = 0$ in the previous example)

$$H(s) = \frac{1}{s} = \frac{1}{\sigma+j\omega} = \frac{\sigma-j\omega}{\sigma^2+\omega^2} \tag{2.218}$$

It can be seen that $H(s) = \frac{1}{s}$ is PR but not SPR.

### 2.18.1 Frequency domain conditions for a transfer function to be SPR

The definition of SPR transfer functions given above are given in terms of conditions in the $s$ complex plane. Such conditions become relatively difficult to be verified as the order of the system increases. The following theorem establishes conditions in the frequency domain $\omega$ for a transfer function to be SPR.

**Theorem 2.11 (Strictly Positive Real)** A rational transfer function $H(s)$ is SPR if

1. $H(s)$ is analytic in $\mathbf{Re}[s] \geq 0$, i.e. the system is asymptotically stable)

2. $\mathbf{Re}[H(j\omega)] > 0$, for all $\omega \in (-\infty, \infty)$ and

3. (a) $\lim_{\omega^2 \to \infty} \omega^2 \mathbf{Re}[H(j\omega)] > 0$ when $n^* = 1$

   (b) $\lim_{\omega^2 \to \infty} \omega \mathbf{Re}[H(j\omega)] > 0, \ \lim_{|\omega| \to \infty} \frac{H(j\omega)}{j\omega} > 0$ when $n^* = -1$

where $n^*$ is the relative degree of the system.

♠♠

**Proof**
The proof has been established by Ioannou and Tao [71] and is not given here for the sake of briefness.

### 2.18.2 Necessary conditions for $H(s)$ to be PR

In general, before checking all the conditions for a specific transfer function to be PR or SPR it is useful to check first that it satisfies a series of necessary conditions. The following are necessary conditions for a system to be PR (SPR)

- $H(s)$ is (asymptotically) stable

- The Nyquist plot of $H(j\omega)$ lies entirely in the (closed) right half complex plane.

- The relative degree of $H(s)$ is either $n^* = 0$ or $n^* = \pm 1$.

- $H(s)$ is (strictly) minimum-phase, *i.e.* the zeros of $H(s)$ lie in $\mathbf{Re}[s] \leq 0$ ($\mathbf{Re}[s] < 0$).

**Remark 2.5** In view of the above necessary conditions it is clear that unstable systems or nonminimum phase systems are not positive real. Furthermore proper transfer functions can be PR only if their relative degree is 0 or 1. This means for instance that a double integrator, *i.e.* $H(s) = \frac{1}{s^2}$ is not PR. This remark will turn out to be important when dealing with passivity of nonlinear systems. In particular for a robot manipulator we will be able to prove passivity from the torque control input to the velocity of the generalized coordinates but not to the position of the generalized coordinates.

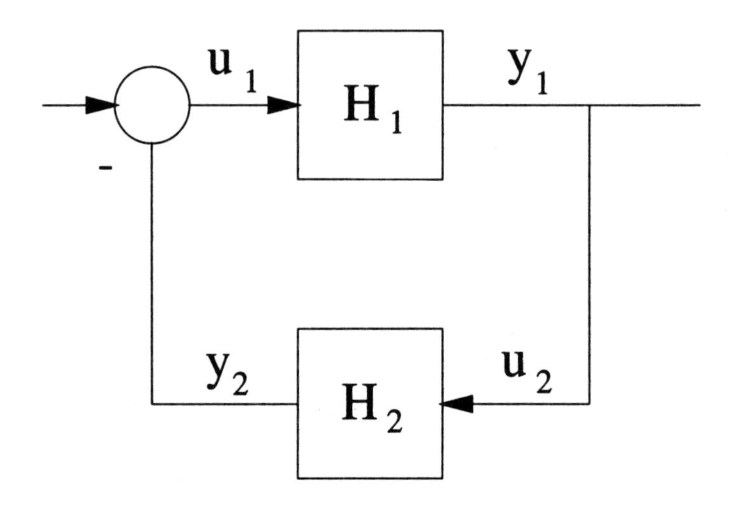


Figure 2.21: Negative feedback interconnection of $H_1$ and $H_2$.

### 2.18.3 Interconnection of positive real systems

One of the important properties of positive real systems is that the inverse of a PR system is also PR. In addition the interconnection of PR systems in parallel or in negative feedback (see figure 2.21) inherit the PR property. More specifically we have the following properties (see [71]):

- $H(s)$ is PR (SPR) $\Leftrightarrow \frac{1}{H(s)}$ is PR (SPR)
- If $H_1(s)$ and $H_2(s)$ are SPR so is $H(s) = \alpha_1 H_1(s) + \alpha_2 H_2(s)$ for $\alpha_1 > 0,\ \alpha_2 > 0$.
- If $H_1(s)$ and $H_2(s)$ are SPR, so is $H(s) = \frac{H_1(s)}{1+H_1(s)H_2(s)}$

**Remark 2.6** Note that a transfer function $H(s)$ need not be proper to be PR or SPR. For instance, the non-proper transfer function $s$ is PR.

### 2.18.4 Special cases of positive real systems

We will now introduce two additional definitions of classes of systems. Both of them are PR systems but one of them is weaker than SPR systems and the other one is stronger. Weak SPR (WSPR) are important because they allow one to extend the KYP lemma presented in chapter 3 for systems other than PR. They are also important because they allow one to relax the conditions for stability of the negative feedback interconnection of a PR system and an SPR system. We will actually show that the negative feedback interconnection between a PR system and a WSPR produces an asymptotically stable system. Both properties will be seen later.

**Remark 2.7** Consider again an electric circuit composed of an inductor in parallel with a capacitor. Such a circuit will exhibit sustained oscillatory behavior. If we have instead a lossy capacitor in parallel with a lossy inductor, it is clear that the energy stored in the system will be dissipated. However it is sufficient that at least one of the two is a lossy element (either a lossy capacitor or a lossy inductor) to guarantee that the oscillatory behavior will asymptotically converge to zero. This example motivates the notion of weakly SPR transfer function defined next.

**Definition 2.5 (Weakly SPR)** A rational function $H(s)$ is weakly SPR (WSPR) if

1. $H(s)$ is analytic in $\mathbf{Re}[s] \geq 0$.
2. $\mathbf{Re}[H(j\omega)] > 0, \quad \text{for all } \omega \in (-\infty, \infty)$.

♠♠

**Definition 2.6 (Strong SPR)** A rational function $H(s)$ is strongly SPR (SSPR) if

1. $H(s)$ is analytic in $\mathbf{Re}[s] \geq 0$.
2. $\mathbf{Re}[H(j\omega)] \geq \delta > 0, \quad \text{for all } \omega \in [-\infty, \infty] \text{ and some } \delta \in I\!R$.

♠♠

Let us now illustrate the various definitions of PR, SPR and WSPR functions on examples.

**Example 2.15** Consider again an asymptotically stable first order system

$$H(s) = \frac{1}{s+\lambda}, \quad \text{with} \quad \lambda > 0 \tag{2.219}$$

Let us check the conditions for $H(s)$ to be SPR

1. $H(s)$ has only poles in $\mathbf{Re}[s] < 0$
2. $H(j\omega)$ is given by

$$H(j\omega) = \frac{1}{\lambda + j\omega} = \frac{\lambda - j\omega}{\lambda^2 + \omega^2} \tag{2.220}$$

Therefore

$$\mathbf{Re}[H(j\omega)] = \frac{\lambda}{\lambda^2 + \omega^2} > 0 \ \forall \omega \in (-\infty, \infty) \tag{2.221}$$

1. $\lim_{\omega^2\to\infty} \omega^2 \mathbf{Re}[H(j\omega)] = \lim_{\omega^2\to\infty} \frac{\omega^2\lambda}{\lambda^2+\omega^2} = \lambda > 0$

Therefore $\frac{1}{s+\lambda}$ is SPR. However $\frac{1}{s+\lambda}$ is not SSPR because there does not exist a $\delta > 0$ such that $\mathbf{Re}[H(j\omega)] > \delta$, for all $\omega \in [-\infty, \infty]$ since $\lim_{\omega^2\to\infty} \frac{\lambda}{\lambda^2+\omega^2} = 0$.

**Example 2.16** Similarily it can be proved that $H(s) = \frac{1}{s}$ and $H(s) = \frac{s}{s^2+\omega^2}$ are PR but they are not WSPR. $H(s) = 1$ and $H(s) = \frac{s+a^2}{s+b^2}$ are both SSPR.

The following is an example of a system that is WSPR but is not SPR.

**Example 2.17** Consider the second order system

$$H(s) = \frac{s+\alpha+\beta}{(s+\alpha)(s+\beta)} \; ; \; \alpha, \; \beta > 0 \tag{2.222}$$

Let us verify the conditions for $H(s)$ to be WSPR. $H(j\omega)$ is given by

$$\begin{aligned} H(j\omega) &= \frac{j\omega+\alpha+\beta}{(j\omega+\alpha)(j\omega+\beta)} \\ &= \frac{(j\omega+\alpha+\beta)(\alpha-j\omega)(\beta-j\omega)}{(\omega^2+\alpha^2)(\omega^2+\beta^2)} \\ &= \frac{(j\omega+\alpha+\beta)(\alpha\beta-j\omega(\alpha+\beta)-\omega^2)}{(\omega^2+\alpha^2)(\omega^2+\beta^2)} \end{aligned} \tag{2.223}$$

Thus

$$\begin{aligned} \mathbf{Re}[H(j\omega)] &= \frac{\omega^2(\alpha+\beta)+(\alpha+\beta)(\alpha\beta-\omega^2)}{(\omega^2+\alpha^2)(\omega^2+\beta^2)} \\ &= \frac{\alpha\beta(\alpha+\beta)}{(\omega^2+\alpha^2)(\omega^2+\beta^2)} > 0, \;\; \text{for all } \omega \in (-\infty, \infty) \end{aligned} \tag{2.224}$$

so $H(s)$ is weakly SPR. However $H(s)$ is not SPR since

$$\lim_{\omega^2\to\infty} \frac{\omega^2\alpha\beta(\alpha+\beta)}{(\omega^2+\alpha^2)(\omega^2+\beta^2)} = 0 \tag{2.225}$$

## 2.19 SPR and adaptive control

The concept of SPR transfer functions is very useful in the design of some type of adaptive control schemes. This will be shown next for the control of an unknown plant in a state space representation and it is due to Parks [153] (See also [77]).

Consider a linear time-invariant system in the following state space representation

$$\begin{aligned} \dot{x} &= Ax + Bu \\ y &= Cx \end{aligned} \tag{2.226}$$

with state $x \in I\!R^n$, input $u \in I\!R$ and output $y \in I\!R$.
Let us assume that there exists a control input

$$u = -L^T x + r(t) \tag{2.227}$$

where $r(t)$ is a reference input and $L \in I\!R^n$, such that the closed loop system behaves as the reference model

$$\begin{aligned} \dot{x}_r &= (A - BL^T)x_r + Br(t) \\ y_r &= Cx_r \end{aligned} \tag{2.228}$$

which has an SPR transfer function. This means that there exists a matrix $P > 0$, a matrix $L$, and a positive constant $\varepsilon$ such that

$$\begin{aligned} A_{cl}^T P + PA_{cl} &= -LL^T - \varepsilon P \\ PB &= C^T \end{aligned} \tag{2.229}$$

where

$$A_{cl} = A - BL^T$$

Since the system parameters are unknown, let us consider the following adaptive control law:

$$\begin{aligned} u &= -\hat{L}^T x + r(t) \\ &= -L^T x + r(t) - \tilde{L}^T x \end{aligned} \tag{2.230}$$

where $\hat{L}$ is the estimate of $L$ and $\tilde{L}$ is the parametric error

$$\tilde{L} = \hat{L} - L$$

Introducing the above control law into the system (2.226) we obtain

$$\dot{x} = (A - BL^T)x + B(r(t) - \tilde{L}^T x) \tag{2.231}$$

Define the state error $\tilde{x} = x - x_r$ and the output error $e = y - y_r$. From the above we obtain

$$\begin{aligned} \frac{d\tilde{x}}{dt} &= A_{cl}\tilde{x} - B\tilde{L}^T x \\ e &= C\tilde{x} \end{aligned} \tag{2.232}$$

Consider the following Lyapunov function candidate

$$V(\tilde{x}, \tilde{L}) = \tilde{x}^T P \tilde{x} + \tilde{L}^T P_L \tilde{L}$$

where $P > 0$ and $P_L > 0$. Therefore

$$\dot{V} = \tilde{x}^T(A_{cl}^T P + PA_{cl})\tilde{x} - 2\tilde{x}^T PB\tilde{L}^T x + 2\tilde{L}^T P_L \frac{d\tilde{L}}{dt}$$

Choosing the following parameter adaptation law

$$\frac{d\hat{L}}{dt} = -P_L^{-1}xe = -P_L^{-1}xC\tilde{x}$$

we obtain

$$\dot{V} = \tilde{x}^T(A_{cl}^T P + PA_{cl})\tilde{x} - 2\tilde{x}^T(PB - C^T)\tilde{L}^T x$$

Introducing (2.229) in the above we get

$$\dot{V} = -\tilde{x}^T(LL^T + \varepsilon P)\tilde{x} \leq 0$$

It follows that $\tilde{x}, x$ and $\tilde{L}$ are bounded. Integrating the above we get

$$\int_0^T \tilde{x}^T(LL^T + \varepsilon P)\tilde{x}dt \leq V(\tilde{x}(0), \tilde{L}(0))$$

Thus $\tilde{x} \in \mathcal{L}_2$. From (2.232) it follows that $\frac{d\tilde{x}}{dt}$ is bounded and we conclude that $\tilde{x}$ converges to zero.

## 2.20 Adaptive output feedback

In the previous section we presented an adaptive control based on the assumption that there exists a state feedback control law such that the resulting closed-loop system is SPR. In this section we present a similar approach but this time we only require output feedback. In the next section we will present the conditions under which there exists an output feedback that renders the closed loop SPR. The material in this section and the next has been presented in [70].
Consider again system (2.226) in the MIMO (multiple-input multiple-output) case, i.e., with state $x \in \mathbb{R}^n$, input $u \in \mathbb{R}^m$ and output $y \in \mathbb{R}^p$. Assume that there exists a constant output feedback control law

$$u = -Ky + r(t) \tag{2.233}$$

such that the closed loop system

$$\begin{array}{rcl} \dot{x} & = & A^*x + Br(t) \\ y & = & Cx \end{array} \tag{2.234}$$

with

$$A^* = A - BKC$$

is SPR, i.e.(2.229) is satisfied. Since the plant parameters are unknown, consider the following adaptive controller for r(t)=0

$$u = -\hat{K}y$$

where $\hat{K}$ is the estimate of $K$ at time $t$. The closed loop system can be written as

$$\begin{aligned} \dot{x} &= A^*x - B(\hat{K} - K)y \\ y &= Cx \end{aligned}$$

Define

$$\tilde{K} = \hat{K} - K$$

and consider the following Lyapunov function candidate

$$V = x^T Px + \text{trace}\left(\tilde{K}^T \Gamma^{-1} \tilde{K}\right)$$

where $P$ satisfies (2.229) and $\Gamma > 0$ is an arbitrary positive definite matrix. The time derivative of $V$ along the system trajectories is given by

$$\dot{V} = x^T(A^{*T}P + PA^*)x - 2x^T PB\tilde{K}y + 2\text{trace}\left(\tilde{K}^T \Gamma^{-1} \frac{d}{dt}\left(\tilde{K}\right)\right)$$

Introducing (2.226) and (2.229) we obtain

$$\dot{V} = x^T(A^{*T}P + PA^*)x - 2\text{trace}\left(\tilde{K}yy^T - \tilde{K}^T \Gamma^{-1} \frac{d}{dt}\left(\tilde{K}\right)\right)$$

Choosing the parameter adaptation law

$$\frac{d}{dt}\left(\hat{K}\right) = \Gamma yy^T$$

and introducing (2.229) we obtain

$$\dot{V} = -x^T(LL^T + \varepsilon P)x \leq 0$$

Therefore $V$ is a Lyapunov function and thus $x$ and $\hat{K}$ are both bounded. Integrating the above equation it follows that $x \in \mathcal{L}_2$. Since $\dot{x}$ is also bounded we conclude that $x(t) \to 0$ as $t \to 0$.
Hence the proposed adaptive control law stabilizes the system as long as the assumption of the existence of a constant output feedback that makes the closed-loop transfer matrix SPR is satisfied. The conditions for the existence of such control law are established in the next section.

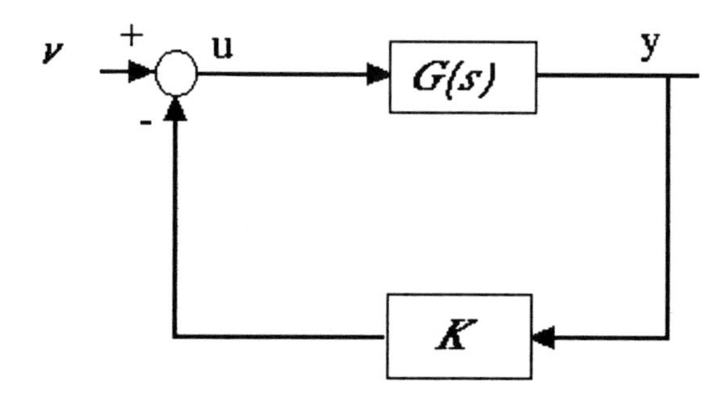


Figure 2.22: Closed-loop system $T(s)$ using constant output feedback.

## 2.21 Design of SPR systems

The adaptive control scheme presented in the previous section motivates the study of constant output feedback control designs such that the resulting closed-loop is SPR. The positive real synthesis problem is important in its own right and has been investigated by [170], [62], [198] and [207]. Necessary and sufficient conditions have been obtained in [70] for a linear system to become SPR under constant output feedback. Furthermore, they show that if no constant feedback can lead to an SPR closed-loop system, then no dynamic feedback with proper feedback transfer matrix can do it neither. Hence, there exists an output feedback such that the closed-loop system is SPR if and only if there exists a constant output feedback rendering the closed-loop system SPR.

Consider again the system (2.226) in the MIMO case, i.e., with state $x \in I\!R^n$, input $u \in I\!R^m$ and output $y \in I\!R^p$ and the constant output feedback in (2.233). The closed loop is represented in figure 2.22 where $G(s)$ is the transfer function of the system (2.226).

The equation of the closed-loop $T(s)$ of figure 2.22 is given in (2.234). We now state the following result where we assume that $B$ and $C$ are full rank.

**Theorem 2.12 (SPR synthesis [70])** There exists a constant matrix $K$ such that the closed-loop transfer function matrix $T(s)$ in figure 2.22 is SPR if and only if

$$B^T C = C^T B > 0$$

and there exists a positive definite matrix $X$ such that

$$C_\perp^T \text{herm}\{B_\perp X B_\perp^T A\} C_\perp < 0$$

When the above conditions hold $K$ is given by

$$K = C^\dagger Z (I - C_\perp (C_\perp^T Z C_\perp)^{-1} C_\perp^T Z) C^{\dagger T} + S$$

where

$$Z = \text{herm}\{PA\}$$

and

$$P = C(B^T C)^{-1} C^T + B_\perp X B_\perp^T$$

and $S$ is an arbitrary positive definite matrix. ♠♠

The notation used above is

| | |
|---|---|
| $X^\dagger$ | The Moore-Penrose pseudo-inverse of $X$ |
| $\text{herm}\{X\}$ | $\frac{1}{2}(X + X^*)$ |
| $X_\perp$ | $X_\perp^T X = 0$ and $X_\perp^T X_\perp = I$ |

In the single-input single-output case, the necessary condition $B^T C > 0$ implies the relative degree of $G(s)$ is one.

# Chapter 3

# Kalman-Yakubovich-Popov Lemma

The Kalman-Yakubovich-Popov lemma is considered to be one of the cornerstones of Control and System Theory due to its applications in Absolute Stability, Hyperstability, Dissipativity, Passivity, Optimal Control, Adaptive Control, Stochastic Control and Filtering. Despite its broad applications the lemma has been motivated by a very specific problem which is called the *Absolute Stability Lur'e problem* [157]. The first results on the Kalman-Yacubovich-Popov lemma are due to Yakubovich [215]. The proof of Kalman [80] was based on *factorization of polynomials*, which were very popular among electrical engineers. They later became the starting point for new development. Using general factorization of matrix polynomials, Popov [156], [158] obtained the lemma in the multivariable case. In the following years the lemma was further extended to the infinite dimensional case (Yakubovich [216], Brusin [37], Likhtarnikov and Yakubovich [102]) and discrete-time case (Szegö and Kalman [199]).

The Kalman-Yakubovich-Popov (KYP) lemma establishes an equivalence between the conditions in the frequency domain for a system to be positive real, an input-output relationship of the system in the time domain, and conditions on the matrices describing the state-space representation of the system. A proof of this lemma in the multivariable case is also due to Anderson [6]. This result is very useful in the stability analysis of dynamical systems and is also extensively used in the analysis of adaptive control schemes. We will use this lemma to prove the Passivity theorem which ensures the stability of a closed loop system composed of two passive systems connected in negative feedback. Both results are extensively used in the analysis and synthesis of dynamical systems and will be presented in detail in this chapter.

## 3.1 The positive real lemma

### 3.1.1 PR functions

Let us consider a controllable multivariable linear time-invariant system described by the following state-space representation

$$\begin{aligned} \dot{x} &= Ax + Bu \\ y &= Cx + Du \end{aligned} \tag{3.1}$$

where $x \in \mathbb{R}^n, u, y, \in \mathbb{R}^m$ with $n \geq m$. The Positive Real lemma can be stated as follows [4].

**Lemma 3.1 (Positive real lemma or KYP lemma)** The transfer function $H(s) = C[sI - A]^{-1}B + D$, with $A \in \mathbb{R}^{n \times n}, B \in \mathbb{R}^{n \times m}, C \in \mathbb{R}^{m \times n}, D \in \mathbb{R}^{m \times m}$ is PR with $H(s) \in \mathbb{R}^{m \times m}$ , if and only if there exists matrices $P > 0$, $P \in \mathbb{R}^{n \times n}$, $L \in \mathbb{R}^{n \times m}$and $W \in \mathbb{R}^{m \times m}$ such that:

$$\begin{aligned} PA + A^T P &= -LL^T \\ PB - C^T &= -LW \\ D + D^T &= W^T W \end{aligned} \tag{3.2}$$

♠♠

The first equation above is known as the Lyapunov equation. Note that $LL^T$ is not required to be positive definite. The third equation above can be interpreted as the factorization of $D + D^T$. For the case $D = 0$, the above set of equations reduces to the first two equations with $W = 0$. Note that (3.2) can also be written as:

$$\begin{bmatrix} -PA - A^T P & C^T - PB \\ C - B^T P & D + D^T \end{bmatrix} = \begin{bmatrix} L \\ W^T \end{bmatrix} \begin{bmatrix} L^T & W \end{bmatrix} \geq 0 \tag{3.3}$$

**Remark 3.1** Provided $D + D^T$ is full-rank, the matrix inequality in (3.3) is equivalent to the following Riccati equation in [55] p.85:

$$-PA - A^T P - (C - B^T P)(D + D^T)^{-1}(C^T - PB) \geq 0 \tag{3.4}$$

which corresponds to the Riccati inequation of an infinite horizon LQ problem whose cost matrix is given by

$$\begin{bmatrix} -PA - A^T P & C \\ C^T & D + D^T \end{bmatrix} \tag{3.5}$$

The equivalence can be established by calculating the determinant of the matrix in (3.3). As we shall see further in the book, this corresponds in the nonlinear case to a partial differential inequation, whose solutions serve as Lyapunov functions candidates. The set of solutions is convex [55] and possesses two extremal solutions (which will be called the available storage and required supply) which satisfy the Riccati equation (or a PDE for the nonlinear case), i.e. 3.4 with equality.

### 3.1.2 Positive real lemma for SPR systems

#### The Lefschetz-Kalman-Yakubovich lemma

We now present the Lefschetz-Kalman-Yakubovich lemma which gives necessary and sufficient conditions for a system in state space representation to be SPR. The proof of the following lemma can be found in [201].

**Lemma 3.2 (Lefschetz-Kalman-Yakubovich)** Consider the system in (3.1). Assume that the rational transfer matrix $H(s) = C(sI - A)^T B + D$ has poles which lie in $\mathbf{Re}[s] < -\gamma$ where $\gamma > 0$ and $(A, B, C, D)$ is a minimal realization of $H(s)$. Then $H(s - \mu)$ for $\mu > 0$ is PR if and only if a matrix $P = P^T > 0$, and matrices $L$ and $W$ exist such that

$$\begin{array}{rcl} PA + A^T P & = & -LL^T - 2\mu P \\ PB - C^T & = & -LW \\ D + D^T & = & W^T W \end{array} \tag{3.6}$$

♠♠

#### Time domain conditions for strict positive realness

The next result is due to J.T. Wen [209] who established different relationships between conditions in the frequency domain and the time domain for SPR systems.

**Lemma 3.3 (KYP lemma for SPR systems)** Consider the LTI, minimal (controllable and observable) system (3.1) whose transfer function is given by

$$H(s) = D + C(sI - A)^{-1} B \tag{3.7}$$

where $B$ is full rank. Assume that the system is exponentially stable. Consider the following statements:

1) $\exists \ P > 0, Q \in I\!R^{n \times n}, \ \lambda_{min}(Q) > 0$ ($\lambda_{min}$ denotes the minimum eigenvalue of $Q$), $L \in I\!R^{m \times n}, W \in I\!R^{m \times m}$, such that

$$\begin{array}{rcl} PA + A^T P & = & -LL^T - Q \\ PB - C^T & = & -LW \\ D + D^T & = & W^T W \end{array} \tag{3.8}$$

1') Same as 1) except $Q$ is related to $P$ by

$$Q = 2\mu P, \ \mu > 0$$

2) (**SSPR**)

$\exists \ \delta > 0$ such that

$$H(j\omega) + H^*(j\omega) \geq \delta I, \ \forall \ \omega \in I\!R$$

(the superscript $*$ denotes complex conjugate transposition)

3) (**WSPR**)

$$H(j\omega) + H^*(j\omega) > 0, \ \forall \ \omega \in I\!R$$

4) (**SPR**)

$$H(j\omega) + H^*(j\omega) > 0, \ \forall \ \omega \in I\!R$$

and

$$\lim_{\omega \to \infty} \omega^2 (H(j\omega) + H^*(j\omega)) > 0$$

5) The system (3.1) can be realized as a driving point impedance of a multi-port dissipative network (composed of resistors, lossy capacitors and lossy inductors).

6) $\exists \ \mu > 0$ such that

$$H(j\omega - \mu) + H^*(j\omega - \mu) \geq 0, \ \forall \ \omega \in I\!R$$

7) $\exists \ \rho \in I\!R^+, \xi(x_0) \in I\!R, \xi(0) = 0$ such that:

$$\int_0^T u^T(t) y(t) dt \geq \xi(x_0) + \rho \int_0^T \|u(t)\|^2 dt, \ \forall \ T \geq 0.$$

These statements are related as follows:

- i) $(1) \Leftarrow (if \ D > 0 \Leftrightarrow)(2) \Leftrightarrow (7)$
- ii) $(1) \Leftarrow (if \ D = 0 \Leftrightarrow)(1') \Leftrightarrow (4) \Leftrightarrow (5) \Leftrightarrow (6)$
- iii) $(i) \Rightarrow (ii) \Rightarrow (3)$ ♠♠

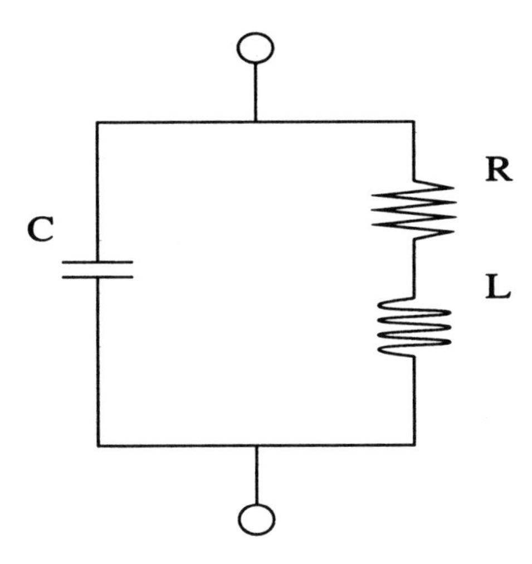


Figure 3.1: An ideal capacitor in parallel with a lossy inductor.

## 3.2 Weakly SPR systems and the KYP lemma

A dissipative network is composed of resistors, lossy inductors and lossy capacitors. Consider the circuit depicted in figure 3.1 of an ideal capacitor in parallel with a lossy inductor. Even though this circuit is not only composed of dissipative elements, the energy stored in the network always decreases. This suggests that the concept of SPR may be unnecessarily restrictive for some control applications. This motivates the study of weakly SPR systems and its relationship with the Kalman-Yakubovich-Popov lemma. Lozano and Joshi [105] proposed the following lemma which establishes equivalent conditions in the frequency and time domain for a system to be weakly SPR (WSPR)

**Lemma 3.4 (weakly SPR)** Consider the minimal (controllable and observable) LTI system (3.1) whose transfer function is given by

$$H(s) = D + C(sI - A)^{-1}B \tag{3.9}$$

Assume that the system is exponentially stable and minimum-phase. Under such conditions the following statements are equivalent:

1. $\exists\ P > 0, P \in \mathbb{R}^{n\times n}, W \in \mathbb{R}^{m\times m}, L \in \mathbb{R}^{n\times m}$

$$\begin{aligned} PA + A^T P &= -LL^T \\ PB - C^T &= -LW \\ D + D^T &= W^T W \end{aligned} \tag{3.10}$$

and such that the quadruplet $(A, B, L, W)$ is a minimal realization whose transfer function: $\overline{H}(s) = W + L^T(sI - A)^{-1}B$ has no zeros in the $j\omega$ axis (i.e. rank $\bar{H}(j\omega) = m, \ \forall \ \omega < \infty$).

2. $H(j\omega) + H^*(j\omega) > 0, \ \forall \ \omega \in I\!R$

3. The following input-output relationship holds

$$\int_0^T u^T(t)y(t)dt + \beta \geq \int_0^T \bar{y}^T(t)\bar{y}dt \ , \forall \ T > 0$$

with $\beta = x(0)^T Px(0), P > 0$ and $\bar{y}(s) = \overline{H}(s)u(s)$ ♠♠

**Proof**
**(1) ⇒ (2)**

Using (3.9) and (3.10) we obtain

$$\begin{aligned}
& H(j\omega) + H^*(j\omega) \\
= \ & D + D^T + C(j\omega I - A)^{-1}B + B^T(-j\omega I - A^T)^{-1}C^T \\
= \ & W^TW + (B^TP + W^TL^T)(j\omega I - A)^{-1}B \\
& +B^T(-j\omega I - A^T)^{-1}(PB + LW) \\
= \ & W^TW + B^T(-j\omega I - A^T)^{-1}[(-j\omega I - A^T)P \\
& +P(j\omega I - A)](j\omega I - A)^{-1}B + W^TL^T(j\omega I - A)^{-1}B \\
& +B^T(-j\omega I - A^T)^{-1}LW \\
= \ & W^TW + B^T(-j\omega I - A^T)^{-1}LL^T(j\omega - A)^{-1}B \\
& +W^TL^T(j\omega I - A)^{-1}B + B^T(-j\omega I - A^T)^{-1}LW \\
= \ & (W + L^T(-j\omega I - A)^{-1}B)^T(W + L^T(j\omega - A)^{-1}B)
\end{aligned}$$

It then follows

$$H(j\omega) + H^*(j\omega) = \overline{H}^*(j\omega)\overline{H}(j\omega) > 0 \tag{3.11}$$

Since $\overline{H}(s)$ has no zeros on the $j\omega$-axis, $\overline{H}(j\omega)$ has full rank and, therefore, the RHS of (3.11) is strictly positive.
**(2) ⇒ (1)**

In view of statement 2, there exists an asymptotically stable transfer function $H(s)$ such that: (see [155] or [55]).

$$H(j\omega) + H^*(j\omega) = \overline{H}^*(j\omega)\overline{H}(j\omega) > 0 \tag{3.12}$$

Without loss of generality let us assume that

$$\overline{H}(s) = W + J(sI - F)^{-1}G \tag{3.13}$$

with $(J, F)$ observable and $\lambda_i(F) < 0 \ \forall \ i$. Therefore, there exists $\bar{P} > 0$ (see [94]) such that

$$\bar{P}F + F^T\bar{P} = -JJ^T \tag{3.14}$$

using (3.13) and (3.14) we have

$$\begin{aligned} \overline{H}^T(-j\omega)\overline{H}(j\omega) &= [W + J(-j\omega I - F)^{-1}G]^T \\ &\quad \times[W + J(j\omega I - F)^{-1}G] \\ &= W^TW + W^TJ(j\omega I - F)^{-1}G \\ &\quad +G^T(-j\omega I - F^T)^{-1}J^TW + X \end{aligned} \tag{3.15}$$

where

$$\begin{aligned} X &= G^T(-j\omega I - F^T)^{-1}J^TJ(j\omega I - F)^{-1}G \\ &= -G^T(-j\omega I - F^T)^{-1}[\bar{P}(F - j\omega I) \\ &\quad +(F^T + j\omega I)\bar{P}](j\omega I - F)^{-1}G \\ &= G^T(-j\omega I - F^T)^{-1}\bar{P}G + G^T\bar{P}(j\omega I - F)^{-1}G \end{aligned} \tag{3.16}$$

Introducing (3.16) into (3.15) and using (3.12)

$$\begin{aligned} \overline{H}^T(-j\omega)\overline{H}(j\omega) &= W^TW + (W^TJ + G^T\bar{P})(j\omega I - F)^{-1}G \\ &\quad +G^T(-j\omega I - F^T)^{-1}(J^TW + \bar{P}G) \\ &= H(j\omega) + H^T(-j\omega) \\ &= D + D^T + C(j\omega I - A)^{-1}B \\ &\quad +B^T(-j\omega I - A^T)^{-1}C^T \end{aligned} \tag{3.17}$$

From (3.17) it follows that $W^TW = D + D^T$. Since $\lambda_i(A) < 0$ and $\lambda_i(F) < 0$, then

$$C(j\omega I - A)^{-1}B = (W^TJ + G^T\bar{P})(j\omega I - F)^{-1}G \tag{3.18}$$

Therefore the various matrices above can be related through a state space transformation, i.e.

$$TAT^{-1} = F \tag{3.19}$$

$$TB = G \tag{3.20}$$

$$CT^{-1} = W^TJ + G^T\bar{P} \tag{3.21}$$

Defining $P = T^T\bar{P}T$ and $L^T = JT$ and using ( 3.14) and ( 3.19) - (3.21)

$$\begin{aligned} -LL^T &= -T^TJ^TJT \\ &= T^T(\bar{P}F + F^T\bar{P})T \\ &= T^T\bar{P}TT^{-1}FT + T^TF^TT^{-T}T^T\bar{P}T \\ &= PA + A^TP \end{aligned}$$

which is the first equation of (3.10). From (3.20) and (3.21) we get

$$\begin{aligned} C &= W^TJT + G^T\bar{P}T \\ &= W^TL^T + G^TT^{-T}T^T\bar{P}T \\ &= W^TL^T + B^TP \end{aligned} \tag{3.22}$$

which is the second equation of (3.10). $\overline{H}(s)$ was defined by the quadruplet $(F, G, J, W)$ in (3.13) which is equivalent, through a state-space transformation, to the quadruplet $(T^{-1}FT, T^{-1}G, JT, W)$. In view of (3.19)-(3.21) and since $L^T = JT$, $\overline{H}(s)$ can also be represented by the quadruplet $(A, B, L^T, W)$ i.e.

$$\overline{H}(s) = W + L^T(sI - A)^{-1}B \tag{3.23}$$

We finally note from (3.12) that $\overline{H}(j\omega)$ has no zeros on the $j\omega$-axis.

**(1) ⇒ (3)**

Consider the following positive definite function

$$V(x) = \frac{1}{2}x^TPx$$

then using (3.10)

$$\begin{aligned}
\dot{V} &= \tfrac{1}{2}x^T(PA + A^TP)x + x^TPBu \\
&= -\tfrac{1}{2}x^TLL^Tx + u^TB^TPx \\
&= -\tfrac{1}{2}x^TLL^Tx + u^T(C - W^TL^T)x \\
&= -\tfrac{1}{2}x^TLL^Tx + u^Ty - \tfrac{1}{2}u^T(D + D^T)u - u^TW^TL^Tx \qquad (3.24) \\
&= -\tfrac{1}{2}x^TLL^Tx + u^Ty - \tfrac{1}{2}u^TW^TWu - u^TW^TL^Tx \\
&= u^Ty - \tfrac{1}{2}\left(L^Tx + Wu\right)^T\left(L^Tx + Wu\right) \\
&= u^Ty - \tfrac{1}{2}\bar{y}^T\bar{y}
\end{aligned}$$

where $\bar{y}$ is given by

$$\begin{aligned}
\dot{x}(t) &= Ax(t) + Bu(t) \\
\bar{y}(t) &= L^Tx(t) + Wu(t)
\end{aligned} \qquad (3.25)$$

and therefore, in view of (3.12)

$$\bar{y}(s) = \overline{H}(s)u(s) \qquad (3.26)$$

with $\overline{H}(s) = W + L^T(sI - A)^{-1}B$. Integrating (3.24) gives

$$\int_0^T u^T(t)y(t)dt + \beta \geq \frac{1}{2}\int_0^T \bar{y}^T(t)\bar{y}\,(t)\,dt \qquad (3.27)$$

with $\beta = V(x(0))$
**(3)** $\Rightarrow$ **(2)**

Without loss of generality, consider an input $u$ such that $\int_0^T u^T(t)u(t)dt < \infty$, $\forall\ T$. Dividing (3.27) by $\int_0^T u^T(t)u(t)dt$, we obtain

$$\frac{\int_0^T u^T(t)y(t)dt + V(x(0))}{\int_0^T u^T(t)u(t)dt} \geq \frac{\int_0^T \bar{y}^T(t)\bar{y}(t)dt}{\int_0^T u^T(t)u(t)dt} \qquad (3.28)$$

This inequality should also hold for $T = \infty$ and $x(0) = 0$, i.e.

$$\frac{\int_0^\infty u^T(t)y(t)dt}{\int_0^\infty u^T(t)u(t)dt} \geq \frac{\int_0^\infty \bar{y}^T(t)\bar{y}(t)dt}{\int_0^\infty u^T(t)u(t)dt} \qquad (3.29)$$

Since $H(s)$ and $\overline{H}(s)$ are asymptotically stable, $u \in \mathcal{L}_2 \Rightarrow y, \bar{y} \in \mathcal{L}_2$ and we can use Plancherel's theorem [167]. From the above equation we obtain

$$\frac{\int_{-\infty}^{\infty} U^*(j\omega)(H(j\omega)+H^*(j\omega))U(j\omega)d\omega}{\int_{-\infty}^{\infty} U^*(j\omega)U(j\omega)d\omega} \geq \frac{\int_{-\infty}^{\infty} U^*(j\omega)\overline{H}^*(j\omega)\overline{H}(j\omega)U(j\omega)d\omega}{\int_{-\infty}^{\infty} U^*(j\omega)U(j\omega)d\omega}$$

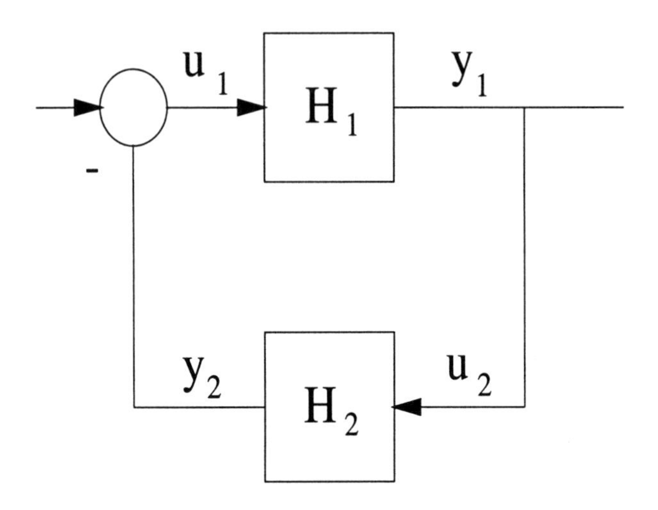


Figure 3.2: Interconnection of $H_1$ and $H_2$.

Since $\overline{H}(s)$ has no zeros on the $j\omega$-axis, the RHS of the above equation is strictly positive and so is the LHS for all nonzero $U(j\omega) \in \mathcal{L}_2$, and thus

$$H(j\omega) + H^*(j\omega) > 0, \;\; \forall \; \omega \in (-\infty, \infty)$$

♠♠

The KYP lemma can also be stated for stabilizable systems. This is done in [43].

## 3.3 Interconnection of PR systems

We will now study the stability properties of positive real or strictly positive real systems when they are connected in negative feedback. We will consider two PR systems $H_1 : u_1 \to y_1$ and $H_2 : u_2 \to y_2$. $H_1$ is in the feedforward path and $H_2$ is in the feedback path(i.e. $u_1 = -y_2$ and $u_2 = y_1$). The stability of the closed loop system is concluded in the following lemma when $H_1$ is PR and $H_2$ is weakly SPR. See figure 3.2.

**Lemma 3.5** Consider a system $H_1 : u_1 \to y_1$ in negative feedback with a system $H_2 : u_2 \to y_2$ as shown in figure 3.2, where $H_1$ is PR and $H_2$ is WSPR. Under those conditions $u_1, u_2, y_1$ and $y_2$ all converge to zero exponentially.♠♠

**Proof**
Let us define the following state-space representation for system $H_1(s)$

$$\begin{array}{rcl} \dot{x}_1 & = & A_1 x_1 + B_1 u_1 \\ y_1 & = & C_1 x_1 + D_1 u_1 \end{array} \tag{3.30}$$

Since $H_1(s)$ is PR there exists matrices $P > 0, P \in I\!R^{n \times n}, W \in I\!R^{m \times m}, L \in I\!R^{n \times m}$ such that

$$\begin{array}{rcl} P_1 A_1 + A_1^T P_1 & = & -L_1 L_1^T \\ P_1 B_1 - C_1^T & = & -L_1 W_1 \\ D_1 + D_1^T & = & W_1^T W_1 \end{array} \tag{3.31}$$

Define the following state-space representation for system $H_2(s)$

$$\begin{array}{rcl} \dot{x}_2 & = & A_2 x_2 + B_2 u_2 \\ y_2 & = & C_2 x_2 + D_2 u_2 \end{array} \tag{3.32}$$

Since $H_2(s)$ is WSPR there exists matrices $P > 0, \in I\!R^{n \times n}, W \in I\!R^{m \times m}, L \in I\!R^{n \times m}$ such that

$$\begin{array}{rcl} P_2 A_2 + A_2^T P_2 & = & -L_2 L_2^T \\ P_2 B_2 - C_2^T & = & -L_2 W_2 \\ D_2 + D_2^T & = & W_2^T W_2 \end{array} \tag{3.33}$$

and

$$\overline{H}_2(s) = W_2 + L_2^T (sI - A_2)^{-1} B_2 \tag{3.34}$$

has no zeros in the $j\omega$-axis. Consider the following positive definite function

$$V_i = x_i^T P_i x_i, \quad i = 1, 2$$

then using (3.31) and (3.33):

$$\begin{array}{rcl} \dot{V}_i & = & (x_i^T A_i^T + u_i^T B_i^T) P_i x_i + x_i^T P_i (A_i x_i + B_i u_i) \\ & = & x_i^T (A_i^T P_i + P_i A_i) x_i + 2 u_i^T B_i^T P_i x_i \\ & = & x_i^T (-L_i L_i^T) x_i + 2 u_i^T B_i^T P_i x_i \\ & = & -x_i^T L_i L_i^T x_i + 2 u_i^T (B_i^T P_i + W_i^T L_i^T) x_i - 2 u_i^T W_i^T L_i^T x_i \\ & = & -x_i^T L_i L_i^T x_i + 2 u_i^T [C_i x_i + D_i u_i] - 2 u_i^T D_i u_i - 2 u_i^T W_i^T L_i^T x_i \\ & = & -x_i^T L_i L_i^T x_i + 2 u_i^T [y_i] - 2 u_i^T D_i u_i - 2 u_i^T W_i^T L_i^T x_i \\ & & -(L_i^T x_i + W_i u_i)^T (L_i^T x_i + W_i u_i) + 2 u_i^T y_i \end{array} \tag{3.35}$$

where we have used the fact that

$$2u_i^T D_i u_i = u_i^T (D_i + D_i^T) u_i = u_i^T W_i^T W_i u_i$$

Define $\bar{y}_i = L_i^T x_i + W_i u_i$ and $V = V_1 + V_2$, then

$$\dot{V} = -\bar{y}_1^T \bar{y}_1 - \bar{y}_2^T \bar{y}_2 + 2(u_1^T y_1 + u_2^T y_2)$$

Since $u_1 = -y_2$ and $u_2 = y_1$ it follows that

$$u_1^T y_1 + u_2^T y_2 = -y_2^T y_1 + y_1^T y_2 = 0$$

Therefore

$$\dot{V} = -\bar{y}_1^T \bar{y}_1 - \bar{y}_2^T \bar{y}_2 \leq -\bar{y}_2^T \bar{y}_2$$

which implies that $V$ is a nondecreasing function and therefore we conclude that $x_i \in \mathcal{L}_\infty$. Integrating the above equation:

$$-V(0) \leq V - V(0) \leq -\int_0^T \bar{y}_2^T \bar{y}_2 dt \tag{3.36}$$

then

$$\int_0^T \bar{y}_2^T \bar{y}_2 dt \leq V(0) \tag{3.37}$$

The feedback interconnection of $H_1$ and $H_2$ is a linear system. Since $x_i \in \mathcal{L}_\infty$, the closed loop is at least stable, i.e. the closed-loop poles are in the left-half plane or in the $jw$-axis. This means that $u_i, y_i$ may have an oscillatory behavior. However the equation above means that $\bar{y}_2 \to 0$. By assumption $\bar{H}_2(s)$ has no zeros on the $j\omega$ axis. Since the state is bounded, $u_2$ can not grown unbounded. It follows that $u_2 \to 0$. This in turn implies that $y_2 \to 0$ since $H_2$ is asymptotically stable. Clearly $u_2 \to 0$ and $y_2 \to 0$. ♠♠

### 3.3.1 On the design of dissipative LQG controllers

The linear-quadratic-Gaussian (LQG) controller has attained considerable maturity since its inception in the fifties and sixties. It has come to be generally regarded as one of the standard design methods. One attribute of LQG-type compensators is that, although they guarantee closed-loop stability, the compensator itself is not necessarily stable. It would be of interest to characterize the class of LQG-type compensators which are *stable*. Going one step further, if the LQG compensator is restricted to be not only *stable,* but also *dissipative,* this would define an important subclass. Since we will be dealing with linear time-invariant systems, we will consider that such systems are dissipative if and only if the corresponding transfer function is strictly positive real (SPR). The importance of such compensators is that they would not only be

dissipative, but would also be optimal with respect to an LQG performance criteria. One reason for considering dissipative compensators is that, when used to control positive-real (PR) plants, they offer excellent robustness to modeling errors as long as the plant is PR. An important application of dissipative compensators would be for vibration suppression in large flexible space structures (LFSS), which are characterized by significant unmodeled dynamics and parameter errors. The linearized elastic-mode dynamics of LFSS [85] with compatible collocated actuators and sensors are PR systems regardless of the unmodeled dynamics or parameter uncertainties can, therefore, be robustly stabilized by an SPR compensator.
The objective of this section is to investigate the conditions under which an LQG-type compensator is SPR, so that one can simultaneously have high performance and robustness to unmodeled dynamics.
Consider a minimal realisation of a PR system expressed by the following state space representation:.

$$\left\{ \begin{array}{rcl} \dot{x} & = & Ax + Bu + v \\ y & = & Cx + w \end{array} \right. \tag{3.38}$$

where $v$ and $w$ are white, zero-mean Gaussian noises. Since the system is PR, we assume, without loss of generality (see remark 3.2 at the end of this section), that the following equations hold for some matrix $Q_a \geq 0$:

$$A + A^T = -Q_A \leq 0 \tag{3.39}$$

and

$$B = C^T \tag{3.40}$$

The above conditions are equivalent to the Kalman-Yakubovich Positive Real lemma. The LQG compensator for the system (3.38), (3.39) and (3.40) is given by (see [5]):

$$u = -u' \tag{3.41}$$

$$\dot{\hat{x}} = \left[A - BR^{-1}B^T P_c - P_f B R_w^{-1} B^T\right] \hat{x} + P_f B R_w^{-1} y \tag{3.42}$$

$$u' = R^{-1}B^T P_c \hat{x} \tag{3.43}$$

where $P_c = P_c^T > 0$ and $P_f = P_f^T > 0$ are the LQ-regulator and the Kalman-Bucy filter Riccati matrices which satisfy the algebraic Riccati equations :

$$P_c A + A^T P_c - P_c B R^{-1} B^T P_c + Q = 0 \tag{3.44}$$

$$P_f A^T + AP_f - P_f BR_w^{-1} B^T P_f + Q_V = 0 \tag{3.45}$$

where $Q$ and $R$ are the usual weighting matrices for the state and input, and $Q_V$ and $R_W$ are the covariance matrices of $v$ and $w$. It is assumed that $Q > 0$ and that the pair $(A, Q_V^{1/2})$ is observable. The main result is stated as follows:

**Theorem 3.1** Consider the PR system in (3.38), (3.39) and (3.40) and the LQG-type controller in (3.41) through (3.45). If $Q$, $R$, $Q_v$ and $R_w$ are such that:

$$Q_v = Q_a + BR^{-1}B^T \tag{3.46}$$

$$R_w = R \tag{3.47}$$

and

$$Q - BR^{-1}B^T \triangleq Q_B > O \tag{3.48}$$

then the controller in (3.42) through (3.43) (described by the transfer function from $y$ to $u^{'}$) is SPR. ♠♠

**Proof**

Introducing (3.39), (3.46), (3.47) into (3.45), it becomes clear that $P_f = I$ is a solution to (3.45). From (3.44) it follows:

$$P_c(A - BR^{-1}B^T P_c - BR^{-1}B^T) + (A - BR^{-1}B^T P_c - BR^{-1}B^T)^T P_c$$

$$= -Q - P_c BR^{-1}B^T P_c - P_c BR^{-1}B^T - BR^{-1}B^T P_c$$

$$= -Q - (P_c + I)BR^{-1}B^T(P_c + I) + BR^{-1}B^T$$

$$= -Q_B - (P_c + I)BR^{-1}B^T(P_c + I) < 0$$

where $Q_B$ is defined in (3.48). In view of equations (3.40 ), (3.47) and the above, it follows that the controller in equations (3.42) and (3.43) is strictly positive real. ♠♠

The above result states that, if the weighting matrices for the regulator and the filters are chosen in a certain manner the resulting LQG-type compensator is SPR. However, it should be noted that this compensator would not be optimal with respect to actual noise covariance matrices. The noise covariance matrices are used herein merely as compensator design parameters and have no

statistical meaning. Condition (3.48) is equivalent to introducing an additional term $y^T R^{-1} y$ in the LQ performance index (since $Q = Q_B + CR^{-1}C^T$) and is not particularly restrictive.
The problem of designing linear-quadratic-Gaussian type compensators for positive real systems was considered. Sufficient conditions were obtained for the compensator to be strictly positive real. The resulting feedback configuration is guaranteed to be stable despite unmodeled plant dynamics and parameter inaccuracies, as long as the plant is positive real. One application of such compensators would be for controlling elastic motion of large flexible space structures using collocated actuators and sensors. The material presented here is based on [107]. Further work on dissipative controller has been carried out in [62] and [74].

**Remark 3.2** Consider a positive real system expressed as:

$$\dot{z} = Dz + Fu \tag{3.49}$$

$$y = Gz \tag{3.50}$$

Then, there exists matrices $P > 0$ and $L$ such that

$$PD + D^T P = -LL^T \tag{3.51}$$

$$PF = G^T \tag{3.52}$$

Define $x = P^{\frac{1}{2}} z$, where $P^{\frac{1}{2}}$ is a symmetric square root of $P$. Introducing this definition in (3.49) and (3.50). We obtain a state space representation as the one in (3.38), but with $A = P^{\frac{1}{2}} D P^{-\frac{1}{2}}$, $B = P^{\frac{1}{2}} F$, $C = GP^{-\frac{1}{2}}$. Multiplying (3.51) on the left and on the right by $P^{-\frac{1}{2}}$ we obtain (3.39) with $Q_A = P^{-\frac{1}{2}} LL^T P^{-\frac{1}{2}}$. Multiplying (3.52) on the left by $P^{-\frac{1}{2}}$ we obtain (3.40).

♠♠

### 3.3.2 The Lur'e problem

In this section we study the stability of an important class of control systems. The Lur'e problem was very popular in the fifties and can be considered as the first steps towards the synthesis of controllers based on passivity. Consider the closed-loop system shown in figure 3.3.2. We are interested in obtaining the conditions on the linear system and on the static nonlinearity such that the closed-loop system is stable. This is what is called the Lur'e problem.

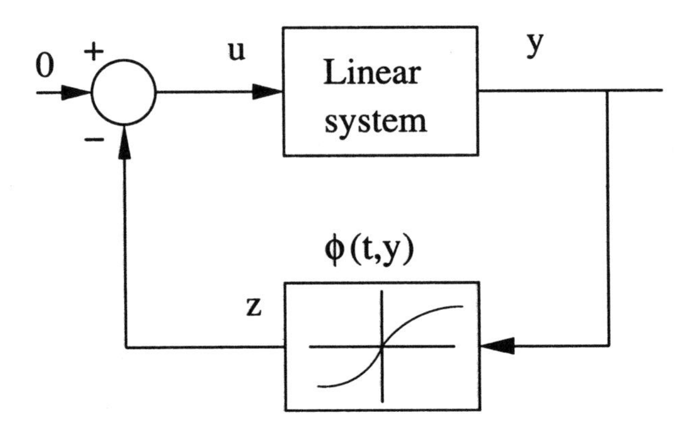


Figure 3.3: The Lur'e problem.

The linear system is given by the following state-space representation:

$$\begin{aligned} \dot{x} &= Ax + Bu \qquad & x \in \mathbb{R}^n, u, y \in \mathbb{R}^m \\ y &= Cx + Du \qquad & m < n \end{aligned} \tag{3.53}$$

The static nonlinearity is possibly time-varying and described by

$$\begin{aligned} z(t) &= \phi(t, y(t)) \\ u(t) &= -z(t) \end{aligned} \tag{3.54}$$

Note that both subsystems have input and output vectors of identical dimensions. The linear system is assumed to be minimal, i.e. controllable and observable which means that:

$$\text{rank}\begin{bmatrix} B & AB & \dots & A^{n-1}B \end{bmatrix} = n$$

and

$$\text{rank}\begin{bmatrix} C \\ CA \\ \vdots \\ CA^{n-1} \end{bmatrix} = n$$

The nonlinearity is assumed to belong to the sector $[a, b]$, i.e.:

i) $\phi(t, 0) = 0 \qquad \forall t$

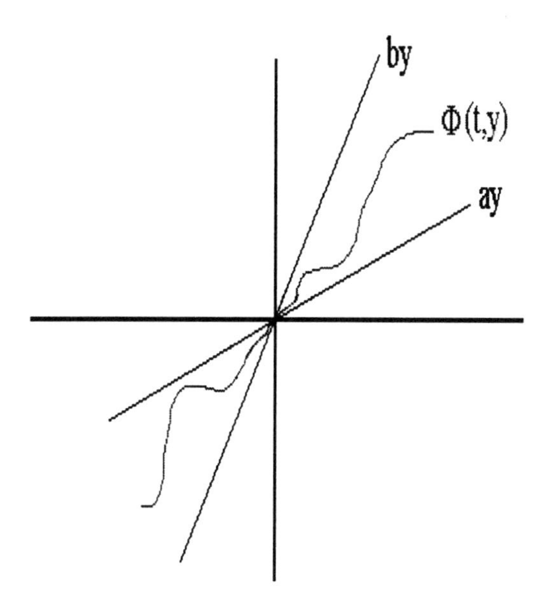


Figure 3.4: Static nonlinearity for $n = 1$.

ii) $[\phi(t,y) - ay]^T [by - \phi(t,y)] \geq 0 \qquad \forall t, \qquad y \in I\!R^m$

In the scalar case ($m = 1$), the static nonlinearity is shown in figure 3.4.

## 3.4 Absolute stability problem

Lur'e problem in figure 3.3 can be stated as follows: Find the conditions on $(A, B, C, D)$ such that the equilibrium point $x = 0$ is globally asymptotically stable for all nonlinearities $\phi$ in the sector $[a, b]$.
Lur'e problem has given rise to a couple of very popular conjectures:

**Conjecture 3.1 (Aizerman's conjecture)** If the linear subsystem with $D = 0$ and $m = 1$ in figure 3.5 is asymptotically stable for all $\phi(y) = ky$, $k \in [a, b]$, then the closed loop system in figure 3.6 with a nonlinearity $\phi$ in the sector $[a, b]$ is also globally asymptotically stable. ♠♠

**Conjecture 3.2 (Kalman's conjecture)** Consider again the system in figure 3.3 with a nonlinearity such that $\phi(t, y) = \phi(y)$ (i.e. a time-invariant and continuously differentiable nonlinearity), $m = 1$, $\phi(0) = 0$ and $a \leq \phi'(y) \leq b$. Then the system in (3.53) with $D = 0$ is globally asymptotically stable if it is globally asymptotically stable for all nonlinearities $\phi(y) = ky$, $k \in [a, b]$. ♠♠

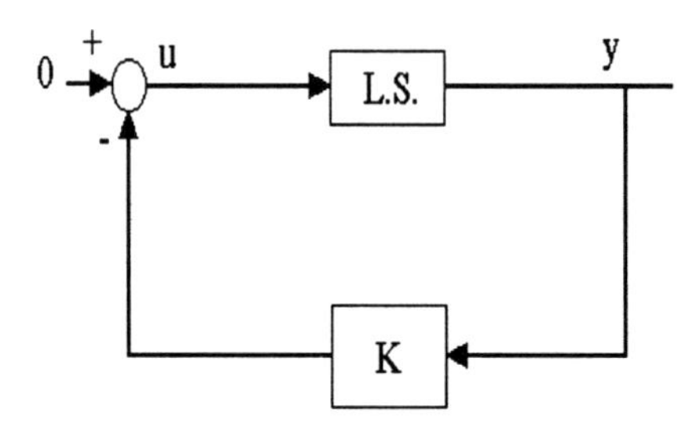


Figure 3.5: Linear system with a constant output feedback.

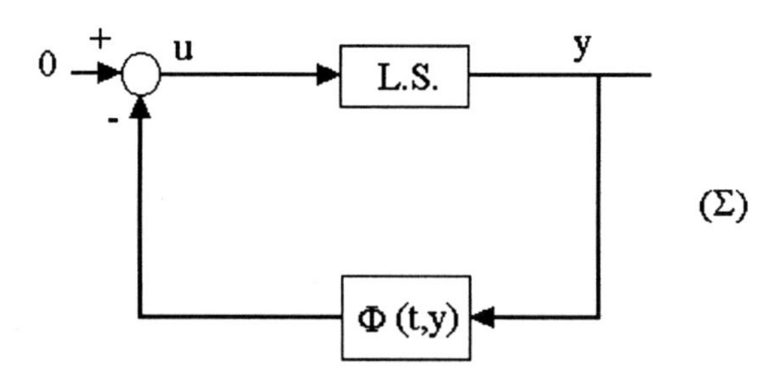


Figure 3.6: Linear system with a sector nonlinearity in negative feedback.

Thus Kalman's conjecture says that if $A - kBC$ is Hurwitz for all $k \in [a, b]$, $x = 0$ should be a globally stable fixed point for (3.53) (3.54) with $\phi$ as described in conjecture 3.2.
However it turns out that both conjectures are false in general. In fact, the absolute stability problem, and consequently Kalman conjecture, may be considered as a particular case of a more general problem known in the Applied Mathematics literature as the Markus-Yamabe conjecture (MYC in short). The MYC can be stated as follows [134]:

**Conjecture 3.3 (Markus-Yamabe's conjecture)** If a $C^1$ map $f : \mathbb{R}^n \to \mathbb{R}^n$ satisfies $f(0) = 0$ and if its Jacobian matrix $\left.\frac{\partial f}{\partial x}\right|_{x_0}$ is stable **for all** $x_0 \in \mathbb{R}^n$, then 0 is a **global attractor** of the system $\dot{x} = f(x)$. ♠♠

In other words, the MYC states that if the Jacobian of a system at any point of the state space has eigenvalues with stricty negative real parts, then the fixed point of the system should be globally stable as well. Although this conjecture seems very sound from an intuitive point of view, it is false for $n \geq 3$. Counterexamples have been given for instance in [42]. It is however true in dimension 2, i.e. $n = 2$. This has been proved in [60]. The proof is highly technical and takes around 40 pages. Since it is, moreover, outside the scope of this monograph dedicated to dissipative systems, it will not be reproduced nor summarized here. This is however one nice example of a result that is apparently quite simple and whose proof is quite complex.
It is clear that the conditions of the Kalman's conjecture with $f(x) = Ax + b\phi(y)$, $\phi(0) = 0$, make it a particular case of the MYC. In short one could say that Kalman's conjecture (as well as Aizerman's conjecture) is a version of MYC for control theory applications. Since, as we shall see in the next subsections, there has been a major interest in developing (sufficient) conditions for Lur'e problem and absolute stability in the Systems and Control community, it is also of significant interest to know the following result:

**Theorem 3.2 ([13] [19])** Kalman's conjecture is true for dimensions $n = 1, 2, 3$. It is false for $n > 3$. ♠♠

In short, Kalman's conjecture states that if the vector field $Ax + b\phi(y)$ is Hurwitz (i.e. its Jacobian satisfies the conditions of the MYC) for all linear characteristic functions $\phi(\cdot)$, then the fixed point $x = 0$ should be globally asymptotically stable for any continuously differentiable $\phi(\cdot)$ whose slope remain bounded inside $[a, b]$.

### The case $n \leq 3$

Since it has been shown in [60] that the MYC is true for $n = 1, 2$, it follows immediately that this is also the case for the Kalman's conjecture. Aizerman's conjecture has been shown to be true for $n = 1, 2$ in [57], proving in a different way that Kalman's conjecture holds for $n = 1, 2$. The following holds for the case $n = 3$:

**Theorem 3.3 ([13])** The system

$$\begin{cases} \dot{x} = Ax + b\phi(y) \\ \\ y = c^T x \end{cases} \tag{3.55}$$

with $x \in \mathbb{R}^3$, $\min_y \frac{d\phi}{dy}(y) = 0$, $\max_y \frac{d\phi}{dy}(y) = k \in (0, +\infty)$, $\phi(0) = 0$, is globally asymptotically stable if the matrices $A + \frac{d\phi}{dy}(y)c^T \in \mathbb{R}^{n\times n}$ are Hurwitz for all $y \in \mathbb{R}$.

♠♠

**Proof**

First of all we need a result from [217] which states that if for all $\omega \geq 0$ one has

$$\mathbf{Re}\left\{Z(i\omega)\left[W(i\omega) + \frac{1}{k}\right]\right\} > 0 \tag{3.56}$$

for some function $Z(\cdot)$, where the set of functions $Z(s)$ is defined as $Z(s) = \tau + \theta s - \kappa s^2$, $\tau \geq 0$, $\kappa \geq 0$, $\tau + \kappa + |\theta| > 0$, and $W(s) = c^T(A - sI)^{-1}b$, then Kalman conjecture holds. One recognizes that $Z(s)$ is a multiplier in the equivalent feedback interconnection of the Lur'e problem, and that the inequality in (3.56) expresses the weak SPRness of the transfer function inside the brackets.

One has $W(s) + \frac{1}{k} = \frac{B(s)}{A(s)}$ where $B(\cdot)$ and $A(\cdot)$ both have degree less or equal to three, and with nonnegative coefficients (note that $A(s) = \det(sI - A)$ is a Hurwitz polynomial). Now notice that the strict inequality in (3.56) can be replaced by a nonstrict one (i.e. with $\geq$) if one allows $W(s) + \frac{1}{k}$ to possess zeroes or poles on the imaginary axis. To understand this property we recall that a WSPR function $H(s)$ is such that $\mathbf{Re}[H(i\omega)] > 0$ for $\omega \in (-\infty, +\infty)$. All its poles and zeroes must lie in the open half left plane. A PR function $H(s)$ may have poles or zeroes on the imaginary axis and it satisfies $\mathbf{Re}[H(i\omega)] \geq 0$ for $\omega \in (-\infty, +\infty)$. So one is only considering that the transfer function between brackets in (3.56) is PR instead of being WSPR. In other words it is argued in [13] that if there exists a $Z(s)$ such that $Z(s)[W(s) + 1/k]$ is PR, then there exists also a $Z(s)$ such that $Z(s)[W(s) + 1/k]$ is WSPR. Now the rest of the proof consists of checking that the nonstrict inequality in (3.56) is satisfied for some function $Z(s)$ as described above. One has [13]:

- If $W(s) + \frac{1}{k} = \frac{1}{k}\frac{s(s^2+\beta s+\gamma)}{(s^2+\alpha)(s+\delta)}$: the basic condition for Kalman conjecture (i.e. the stability of matrices $A + b\mu c^T$ for all $\mu \in (0, k)$) is satisfied if $\beta\alpha + \gamma\delta \geq \alpha\delta$.

- One can choose $\tau = 0$, $\theta = 1$, $\kappa = 0$ and (3.56) is satisfied if $\beta\alpha + \gamma\delta = \alpha\delta$.
- In case $\beta\alpha + \alpha\delta > \gamma\delta$, one can choose $\tau = 1$, $\theta = \frac{\gamma-\beta\delta-\alpha}{\beta\alpha+\delta\gamma-\delta\alpha}$, $\kappa = 1$ and (3.56) is satisfied if $\gamma \geq \beta\delta + \alpha$.
- In case $\beta\alpha + \alpha\delta > \gamma\delta$, one can choose $\tau = 0$, $\theta = \frac{\alpha(\gamma-\beta\delta-\alpha)}{\beta\alpha+\delta\gamma-\delta\alpha}$, $\kappa = 1$ and (3.56) is satisfied if $\gamma < \beta\delta + \alpha$.

- If $W(s)+\frac{1}{k} = \frac{1}{k}\frac{(s^2+\alpha_1)(s+\gamma)}{(s^2+\alpha_2)(s+\delta)}$: the basic condition for Kalman's conjecture is satisfied if $\alpha_2(\delta-\gamma) < \alpha_1(\delta-\gamma)$. Moreover under this condition (3.56) is satisfied provided $\tau = \alpha_1\alpha_2$, $\theta = \frac{\delta\gamma-\alpha_2)(\delta\gamma-\alpha_1)}{\gamma-\delta}$, $\kappa = \delta\gamma$.

Therefore one is always able to find a function $Z(\cdot)$ such that the inequality (WSPR condition) in (3.56) is satisfied. This ends the proof. ♠♠

**Bernat and Llibre's counterexample when $n \geq 4$**

This counterexample was inspired by a work from Barabanov [13] and suitably modified in [19]. Let us consider the following fourth order closed-loop system that will serve as a pre-counterexample:

$$(\Sigma)\left\{\begin{array}{l} \dot{x}_1 = x_2 \\ \\ \dot{x}_2 = -x_4 \\ \\ \dot{x}_3 = x_1 - 2x_4 - \frac{9131}{900}\phi(x_4) \\ \\ \dot{x}_4 = x_1 + x_3 - x_4 - \frac{1837}{180}\phi(x_4) \\ \end{array}\right. \tag{3.57}$$

$$\phi(y) = \left\{\begin{array}{ll} -\frac{900}{9185} & \text{if } \ y < -\frac{900}{9185} \\ \\ x & \text{if } \ |y| \leq \frac{900}{9185} \\ \\ \frac{900}{9185} & \text{if } \ y > \frac{900}{9185} \end{array}\right. \tag{3.58}$$

As a first step let us check that the system $(\Sigma)$ satisfies the assumption of Kalman's conjecture, i.e. it is globally asymptotically stable for any $\phi(y) = ky$ with $k \in [0, 1]$. Notice that the characteristic function in (3.58) satisfies $\phi \in [0, 1]$. The tangent linearization of the vector field of $(\Sigma)$ is given by the Jacobian:

$$\begin{pmatrix} 0 & 1 & 0 & 0 \\ 0 & 0 & 0 & -1 \\ 1 & 0 & 0 & -k_1 \\ 1 & 0 & 1 & -k_2 \end{pmatrix} \tag{3.59}$$

with $k_1 = 2 + \frac{9131}{900}\dot{\phi}(x_4)$ and $k_2 = 1 + \frac{1837}{180}\dot{\phi}(x_4)$. The proof is based on the application of the Routh-Hurwitz criterion. One finds that a necessary and sufficient condition such that this Jacobian is Hurwitz for all $x_4 \in \mathbb{R}$ is that $0 < \dot{\phi} < \frac{91310}{5511}$. Notice that the characteristic function in (3.58) satisfies $0 \leq \dot{\phi}$, and not the strict inequality, so it is not yet a suitable nonlinearity. This will not preclude the proof from working, as we shall see later, because one will show the existence of a characteristic function that is close to this one and which satisfies the Hurwitz condition. Actually the $\phi$ in (3.58) will be useful to show that $(\Sigma)$ possesses a stable periodic orbit, and that there exists a slightly perturbed system (that is a system very close to the one in (3.58)) which also possesses such an orbit and which satisfies the assumption of Kalman's conjecture.

**Remark 3.3** The reader may wonder how such a counterexample has been discovered and worked out. Actually the authors in [19] started from another counterexample provided in [13] (but the arguments therein appeared to be incomplete) and employed a numerical simulation procedure to locate a periodic trajectory by trying different parameters. This explains the somewhat surprizing and *ad hoc* values of the parameters in (3.58).

**Construction of a periodic orbit $\Omega(t)$ for (3.58):** The construction of a periodic orbit relies on the explicit integration of the trajectories of (3.58) in the domains where $\phi(\cdot)$ is linear, i.e. $D_0 = \{x : |x_4| \leq \frac{900}{9185}\}$ and $D_- = \{x : x_4 < -\frac{900}{9185}\}$ respectively. Actually since the system is symmetric with respect to the origin (check that the vector field satisfies $f(-x) = -f(x)$) it is not worth considering the domain $D_+ = \{x : x_4 > \frac{900}{9185}\}$ in the limit cycle construction. These domains are separated by the hyperplanes $\Gamma_\pm = \{x \in \mathbb{R}^4 : x_4 = \pm\frac{900}{9185}\}$. These planes will serve later as Poincaré sections for the definition of a Poincaré map and the stability study. See figure 3.7.
The procedure is as follows: let us consider an initial point $x_0 \in \mathbb{R}^4$ in the state space, and choose for instance $x_0 \in \Gamma^-$. The periodic orbit, if it exists, may consist of the concatenation of solutions of the system as $x$ evolves through the 3 domains $D_0$, $D_\pm$ within which the vector field is linear. In each domain one can explicitly obtain the solutions. Then the existence of such a periodic orbit simply means that the state $\bar{x}$ attained by the system after having integrated it sequentially through $D_-$, $D_0$, $D_+$ and $D_0$ again, satisfies $\bar{x} = x_0$. From the very basic properties of solutions to linear differential equations like continuous dependence on initial data, this gives rise to a nonlinear system $g(z) = 0$, where $z$ contains not only the state $x_0$ but the period of the searched orbit as well. In other words, the existence proof is transformed into the existence of the zero of a certain nonlinear system.

**Remark 3.4** Such a manner of proving the existence of periodic orbits has also been widely used in vibro-impact mechanical systems, and is known in that field as the Kobrinskii's method [27] [12].

Let us now investigate the proof in more detail. Due to the above mentioned symmetry, it happens to be sufficient to apply the described concatenation method in the domains $D_0$ and $D_-$. In these domains the system in (3.58) becomes:

$$(\Sigma_0)\begin{cases} \dot{x}_1 = x_2 \\ \dot{x}_2 = -x_4 \\ \dot{x}_3 = x_1 - \frac{10931}{900}x_4 \\ \dot{x}_4 = x_1 + x_3 - \frac{2017}{180}x_4 \end{cases} \tag{3.60}$$

and

$$(\Sigma_-)\begin{cases} \dot{x}_1 = x_2 \\ \dot{x}_2 = -x_4 \\ \dot{x}_3 = x_1 - 2x_4 + \frac{9131}{9185} \\ \dot{x}_4 = x_1 + x_3 - x_4 + 1 \end{cases} \tag{3.61}$$

respectively. For the sake of briefness we will not provide here the whole expressions of the solutions of (3.60) and (3.61), but only those of $x_1$. In case of the system in (3.60) it is given by:

$$\begin{aligned} x_1(t) = &\\ +a_1 &\left(\frac{81}{79222\exp(10t)} - \frac{25}{2002\exp(\frac{6}{5t})} + \frac{25496\cos(\frac{\sqrt{10799}t}{360})}{25207\exp(\frac{t}{360})} - \frac{17704\sin\left(\frac{\sqrt{10799}t}{360}\right)}{25207\sqrt{10799}\exp(\frac{t}{360})}\right)\\ +a_2 &\left(-\frac{81}{792220\exp(10t)} + \frac{125}{12012\exp(\frac{6t}{5})} - \frac{3896\cos\left(\frac{\sqrt{10799}t}{360}\right)}{378105\exp(\frac{t}{360})} + \frac{137674504\sin\left(\frac{\sqrt{10799}t)}{360}\right)}{378105\sqrt{10799}\exp(\frac{t}{360})}\right)\\ +a_3 &\left(\frac{45}{39611\exp(10t)} - \frac{75}{1001\exp(\frac{6t}{5})} + \frac{1860\cos\left(\sqrt{10799}t\right)}{25207\exp(\frac{t}{360})} - \frac{710940\sin\left(\frac{\sqrt{10799}t}{360}\right)}{25207\sqrt{10799}\exp(\frac{t}{360})}\right)\\ +a_4 &\left(-\frac{450}{39611\exp(10t)} + \frac{90}{1001\exp(\frac{6t}{5})} - \frac{1980\cos\left(\frac{\sqrt{10799}t}{360}\right)}{25207\exp(\frac{t}{360})} - \frac{4140\sin\left(\frac{\sqrt{10799}t}{360}\right)}{1939\sqrt{10799}\exp(\frac{t}{360})}\right) \end{aligned} \tag{3.62}$$

and in case of (3.61) one finds:

$$
\begin{aligned}
x_1(t) &= a_1\left(\cos(t)+\sin(t)-\frac{2\sin\left(\frac{\sqrt{3}t}{2}\right)}{\sqrt{3}\exp(\frac{t}{2})}\right)\\
&+a_2\left(-\cos(t)+\sin(t)+\frac{\sin\left(\frac{\sqrt{3}t}{2}\right)}{\sqrt{3}\exp(\frac{t}{2})}+\frac{\cos\left(\frac{\sqrt{3}t}{2}\right)}{\exp(\frac{t}{2})}\right)\\
&+a_3\left(\cos(t)-\frac{\sin\left(\frac{\sqrt{3}t}{2}\right)}{\sqrt{3}\exp(\frac{t}{2})}-\frac{\cos(\left(\frac{\sqrt{3}t}{2}\right)}{\exp(\frac{t}{2})}\right)\\
&+a_4\left(-\sin(t)+\frac{2\sin\left(\frac{\sqrt{3}t}{2}\right)}{\sqrt{3}\exp(\frac{t}{2})}\right)\\
&+\cos(t)-\frac{9131}{9185}+\frac{9131}{9185}\sin(t)-\frac{18316\sin\left(\frac{\sqrt{3}t}{2}\right)}{9185\sqrt{3}\exp(\frac{t}{2})}-\frac{54\cos\left(\frac{\sqrt{3}t}{2}\right)}{9185\exp(\frac{t}{2})}
\end{aligned}
\tag{3.63}
$$

where the initial condition for the integration is $x(0) = (a_1, a_2, a_3, a_4)$. The expressions for the other components of the state are quite similar.
Let us now consider the construction of the nonlinear system $g(z) = 0$, $g \in I\!R^5$. The initial point from which the periodic solution is built is chosen in [19] as $(a_1, a_2, a_3, -\frac{900}{9185})$, i.e. it belongs to the boundary $\Gamma_-$ defined above. Due to the symmetry of the system in (3.58) it is sufficient in fact to construct only one half of this trajectory. In other words the existence can be checked as follows:

- Calculate the time $T > 0$ that the solution of system (3.61) needs, in forwards time, to go from the state $(a_1, a_2, a_3, -\frac{900}{9185})$ to the hyperplane $\Gamma_-$; i.e. $T = \min\{t : t > 0, \phi_2(t; 0, x(0)) \in \Gamma_-\}$.
- Calculate the time $-\tau < 0$ that the solution of system (3.60) needs, in backwards time, to go from the state $(-a_1, -a_2, -a_3, \frac{900}{9185}) \in \Gamma_+$, to the hyperplane $\Gamma_-$; i.e. $-\tau = \max\{\bar{\tau} : \bar{\tau} < 0, \phi_1(\bar{\tau}; 0, -x(0)) \in \Gamma_-\}$.
- Check that $\phi_2(T; 0, x(0)) = \phi_1(-\tau; 0, -x(0))$, i.e. both portions of trajectories coincide when attaining $\Gamma_-$.

This is depicted in figure 3.7, where one half of the searched orbit is drawn. We have denoted the solution of system (3.60) as $\phi_1$ and that of (3.61) as $\phi_2$. Actually the third item represents the nonlinear system $g(z) = 0$, with $z^T = (\tau, T, a_1, a_2, a_3)$. One gets:

$$
\begin{aligned}
g_1(z) &= \phi_{2,1}(T; 0, x(0)) - \phi_{1,1}(\tau; 0, -x(0))\\
g_2(z) &= \phi_{2,2}(T; 0, x(0)) - \phi_{1,2}(\tau; 0, -x(0))\\
g_3(z) &= \phi_{2,3}(T; 0, x(0)) - \phi_{1,3}(\tau; 0, -x(0))\\
g_4(z) &= \phi_{2,4}(T; 0, x(0)) + \frac{900}{9185}\\
g_5(z) &= \phi_{2,4}(\tau; 0, -x(0)) + \frac{900}{9185}
\end{aligned}
\tag{3.64}
$$

For instance, one has:

$$
\begin{aligned}
&g_1(z) = \\
&-\frac{9131}{9185} + \cos(T) - \frac{54\cos\left(\frac{\sqrt{3}t}{2}\right)}{9185\exp(\frac{T}{2})} + \frac{9131\sin(T)}{9185} - \frac{18316\sin\left(\frac{\sqrt{3}T}{2}\right)}{9185\sqrt{3}\exp(\frac{T}{2})} \\
&+a_1\left(\cos(t) + \sin(T) - \frac{2\sin\left(\frac{\sqrt{3}T}{2}\right)}{\sqrt{3}\exp(\frac{T}{2})} - \frac{25\exp(\frac{6\tau}{5})}{2002} + \frac{81\exp(10\tau)}{79222}\right. \\
&\left.+\frac{25496\exp(\frac{\tau}{360})\cos(\frac{\sqrt{10799}\tau}{360})}{25207} + \frac{17704\exp(\frac{\tau}{360})\sin\left(\frac{\sqrt{10799}\tau}{360}\right)}{25207\sqrt{10799}}\right) \\
&+a_2\left(-\cos(T) + \sin(T) + \frac{\cos\left(\frac{\sqrt{3}T}{2}\right)}{\exp(\frac{T}{2})} + \frac{\sin\left(\frac{\sqrt{3}T}{2}\right)}{\sqrt{3}\exp(\frac{T}{2})} + \frac{125\exp(\frac{6\tau}{5})}{12012}\right. \\
&\left.-\frac{81\exp(10\tau)}{792220} - \frac{3896\exp(\frac{\tau}{360})\cos\left(\frac{\sqrt{10799}\tau}{360}\right)}{378105} - \frac{137674504\exp(\frac{\tau}{360})\sin\left(\frac{\sqrt{10799}\tau}{360}\right)}{378105\sqrt{10799}}\right) \\
&+a_3\left(\cos(T) - \frac{\cos\left(\frac{\sqrt{3}T}{2}\right)}{\exp(\frac{T}{2})} - \frac{\sin\left(\frac{\sqrt{3}T}{2}\right)}{\sqrt{3}\exp(\frac{T}{2})} - \frac{75\exp(\frac{6\tau}{5})}{1001} + \frac{45\exp(10\tau)}{39611}\right. \\
&\left.+\frac{1860\exp(\frac{\tau}{360})\cos\left(\frac{\sqrt{10799}\tau}{360}\right)}{25207} + \frac{710940\exp(\frac{\tau}{360})\sin\left(\frac{\sqrt{10799}\tau}{360}\right)}{25207\sqrt{10799}}\right) \\
&+\frac{900}{9185}\left(\sin(T) - \frac{2\sin\left(\frac{\sqrt{3}T}{2}\right)}{\sqrt{3}\exp(\frac{T}{2})} - \frac{90\exp(\frac{6\tau}{5})}{1001} + \frac{450\exp(10\tau)}{39611}\right. \\
&\left.+\frac{1980\exp(\frac{\tau}{360})\cos\left(\frac{\sqrt{10799}\tau}{360}\right)}{25207} - \frac{4140\exp(\frac{\tau}{360})\sin\left(\frac{\sqrt{10799}\tau}{360}\right)}{1939\sqrt{10799}}\right)
\end{aligned}
\tag{3.65}
$$

As we announced above we will not write down the whole vector $g$ here, the rest of the entries having similar expressions. The next step is therefore to find a zero of the system $g(z) = 0$. Actually there exists many different results in the Applied Mathematics literature – see [83] chapter XVIII – that provide conditions assuring the existence of a zero and a way to compute it. However they are in general of local nature, i.e. the iterative mapping that is proposed (Newton-like) converges towards the zero (which is a fixed point of this mapping) only locally. In order to cope with this feature, Bernat and Llibre first locate numerically a periodic orbit for (3.58) and notice that it passes close to the point $(a_1, a_2, a_3, -\frac{900}{9185})$ with $a_1 = 0.2227501959407$, $a_2 = -2.13366751019745$, $a_3 = -1.3951391555710$, whereas $T = 0.4317679732343$, $\tau = 4.1523442055633$. Obviously these are approximate values. The value of $g(z)$ at this point is equal to $(3.91 \cdot 10^{-14}, 4.95 \cdot 10^{-11}, 5.73 \cdot 10^{-10}, -1.67 \cdot 10^{12}, -4.84 \cdot 10^{-10})$, that is quite close to zero indeed. The so-called Newton-Kantorovich theorem is used to prove that in a neighbourhood of this point

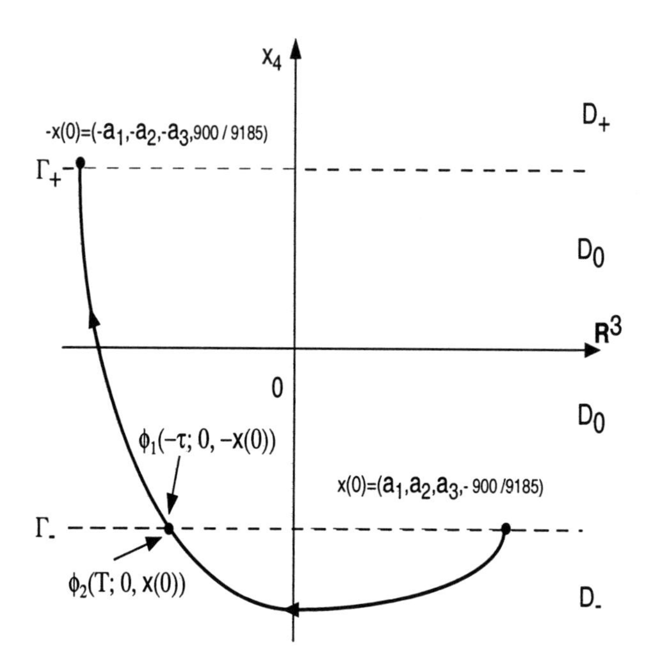


Figure 3.7: Construction of the periodic orbit.

there exists a zero. Let $B_r(x_0)$ be the open ball of radius $r$ centered at $x_0$, and $\bar{B}_r(x_0)$ its closure (the closed ball).

**Theorem 3.4 (Newton-Kantorovich)** Given a $C^1$ function $g : C \subset \mathbb{R}^n \longrightarrow \mathbb{R}^n$ and a convex set $C_0 \subset C$, assume that the following assumptions hold:

- $|Dg(z) - Dg(y)| \leq \gamma |z - y|$ for all $z, y \in C_0$,
- $|Dg(z_0)^{-1} g(z_0)| \leq \alpha$,
- $|Dg(z_0)^{-1}| \leq \beta$,

for some $z_0 \in C_0$. Consider $h = \alpha\beta\gamma$, $r_{1,2} = \frac{1 \pm \sqrt{1-2h}}{h}\alpha$. If $h \leq \frac{1}{2}$ and $\bar{B}_{r_1}(z_0) \subset C_0$, then the sequence $\{z_k\}_{k \geq 0}$ defined as:

$$z_{k+1} = z_k - Dg(z_k^{-1}) g(z_k) \tag{3.66}$$

is contained in the ball $B_{r_1}(z_0)$ and converges towards the unique zero of $g(z)$ that is inside the set $C_0 \subset B_{r_2}(z_0)$. ♠♠

$Dg(x)$ denotes the Jacobian of $g(\cdot)$ computed at $x$. The authors in [19] choose

$$C_0 = [0.4, 0.5] \times [4.1, 4.2] \times [0.17, 0.27] \times [-2.1, 2.2] \times [-1.33, -1.45]$$

$$z_0 = (0.4317679732343, 4.1523442055633, 0.2227501959407,$$

$$-2.13366751019745, -1.3951391555710)$$

and take the $\|\cdot\|_\infty$ matrix norm. As one can see the application of the theorem requires the computation of Jacobians and bounds on their norms. The whole thing takes 16 journal pages in [19], and is omitted here for obvious reasons. All the computations are made with an accuracy of $10^{-20}$ and the numerical errors are monitored. All the parameters appearing in theorem 3.4 are calculated and the conditions are fulfilled. So the existence of a zero $\bar{z}_0$ is shown, consequently the system (3.58) possesses a periodic orbit $\Omega(t)$ that passes through $\bar{x}_0$, where $\bar{z}_0 = (\bar{T}_0, \bar{\tau}_0, \bar{x}_0)$.

**Stability of the periodic orbit $\Omega(t)$:** The study of the stability of periodic trajectories can be classically attacked with Poincaré maps.
Due to the way the trajectory $\Omega(t)$ has been built, one suspects that the Poincaré section will be chosen to be $\Gamma_-$, whereas the Poincaré map will be the concatenation of four maps:

$$\begin{aligned} P_1 : &\quad B_r(\bar{x}_0) \cap \Gamma_- \longrightarrow \Gamma_- \\ P_2 : &\quad B_r(\bar{x}_1) \cap \Gamma_- \longrightarrow \Gamma_+ \\ P_3 : &\quad B_r(-\bar{x}_0) \cap \Gamma_+ \longrightarrow \Gamma_+ \\ P_4 : &\quad B_r(-\bar{x}_1) \cap \Gamma_+ \longrightarrow \Gamma_- \end{aligned} \tag{3.67}$$

where obviously $\bar{x}_1 \in \Gamma_-$ is a point that belongs to $\Omega(t)$. In a neighbourhood of $\Omega(0) = \bar{x}_0$ the Poincaré map is defined as $P = P_4 \circ P_3 \circ P_2 \circ P_1 : B_r(\bar{x}_0) \cap \Gamma_- \longrightarrow \Gamma_-$. The local stability analysis consists of studying the eigenvalues of the Jacobian $DP(\bar{x}_0)$. The chain rule yields $DP(\bar{x}_0) = DP_4(P_3 \circ P_2 \circ P_1(\bar{x}_0)).DP_3(P_2 \circ P_1(\bar{x}_0)).DP_2(P_1(\bar{x}_0)).DP_1(\bar{x}_0) = DP_4(-\bar{x}_1).DP_3(-\bar{x}_0).DP_2(\bar{x}_1).DP_1(\bar{x}_0)$. The solution of system (3.61) that passes at $t = 0$ through the point $x \in B_r(\bar{x}_0) \cap \Gamma_-$ is denoted as $\phi_2(t; x)$. If $\bar{T} > 0$ is the smallest time such that $\phi_2(\bar{T}; x) \in \Gamma_-$, then $P_1(x) = E\phi_2(\bar{T}; x)$, where $E \in I\!R^{4\times 4}$ is equal to the identity matrix except for its last row whose entries are all zeros (recall that the system we deal with is an autonomous 4-dimensional system with a codimension 1 Poincaré section, so that the Poincaré map has dimension 3). Hence:

$$DP_1(\bar{x}_0) = \left( \frac{\partial \phi_{2,i}(\bar{T}; x)}{\partial x_j} \right)_{x=\bar{x}_0} \in I\!R^{3\times 3} \tag{3.68}$$

for $1 \leq i, j \leq 3$. One has:

$$\begin{aligned} \frac{\partial \phi_{2,i}(\bar{T};x)}{\partial x_j} &= \frac{\partial \phi_{2,i}}{\partial \bar{T}}\frac{\partial \bar{T}}{\partial x_j} + \frac{\partial \phi_{2,i}}{\partial x}\frac{\partial x}{\partial x_j} \\ &= \frac{\partial \phi_{2,i}}{\partial \bar{T}}\frac{\partial \bar{T}}{\partial x_j} + \frac{\partial \phi_{2,i}}{\partial x_j} \end{aligned} \tag{3.69}$$

Since the expressions for the solutions are known, the partial derivatives of $\phi_{2,i}$ can be calculated. The term $\frac{\partial \bar{T}}{\partial x_j}$ can be obtained from $\phi_{2,4}(\bar{T};x) = -\frac{900}{9185}$. Plugging this into (3.69) yields:

$$\frac{\partial \bar{T}}{\partial x_j} = -\frac{\frac{\partial \phi_{2,4}}{\partial x_j}}{\frac{\partial \phi_{2,4}}{\partial \bar{T}}} \tag{3.70}$$

At this stage one should recall that the zero $\bar{z}_0$ of $g(z) = 0$ is not known exactly, only its existence in a neighbourhood of a known point has been established. So one is led to make the computations with the numerical approximation, and to monitor the numerical error afterwards. The computation of the Jacobian is therefore done with the values computed above, i.e. the first three components of $x_0$ equal to $(0.2227501959407, -2.13366751019745, -1.3951391555710)$, whereas the time $T$ is taken as $T_0 = 4.1523442055633$ s. The other Jacobians are computed in an analogous way, and one finally finds that the three eigenvalues of $DP(x_0)$ are equal to $0.305, 0.006, 9.1\,10^{-6}$. Then one concludes that the eigenvalues of $DP(\bar{x}_0)$ also are smaller than 1, using a result on the characterization of the error in the calculation of eigenvalues of diagonalizable matrices.

**Summary:** The system in (3.58) has been proved to possess a periodic orbit *via* a classical method that consists of constructing *a priori* an orbit $\Omega$ and then of proving that it does exist by showing the existence of the zero of some nonlinear system. Since the main problem associated to such a method is to "guess" that the constructed orbit has some chance to exist, a preliminary numerical study has been done to locate an approximation of the orbit. Then investigating the local stability of $\Omega(t)$ by computing the Jacobian of its Poincaré map (at this stage the reader should remark that we do not care about the explicit knowledge of the Poincaré map itself: the essential point is that we are able to calculate, in an approximate way here, its Jacobian and consequently the eigenvalues of the Jacobian). The system in (3.58) does not exactly fit within the Kalman's conjecture assumptions since it does not satisfy $\dot{\phi} > 0$. The next step thus completes the counterexample by using a property of structural stability of the perturbed vector field in (3.58).

**The counterexample:** Let us denote $T_\mu$ the Poincaré map with Poincaré section $\Gamma_-$, defined from the flow of the system in (3.58), with characteristic function $\mu(\cdot)$, in the vicinity of $x_0$. Then the following is true:

**Lemma 3.6 ([19])** There exists a characteristic function $\Psi$ such that:

- $\Psi$ is $C^1$,
- $0 < \dot{\Psi}(y) < 10$ for all $y \in \mathbb{R}$,
- $\Psi$ is sufficiently close to $\phi$ in (3.58), with the $C^0$ topology in $\bar{B}_r(0)$ ([1]),
- $T_\Psi$ has a stable fixed point near $\bar{x}_0$.

where one assumes that the periodic orbit $\Omega(t) \subset B_r(0)$, $r > 0$. ♠♠

The proof is as follows: we know that $T_\phi$ has a stable fixed point $\bar{x}_0$. Due to the stability there exists a ball $\bar{B}_{r'}(\bar{x}_0) \subset D_-$ such that $T_\phi\left(\bar{B}_{r'}(\bar{x}_0)\right) \subset \bar{B}_{r'}(x_0)$. Then using the theorem of continuous dependence on initial conditions and parameters for ordinary differential equations with $C^0$ and Lipschitz continuous – in the variable $x$ – vector field, there exists a function $\Psi$ satisfying the first three items in lemma 3.6, and such that $T_\Psi\left(\bar{B}_{r'}(\bar{x}_0)\right) \subset \bar{B}_{r'}(x_0)$. In other words a slight enough perturbation of the vector field in (3.58) allows one to transform the characteristic function $\phi(\cdot)$, hence the whole vector field, into a $C^1$ function, assuring that the system is Hurwitz (its Jacobian is Hurwitz at any point of the state space). The so-called Brouwer's fixed point theorem guarantees then the existence of a fixed point for $T_\mu$ inside $B_{r'}(\bar{x}_0)$ (let us recall that Brouwer's theorem states that a continuous function $g(\cdot)$ that maps a closed ball to itself, satisfies $g(y) = y$ for some $y$). The fixed point of $T_\Psi$ corresponds to a periodic orbit of the system in (3.58) with $\Psi$ as a characteristic function. The second item in lemma 3.6 assures that such a system is Hurwitz, see [19] proposition 8.1. This system therefore consitutes a counterexample to the Kalman conjecture in dimension 4. As shown in [19] it is easily extended to the dimensions $n > 4$, by adding subsystems $\dot{x}_i = -x_i$, $i \geq 5$.

## 3.5 The circle criterion

Consider again the observable and controllable system in (3.53). Its transfer function $H(s)$

$$H(s) = C(sI - A)^{-1} B + D \tag{3.71}$$

Assume that the transfe. function $H(s)$ is SPR and is connected in negative feedback with a nonlinearity $\phi$ as illustrated in figure 3.8. The conditions for stability of such a scheme are stated in the following theorem.

[1]i.e. the distance between $\Psi$ and $\phi$ is defined from the norm of the uniform convergence as $\sup_{y \in \bar{B}_r(0)} ||\Psi(y) - \phi(y)||$.

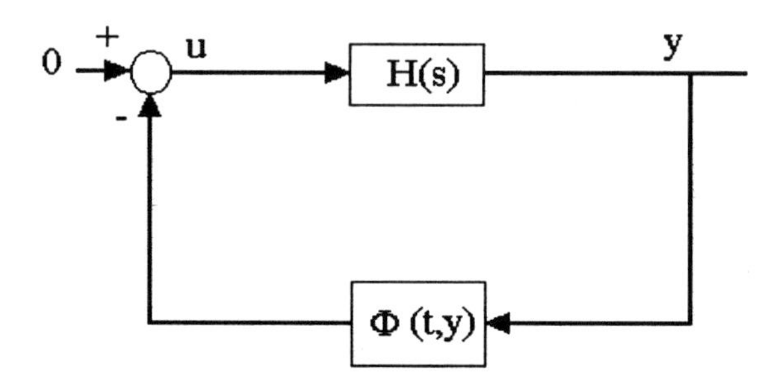


Figure 3.8: Linear system with a sector nonlinearity in negative feedback.

**Theorem 3.5** Consider the system in figure 3.8. If $H(s)$ in (3.71) is SPR and if $\phi(t,y)$ is in the sector $[0,\infty)$, i.e.:
i) $\phi(t,0) = 0 \quad \forall t \geq 0$
ii) $y^T\phi(t,y) \geq 0 \quad \forall t \geq 0, \quad \forall y \in I\!R^m$

then the origin is a globally exponentially stable equilibrium point. ♠♠

**Proof**
Since $H(s) = C(sI - A)^{-1}B + D$ is SPR, then there exist $P > 0$, $Q$ and $W$, $\epsilon > 0$ such that

$$\begin{array}{ll} A^TP + PA & = -\epsilon P - Q^TQ \\ B^TP + W^TQ & = C \\ W^TW & = D + D^T \end{array} \tag{3.72}$$

Define the Lyapunov function candidate:

$$V = x^TPx$$

then

$$\begin{aligned} \dot{V}(x(t)) &= \dot{x}^TPx + x^TP\dot{x} \\ &= [Ax - B\phi]^T Px + x^TP[Ax - B\phi] \\ &= x^T\left(A^TP + PA\right)x - \phi^TB^TPx - x^TPB\phi \end{aligned} \tag{3.73}$$

Note that:

$$B^TP = C - W^TQ$$

Hence, using the above, (3.53) and the control $u = -\phi(t,y)$, we get:

$$\begin{aligned} x^T P B \phi &= \phi^T B^T P x \\ &= \phi^T C x - \phi^T W^T Q x \\ &= \phi^T [y - Du] - \phi^T W^T Q x \\ &= \phi^T [y + D\phi] - \phi^T W^T Q x \end{aligned}$$

Substituting the above into (3.73) we get:

$$\begin{aligned} \dot{V} &= -\epsilon x^T P x - x^T Q^T Q x - \phi^T \left(D + D^T\right) \phi \\ & \quad -\phi^T W^T Q x - x^T Q^T W \phi - \phi^T y - y^T \phi \end{aligned}$$

Using (3.72) and the fact that $y^T \phi \geq 0$ we have

$$\begin{aligned} \dot{V} &\leq -\epsilon x^T P x - x^T Q^T Q x - \phi^T W^T W \phi - \phi^T W^T Q x - x^T Q^T W \phi \\ &= -\epsilon x^T P x - [Qx + W\phi]^T [Qx + W\phi] \\ &\leq -\epsilon x^T P x \end{aligned}$$

Define $\bar{z} \triangleq \dot{V} + \epsilon V \leq 0$ which can also be rewritten as $\dot{V} = -\epsilon V + \bar{z}$
Thus

$$\begin{aligned} V(x(t)) &= e^{-\epsilon t} V(0) + \int_0^t e^{-\epsilon(t-\tau)} \bar{z}(\tau)\, d\tau \\ &\leq e^{-\epsilon t} V(0) \end{aligned}$$

Finally the fixed point $x = 0$ is globally exponentially stable. ♠♠

### 3.5.1 Loop transformations

The above theorem applies when $\phi$ belongs to the sector $[0, \infty)$. In order to use the above result when $\phi$ belongs to the sector $[a, b]$ we have to make some loop transformations which are given next.

1) If $\phi \in [a, b]$ then $\phi_1 \doteq \phi(t, y) - a \in [0, b - a]$. This is illustrated in figure 3.9

2) If $\phi_1 \in [0, c]$ with $c = b - a$ then we can make the transformation indicated in figure 3.10 where $\bar{y} = \phi_2(t, \bar{u})$ and $\delta > 0$ is arbitrarily small number. Therefore, as is shown next, $\phi_2 \in [0, \infty)$.

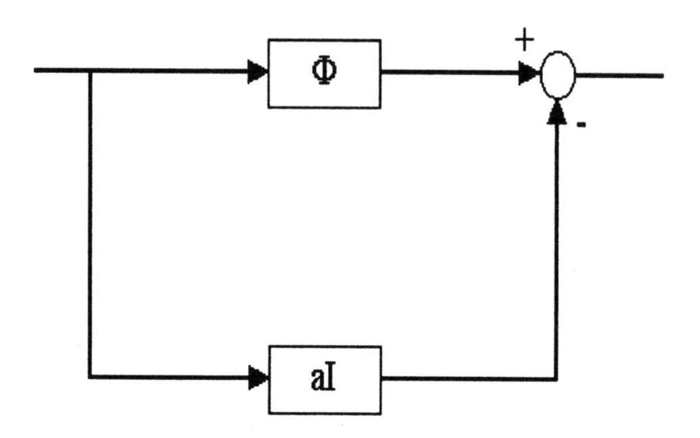


Figure 3.9: Loop transformations.

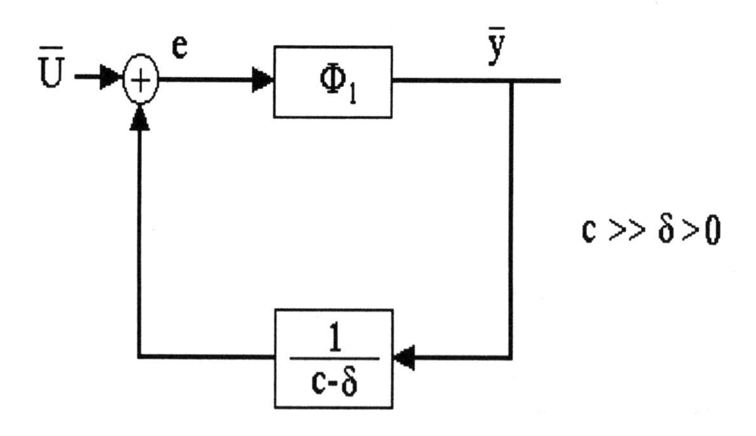


Figure 3.10: Loop transformations.

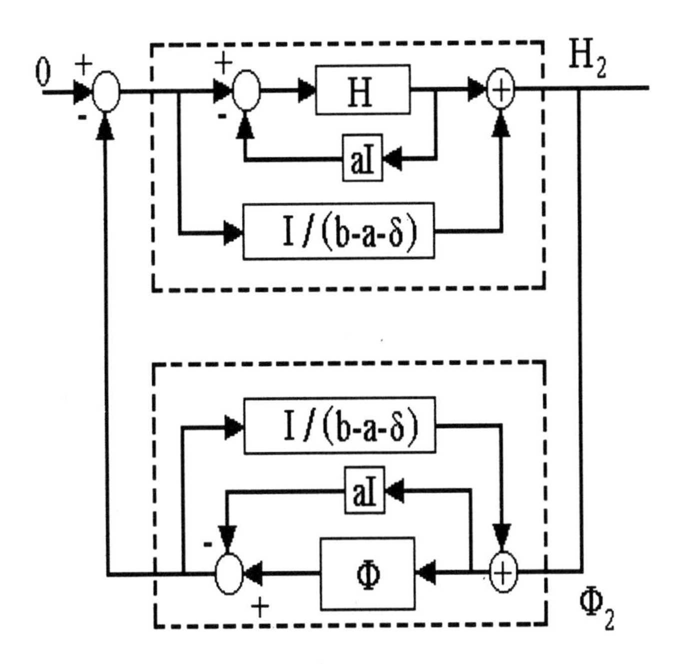


Figure 3.11: Loop transformations.

Note that if $\phi_1 = \bar{c}$, then

$$\bar{y} = \frac{\bar{c}}{1 - \frac{\bar{c}}{c-\delta}} = \frac{\bar{c}\,(c-\delta)}{c-\bar{c}-\delta}\bar{u}$$

Therefore:

1. if $\bar{c} = c$, $\quad \lim_{\delta \to 0} \frac{\bar{y}}{\bar{u}} = \infty$

2. if $\bar{c} = 0$, $\quad \frac{\bar{y}}{\bar{u}} = 0$

Using the two transformations described above, the system in figure 3.8 can be transformed into the system in figure 3.11. We then have the following corollary:

**Corollary 3.1** If $H_2$ in figure 3.11 is SPR and the nonlinearity $\phi \in [0, \infty)$ then the closed-loop system is globally exponentially stable. ♠♠

Note that $H_2$ is SPR if and only if

$$H_1(j\omega) + H_1^*(j\omega) + \frac{2I}{b-a-\delta} > 0$$

with $H_1(s) = H(s)[I + aH(s)]^{-1}$ and $\delta << 1$. ♠♠
For $m = 1$ the above result has a graphical interpretation which leads to the **circle criterion.**
Suppose $z = x + jy$ is a complex number and $a, b \in I\!R$ with $a < b$, $a \neq 0$. Consider the condition

$$\eta = \mathbf{Re}\left\{\frac{z}{1+az} + \frac{1}{b-a}\right\} > 0$$

Now one has

$$\begin{aligned}\frac{z}{1+az} + \frac{1}{b-a} &= \frac{x+jy}{1+a(x+jy)} + \frac{1}{b-a}\\ &= \frac{x+jy[1+ax-jay]}{(1+ax)^2+y^2a^2} + \frac{1}{b-a}\end{aligned}$$

Therefore:

$$\eta = \frac{x(1+ax)+ay^2}{(1+ax)^2+y^2a^2} + \frac{1}{b-a} > 0$$

or equivalently

$$\begin{aligned}&0 < (b-a)\left\{x(1+ax) + ay^2\right\} + (1+ax)^2 + y^2a^2\\ &= (b-a)\left\{x + ax^2 + ay^2\right\} + 1 + 2ax + a^2x^2 + a^2y^2 \qquad (3.74)\\ &= ba\left\{x^2+y^2\right\} + x(b+a) + 1\end{aligned}$$

which implies

$$bay^2 + ba\left(x + \frac{a+b}{2ab}\right)^2 + 1 - \frac{(a+b)^2}{4ab} > 0$$

Note that

$$1 - \frac{(a+b)^2}{4a^2b^2} = \frac{4ab - a^2 - 2ab - b^2}{4ab} = -\frac{(a-b)^2}{4ab}$$

Introducing the above into (3.74) we get

$$bay^2 + ba\left(x + \frac{a+b}{2ab}\right)^2 > \frac{(a-b)^2}{4ab}$$

If $ab > 0$ this can be written as

$$y^2 + ba\left(x + \frac{a+b}{2ab}\right)^2 > \frac{(a-b)^2}{4a^2b^2}$$

or

$$\left|z + \frac{a+b}{2ab}\right| > \frac{|a-b|}{2|ab|}$$

If $ab < 0$ then

$$\left|z + \frac{a+b}{2ab}\right| < \frac{|a-b|}{2|ab|}$$

Let $D(a,b)$ denote the closed disc in the complex plane centered at $\frac{a+b}{2ab}$ and with radius $\frac{|a-b|}{2|ab|}$. Then

$$\mathbf{Re}\left\{\frac{z}{1+az} + \frac{1}{b-a}\right\} > 0$$

if and only if

$$\left|z + \frac{a+b}{2ab}\right| > \frac{|a-b|}{2|ab|}, \qquad ab > 0$$

In other words, the complex number $z$ lies outside the disc $D(a,b)$ in case $ab > 0$ and lies in the interior of the disc $D(a,b)$ in case $ab < 0$. We therefore have the following important result.

**Theorem 3.6 (Circle criterion)** Consider again the system for m=1 in figure 3.11. The closed loop system is globally exponentially stable if:

**(i)** $0 < a < b$ : The plot of $h(j\omega)$ lies outside and is bounded away from the disc $D(a,b)$. Moreover the plot encircles $D(a,b)$ exactly $\nu$ times in the counter-clockwise direction, where $\nu$ is the number of eigenvalues of $A$ with positive real part.

**(ii)** $0 = a < b$ : $A$ is a Hurwitz matrix and

$$\mathbf{Re}\left\{H(j\omega) + \frac{1}{b}\right\} > 0 \tag{3.75}$$

**(iii)** $a < 0 < b$ : $A$ is a Hurwitz matrix; the plot of $h(j\omega)$ lies in the interior of the disc $D(a,b)$ and is bounded away from the circumference of $D(a,b)$.

**(iv)** $a < b \leq 0$ : Replace $h(.)$ by $-h(.)$, $a$ by $-b$, $b$ by $-a$ and apply (i) or (ii) as appropriate. ♠♠

**Remark 3.5** If $b - a \to 0$ the "critical disc" $D(a,b)$ in case (i) shrinks to the "critical point"0 $-1/a$ of the Nyquist criterion. The circle criterion is applicable to time-varying and/or nonlinear systems, whereas the Nyquist criterion is only applicable to linear time invariant systems.

## 3.6 The Popov criterion

Unlike the circle criterion, the Popov criterion [155], [156], [157] is applicable only to autonomous SISO systems:

$$\begin{aligned} \dot{x} &= Ax + bu \\ \dot{\xi} &= u \\ y &= cx + d\xi \\ u &= -\phi(y) \end{aligned}$$

where $u, y \in I\!R$, $y \in \phi : I\!R \longrightarrow I\!R$ is a time-invariant nonlinearity belonging to the open sector $(0, \infty)$ ; i.e.

$$\phi(0) = 0,\ y\phi(y) > 0\ ;\ \forall y \neq 0$$

The linear part can also be written as:

$$\begin{aligned} \begin{bmatrix} \dot{x} \\ \dot{\xi} \end{bmatrix} &= \begin{bmatrix} A & 0 \\ 0 & 0 \end{bmatrix} \begin{bmatrix} x \\ \xi \end{bmatrix} + \begin{bmatrix} b \\ 1 \end{bmatrix} u \\ y &= \begin{bmatrix} c & d \end{bmatrix} \begin{bmatrix} x \\ \xi \end{bmatrix} \end{aligned} \tag{3.76}$$

Hence the transfer function is

$$h(s) = \frac{d}{s} + c(sI - A)^{-1}b$$

which has a pole at the origin. We can now state the following result:

**Theorem 3.7 (Popov's criterion)** Consider the system in (3.76). Assume that

1. $A$ is Hurwitz
2. $(A, b)$ is controllable
3. $(c, A)$ is observable
4. $d > 0$
5. $\phi \in (0, \infty)$

Then the system is globally asymptotically stable if there exixts $r > 0$ such that: $\mathbf{Re}[(1 + j\omega r)h(j\omega)] > 0 \quad \forall\ \omega \in I\!R$. ♠♠

**Remark 3.6** Contrary to Popov's criterion, the circle criterion does not apply to systems with a pole at $s = 0$ and $\phi \in (0, \infty)$ ♠♠

**Proof**
Note that

$$\begin{aligned} s(sI - A)^{-1} &= (sI - A + A)(sI - A)^{-1} \\ &= I + A(sI - A)^{-1} \end{aligned}$$

Hence

$$\begin{aligned} (1 + rs)h(s) &= (1 + rs)\left[\tfrac{d}{s} + c\,(sI - A)^{-1}\,b\right] \\ &= \tfrac{d}{s} + rd + c(sI - A)^{-1}b \\ &\quad +rcb + rcA(sI - A)^{-1}b \end{aligned}$$

Note that $\frac{d}{j\omega}$ is purely imaginary. From the above and by assumption we have

$$\mathbf{Re}\left[(1 + j\omega r)h(j\omega)\right] = \mathbf{Re}\left[r\,(d + cb) + c\,(I + rA)\,(j\omega - A)^{-1}b\right] > 0$$

Define the transfer function

$$g\,(s) = r\,(d + cb) + c\,(I + rA)\,(sI - A)^{-1}\,b$$

i.e. $\{A, b, c\,(I + rA)\,, r\,(d + cb)\}$ is a minimal realization of $g(s)$. If $\mathbf{Re}\,[g(\omega)] > 0$ then there exists $P > 0, q$ and $\omega$ and $\epsilon > 0$ such that

$$\begin{aligned} A^T P + PA &= -\epsilon P - q^T q \\ b^T P + \omega q &= c(I + rA) \\ \omega^2 &= 2r(d + cb) \end{aligned}$$

Choose the Lyapunov function candidate

$$V(x, \xi) = x^T P x + d\xi^2 + 2r\int_0^y \phi\,(\sigma)\,d\sigma$$

Given that $\phi \in [0, \infty)$ it then follows that $\int \phi\,(\sigma)\,d\sigma \geq 0$. Hence $V$ is positive definite and radially unbounded

$$\begin{aligned} \dot{V} &= \dot{x}^T\,Px + x^T P\,\dot{x} + 2d\xi\,\dot{\xi} + 2r\phi(y)\,\dot{y} \\ &= (Ax - b\phi)^T Px + x^T P(Ax - b\phi) \\ &\quad -2d\xi\phi + 2r\phi\,[c\,(Ax - b\phi) - d\phi] \end{aligned}$$

Note from (3.76) that $d\xi = y - cx$, thus

$$\begin{aligned}
\dot{V} &= x^T(A^TP+PA)x - 2\phi b^TPx \\
&\quad +2\phi c(I+rA)x - 2r\,(d+cb)\,\phi^2 - 2y\phi \\
&= -\epsilon x^TPx - (qx-\omega\phi)^2 \\
&\quad -r\,(d+cb)\,\phi^2 - 2y\phi
\end{aligned}$$

Since $g(j\omega) \to r(d+cb)$ as $\omega \to \infty$ it follows that $r(d+cb) > 0$. Hence

$$\dot{V} \leq -\epsilon x^TPx - 2y\phi \leq 0 \;\; \forall x, \epsilon$$

We now show that $\dot{V}\,(x,\xi) < 0$ if $(x,\xi) \neq (0,0)$. If $x \neq 0$ then $\dot{V} < 0$ since $P > 0$. If $x = 0$ but $\xi \neq 0$, then $y = d\xi \neq 0$, and $\phi y > 0$ since $\phi \in [0,\infty)$. Therefore the system (3.76) is globally asymptotically stable. ♠♠

**Corollary 3.2** Suppose now that $\phi \in (0,k)\,, k > 0$. Then the system is globally asymptotically stable if there exists $r > 0$ such that

$$\mathbf{Re}\left[(1+j\omega r)h(j\omega)\right] + \frac{1}{k} > 0$$

**Proof**

It follows from the loop transformation in figure 3.12, where:

$$\begin{aligned}
\phi_1 &= \phi\left[I - \tfrac{1}{k}\phi\right]^{-1} \\
g_1 &= g(s) + \tfrac{1}{k} \\
&= (1+j\omega r)(hj\omega) + \tfrac{1}{k} \\
\mathbf{Re}(g_1) &= \mathbf{Re}[h(j\omega)] + r\omega\mathbf{Im}[h(j\omega)] + \tfrac{1}{k} > 0
\end{aligned}$$

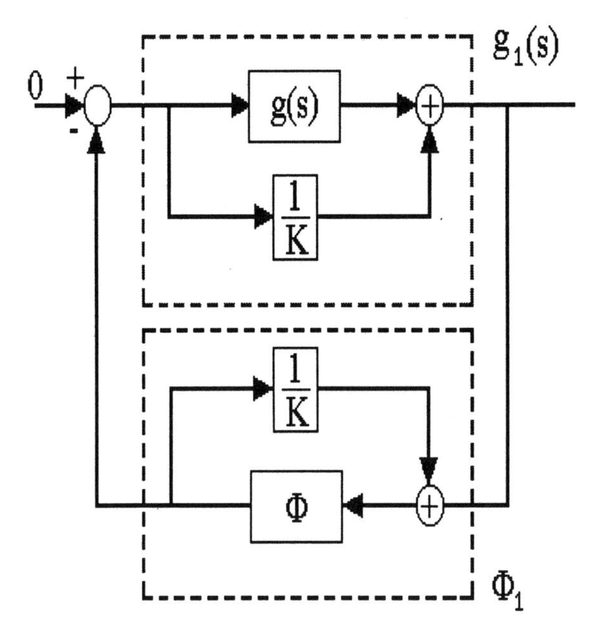


Figure 3.12: Loop transformations.

# Chapter 4

# Dissipative Systems

In this chapter we will further study the concept of dissipative system which is a very useful tool in the analysis and synthesis of control laws for linear and nonlinear dynamical systems. One of the key properties of a dissipative dynamical system is that the total energy stored in the system decreases with time. Dissipativeness can be considered as an extension to the nonlinear case of the definition of PR systems. Some relationships between Positive Real and Passive systems have been established in chapter 2. There exists several important subclasses of dissipative nonlinear systems with slightly different properties which are important in the analysis. Dissipativity is useful in stabilizing mechanical systems like fully actuated robots manipulators, robots with flexible joints, underactuated robot manipulators, electric motors, fully actuated and underactuated satellites [49], combustion engines [61], etc.. Some of these examples will be presented in the following chapters.
Dissipative system theory is intimately related to Lyapunov stability theory. There exists tools from the dissipativity approach that can be used to generate Lyapunov functions. A difference between the two approaches is that the state of the system and the equilibrium point are notions that are required in the Lyapunov approach while the dissipative approach is rather based on input-output behavior of the plant. The input-output properties of a closed loop system can be studied using $\mathcal{L}_p$ stability analysis. The properties of $\mathcal{L}_p$ signals can then be used to analyze the stability of a closed loop control system. $\mathcal{L}_p$ stability analysis has been studied by Desoer and Vidyasagar [45]. A clear presentation of this notions will also be given in this book since they are very useful in the stability analysis of control systems and in particular in the control of robot manipulators.
Dissipativeness of dynamical systems has been introduced by Willems [212] [213]. Hill and Moylan [65] [66] carried out an extension of the Kalman-Yakubovich-Popov (KYP) lemma to the case of nonlinear systems with state space representations that are affine in the input. Byrnes et al [38] further

developed the concept of dissipative systems and characterized the class of dissipative systems by obtaining some necessary conditions for a nonlinear system to be dissipative and studied the stabilization of dissipative systems. Before presenting the definitions of dissipative systems we will study some properties of $\mathcal{L}_p$ signals which will be useful in studying the stability of closed loop control systems.

## 4.1 Normed spaces

We will briefly review next the notations and definitions of normed spaces, $\mathcal{L}_p$ norms and properties of $\mathcal{L}_p$ signals. For a more complete presentation the reader is referred to [45] or any monograph on mathematical analysis.
Let $E$ be a linear space over the field $K$ (typically $K$ is $\mathbb{R}$ or the complex field $\mathbb{C}$). The function $\rho(.)$, $\rho : E \to \mathbb{R}^+$ is a norm on $E$ if and only if:

1. $x \in E$ and $x \neq 0 \Rightarrow \rho(x) > 0$, $\rho(0) = 0$.
2. $\rho(\alpha x) = |\alpha|\rho(x), \forall \alpha \in K, \forall x \in E$
3. $\rho(x + y) \leq \rho(x) + \rho(y)$, $\forall x, y \in E$ (triangle inequality)

## 4.2 $\mathcal{L}_p$ norms

Let $x(\cdot)$ be a function $\mathbb{R} \to \mathbb{R}$, and let $|\cdot|$ denote the absolute value. The most common signal norms are the $\mathcal{L}_1, \mathcal{L}_2, \mathcal{L}_p$ and $\mathcal{L}_\infty$ norms which are respectively defined as

$$\| x \|_1 \triangleq \int |x(t)|\, dt$$

$$\| x \|_2 \triangleq \left(\int |x(t)|^2 dt\right)^{\frac{1}{2}}$$

$$\| x \|_p \triangleq \left(\int |x(t)|^p dt\right)^{\frac{1}{p}}$$

$$\begin{aligned} \| x \|_\infty &\triangleq \operatorname*{ess\,sup}_{t \in \mathbb{R}} |x(t)| dt \\ &= \inf\{a|\ |f(t)| < a,\ a.e.\} \\ &= \sup_{t>0} |x(t)| \end{aligned}$$

We say that $f$ belongs to $\mathcal{L}_p$ if and only if $f$ is locally Lebesgue integrable (i.e. $|\int_a^b f dt| < \infty$) and $\|f(t)\|_p < \infty$.

**Proposition 4.1** If $f \in \mathcal{L}_1 \bigcap \mathcal{L}_\infty$ then $f \in \mathcal{L}_p, p \in [1, \infty]$. ♠♠

**Proof**

Since $f \in \mathcal{L}_1$, the set $A \stackrel{\triangle}{=} \{t|\ |f(t)| \geq 1\}$ has finite Lebesgue measure. Therefore, since $f \in l_\infty$

$$\int_A |f(t)|^p dt < \infty, \ \forall p \in [1, \infty)$$

Define the set $B \stackrel{\triangle}{=} \{t|\ |f(t)| < 1\}$. Then we have

$$\int_B |f(t)|^p dt \leq \int_B |f(t)| dt < \int |f(t)| dt < \infty, \forall p \in [1, \infty)$$

Finally

$$\int |f(t)|^p dt = \int_A |f(t)|^p dt + \int_B |f(t)|^p dt < \infty$$

♠♠

### 4.2.1 Relationships between $\mathcal{L}_1$, $\mathcal{L}_2$ and $\mathcal{L}_\infty$ spaces.

In order to understand the relationship between $\mathcal{L}_1$, $\mathcal{L}_2$ and $\mathcal{L}_\infty$ spaces let us consider the following examples that have been introduced in [45]:

Examples:

- $f_1(t) = 1$
- $f_2(t) = \frac{1}{1+t}$
- $f_3(t) = \frac{1}{1+t} \frac{1+t^{\frac{1}{4}}}{t^{\frac{1}{4}}}$
- $f_4(t) = e^{-t}$
- $f_5(t) = \frac{1}{1+t^2} \frac{1+t^{\frac{1}{4}}}{t^{\frac{1}{4}}}$
- $f_6(t) = \frac{1}{1+t^2} \frac{1+t^{\frac{1}{2}}}{t^{\frac{1}{2}}}$

  It can be shown that (see figure 4.1):

- $f_1 \notin \mathcal{L}_1$ , $f_1 \notin \mathcal{L}_2$ and $f_1 \in \mathcal{L}_\infty$
- $f_2 \notin \mathcal{L}_1$ , $f_2 \in \mathcal{L}_2$ and $f_2 \in \mathcal{L}_\infty$
- $f_3 \notin \mathcal{L}_1$ , $f_3 \in \mathcal{L}_2$ and $f_3 \notin \mathcal{L}_\infty$
- $f_4 \in \mathcal{L}_1$ , $f_4 \in \mathcal{L}_2$ and $f_4 \in \mathcal{L}_\infty$
- $f_5 \in \mathcal{L}_1$ , $f_5 \in \mathcal{L}_2$ and $f_5 \notin \mathcal{L}_\infty$
- $f_6 \in \mathcal{L}_1$ , $f_6 \notin \mathcal{L}_2$ and $f_6 \notin \mathcal{L}_\infty$

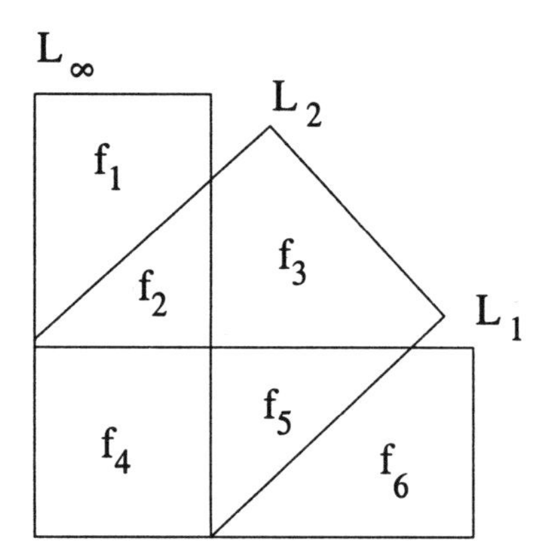


Figure 4.1: Relationships between $L_1$, $L_2$ and $L_\infty$.

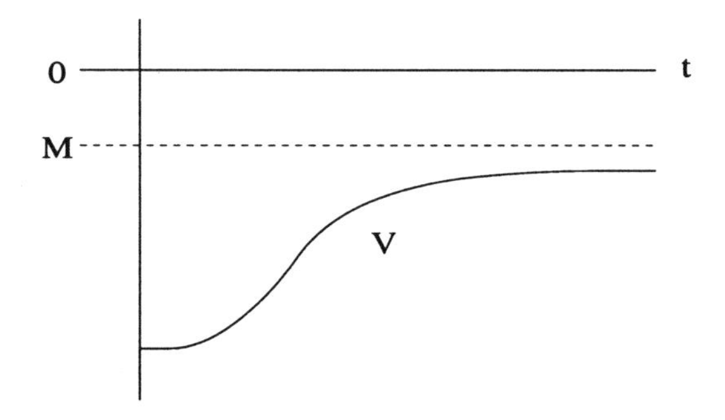


Figure 4.2: A nondecreasing function $V$.

## 4.3 Review of some properties of $\mathcal{L}_p$ signals

The following facts are very useful to prove convergence of signals under different conditions.

**Fact 1**: If $V$ is a non-decreasing function (see figure 4.2) and if $V \leq M$ for some $M \in I\!R$, then $V$ converges.

**Proof**

Since V is non-decreasing, then V can only either increase or remain constant. Assume that V does not converge to a constant limit. Then $V$ has to diverge to infinity since it cannot oscillate. In other words there exists a strictly increasing sequence of time instants $t_1, t_2, t_3$ ... and a $\delta > 0$ such that $V(t_i)+\delta < V(t_{i+1})$. However this leads to a contradiction since $V$ has upper-bound $M$. Therefore, the sequence $V(t_i)$ has a limit. ♠♠

Examples:

- $\int_0^t |s(\tau)|d\tau < \infty \;\Rightarrow\; \int_0^t |s(\tau)|d\tau$ converges
- $V \geq 0$ and $\dot{V} \leq 0 \;\Rightarrow\; V$ converges.

**Fact 2:** If $\int_0^t |f|dt$ converges then $\int_0^t f dt$ converges.

**Proof** In view of the assumption we have

$$\infty > \int_0^t |f|dt = \int_{f>0} |f|dt + \int_{f\leq 0} |f|dt$$

then both integrals in the right-hand side above converge. We also have

$$\int_0^t f dt = \int_{f>0} |f|dt - \int_{f\leq 0} |f|dt$$

then $\int_0^t f d\tau$ converges too. ♠♠

**Fact 3:** $\dot{f} \in \mathcal{L}_1$ implies that $f$ has a limit.

**Proof**

By assumption we have

$$|f(t) - f(0)| = |\int_0^t \dot{f}dt| \leq \int_0^t |\dot{f}|dt < \infty$$

Using Fact1 it follows that $\int_0^t |\dot{f}|dt$ converges. This implies that $\int_0^t \dot{f}dt$ converges which in turn implies that $f$ converges too. ♠♠

**Fact 4:** If $f \in \mathcal{L}_2$ and $\dot{f} \in \mathcal{L}_2$ then $f \to 0$ and $f \in \mathcal{L}_\infty$.

**Proof**

Using the assumptions

$$\begin{aligned} |f^2(t) - f^2(0)| &= |\textstyle\int_0^t \frac{d}{dt}[f^2]dt| \\ &\leq \textstyle\int_0^t |\frac{d}{dt}[f^2]|dt \\ &= 2\textstyle\int_0^t |f\dot{f}|dt \qquad (4.1) \\ &\leq \textstyle\int_0^t f^2 dt + \int_0^t \dot{f}^2 dt \\ &< \infty \end{aligned}$$

In view of Fact 3 it follows that $|\frac{d}{dt}[f^2]| \in \mathcal{L}_1$ which implies that $\int_0^t \frac{d}{dt}[f^2]dt$ converges which in turn implies that $f^2$ converges. But by assumption $\int_0^t f^2 dt < \infty$, then $f$ has to converge to zero. Clearly $f \in \mathcal{L}_\infty$. ♠♠

**Fact 5:** $f \in \mathcal{L}_1$ and $\dot{f} \in \mathcal{L}_1 \Rightarrow f \to 0$

**Proof**
Using fact 3 it follows that $\dot{f} \in \mathcal{L}_1 \Rightarrow f$ has a limit. Since in addition we have $\int_0^t |f|dt < \infty$ then $f$ has to converge to zero. ♠♠
Before presenting further results of $\mathcal{L}_p$ functions, some definitions are in order.

**Definition 4.1** $f(t,x)$ is said to be globally Lipschitz (with respect to $x$) if there exists a bounded $k$ such that :

$$|f(t,x) - f(t,x')| \leq k|x - x'|, \ \forall x, x' \in I\!R^n, t \in I\!R^+$$

**Definition 4.2** $f(t,x)$ is said to be Lipschitz with respect to time if there exists a bounded $k$ such that:

$$|f(t,x) - f(t',x)| \leq k|t - t'|, \ \forall \ x \in I\!R^n, t, t' \in I\!R^+$$

**Definition 4.3** $f$ is uniformly continuous in a set $\mathcal{A}$ if for all $\epsilon > 0$, there exists $\delta(\epsilon) > 0$:

$$|t - t'| < \delta \Rightarrow |f(t) - f(t')| < \epsilon, \ \forall \ t, t' \ \in \mathcal{A}$$

**Remark 4.1** Uniform continuity and Lipschitz continuity are two different notions. Any Lipschitz function is uniformly continuous. However the inverse implication is not true. For instance the function $x \mapsto \sqrt{x}$ is uniformly continuous on $[0,1]$, but it is not Lipschitz on $[0,1]$. This may be easily checked from the definitions. The criterion in Fact 6 is clearly a sufficient condition only ("very sufficient", one should say!) to assure uniform continuity of a function. Furthermore, uniform continuity has a meaning on a set. Asking whether a function is uniformly continuous at a point is meaningless [166].

**Fact 6:** $\dot{f}(t) \in \mathcal{L}_\infty \Rightarrow f(t)$ is uniformly continuous.
**Proof**
$\dot{f} \in \mathcal{L}_\infty$ implies that $f$ is Lipschitz with respect to time $t$ and that $f(t)$ is uniformly continuous.
**Fact 7** If $f \in \mathcal{L}_2$ and is Lipschitz with respect to time then $f \to 0$.
**Proof**
By assumption: $\int_0^t f^2 dt < \infty$ and $|f(t) - f(t')| \leq k|t - t'|, \ \forall t, t'$.
Assume that

$$|f(t_1)| \geq \epsilon \text{ for some } \ t_1, \epsilon > 0$$

and

$$|f(t_2)| = 0 \text{ for some } \ t_2 \geq t_1$$

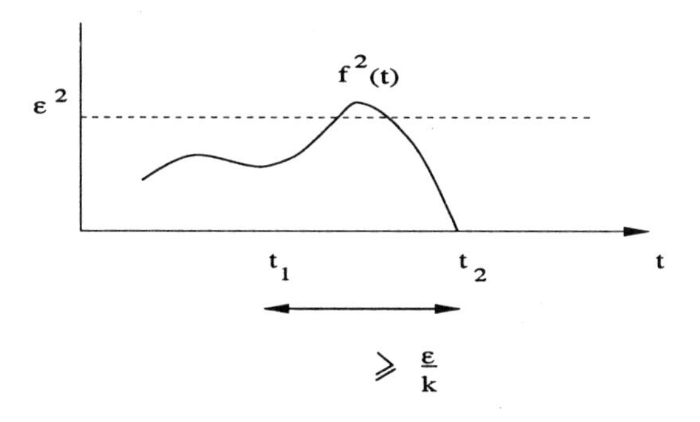


Figure 4.3: Proof of fact 7.

then

$$\epsilon \leq |f(t_1) - f(t_2)| \leq k|t_1 - t_2|$$

i.e. $|t_1 - t_2| \geq \frac{\epsilon}{k}$. We are now interested in computing the smallest upperbound for $\int_{t1}^{t2} f^2 dt$ . We will therefore assume that in the interval of time $(t_1, t_2)$ the function $f(t)$ decreases at maximum rate which is given by $k$ in the equation above. We therefore have (see figure 4.3):

$$\int_{t1}^{t2} f^2 dt \geq \frac{\epsilon^2 \frac{\epsilon}{k}}{2} = \frac{\epsilon^3}{2k}$$

Since $f \in \mathcal{L}_2$, it is clear that the number of times $|f|$ can go from 0 to $\epsilon$ is finite. Since $\epsilon > 0$ is arbitrary, we conclude that $f \to 0$ as $t \to \infty$. ♠♠

**Fact 8:** If $f \in \mathcal{L}_p (1 \leq p \leq \infty)$ and if $f$ is uniformly continuous, then $f \to 0$.

**Proof**

This result can be proved by contradiction following the proof of Fact 7. ♠♠

**Fact 9:** If $f_1 \in \mathcal{L}_2$ and $f_2 \in \mathcal{L}_2$, then $f_1 + f_2 \in \mathcal{L}_2$.

**Proof**

The result follows from

$$\int (f_1 + f_2)^2 \, dt = \int (f_1^2 + f_2^2 + 2 f_1 f_2) dt \leq 2 \int (f_1^2 + f_2^2) dt < \infty$$

♠♠

The following lemma describes the behavior of an asymptotically stable linear system when its input is $\mathcal{L}_2$.

**Lemma 4.1** Consider the state space representation of a linear system

$$\dot{x} = Ax + Bu \tag{4.2}$$

with $u \in I\!R^m, x \in I\!R^n$ and $A$ exponentially stable. If $u \in \mathcal{L}_2$ then $x \in \mathcal{L}_2 \cap \mathcal{L}_\infty, \dot{x} \in \mathcal{L}_2$ and $x \to 0$. ♠♠

**Remark 4.2** The system above with $u \in \mathcal{L}_2$ does not necessarily have an equilibrium point. Therefore, we can not use the Lyapunov approach to study the stability of the system.

**Proof**
Since $A$ is exponentially stable then there exists $P, Q > 0$ such that:

$$PA + A^T P = -Q$$

which is the well known Lyapunov equation. Consider the following positive definite function

$$V = x^T P x + k \int_t^\infty u^T u dt$$

where $k$ is a constant to be defined later. $V$ is not a Lyapunov function since the system may not have an equilibrium point. Note that since $u \in \mathcal{L}_2$, there exists a constant $k'$ such that

$$\int_0^t u^T u dt + \int_t^\infty u^T u dt = k' < \infty$$

taking the derivative with respect to time we obtain

$$u^T u + \frac{d}{dt}\left[\int_t^\infty u^T u dt\right] = 0$$

Using the above equations we get

$$\begin{aligned} \dot{V} &= \dot{x} P x + x^T P \dot{x} - k u^T u \\ &= (x^T A^T + u^T B^T) P x + x^T P (Ax + Bu) - k u^T u \\ &= x^T (A^T P + PA) x + 2 u^T B^T P x - k u^T u \\ &= -x^T Q x + 2 u^T B^T P x - k u^T u \end{aligned} \tag{4.3}$$

Note that

$$\begin{aligned} 2u^T B^T P x &\leq 2|u^T B^T P x| \\ &\leq 2\|u\| \|B^T P\| \|x\| \\ &\leq 2\|u\| \|B^T P\| \left[\frac{2}{\lambda_{min} Q}\right]^{\frac{1}{2}} \left[\frac{\lambda_{min} Q}{2}\right]^{\frac{1}{2}} \|x\| \\ &\leq \|u\|^2 \|B^T P\|^2 \frac{2}{\lambda_{min} Q} + \frac{\lambda_{min} Q}{2} \|x\|^2 \end{aligned} \tag{4.4}$$

where we have used the inequality $2ab \le a^2 + b^2$, for all $a, b \in I\!R$ and $\lambda_{min}Q$ denotes the minimum eigenvalue of $Q$. Choosing $k = ||B^T P||^2 \frac{2}{\lambda_{min}Q}$ we get

$$\dot{V} \le -\frac{\lambda_{min}Q}{2}||x||^2$$

Therefore $V$ is a non-increasing function and thus $V \in \mathcal{L}_\infty$ which implies that $x \in \mathcal{L}_\infty$. Integrating the above equation we conclude that $x \in \mathcal{L}_2$. From the system equation we conclude that $\dot{x} \in \mathcal{L}_2$ (see also fact 9). Finally $x, \dot{x} \in \mathcal{L}_2 \Rightarrow x \to 0$ (see fact 4).
A more general result is stated in the following theorem which can be found in [45, p.59] where $*$ denotes the convolution product.

**Theorem 4.1** Consider the exponentially stable and strictly proper system.

$$\begin{aligned} \dot{x} &= Ax + Bu \\ y &= C^T x \end{aligned} \tag{4.5}$$

and

$$H(s) = C(sI - A)^{-1}B$$

If $u \in \mathcal{L}_p$, then $y = h * u \in \mathcal{L}_p \cap \mathcal{L}_\infty$, $\dot{y} \in \mathcal{L}_p$ for $p = 1, 2$ and $\infty$. For $p = 1, 2$, $y \to 0$.

### 4.3.1 Example of applications of the properties of $\mathcal{L}_p$ functions in adaptive control

Let us first briefly review the Gradient type Parameter Estimation Algorithm. Let $y \in I\!R, \phi \in I\!R^n$ be measurable functions [1] which satisfy the following linear relation:

$$y = \theta^T \phi$$

where $\theta \in I\!R^n$ is an unknown constant vector. Define $\hat{y} = \phi^T \hat{\theta}$ and $e = \hat{y} - y$ then

$$e = (\hat{\theta}^T - \theta^T)\phi = \tilde{\theta}^T \phi$$

Let $V = \frac{1}{2}e^2$, then $\dot{V} = \frac{\partial V}{\partial \tilde{\theta}} \frac{d\tilde{\theta}}{dt}$. Note that $\frac{d\tilde{\theta}}{dt} = \frac{d\hat{\theta}}{dt}$. Let us choose

$$\frac{d\hat{\theta}}{dt} = \frac{d\tilde{\theta}}{dt} = -\left(\frac{\partial V}{\partial \tilde{\theta}}\right)^T = -e\left(\frac{\partial e}{\partial \tilde{\theta}}\right)^T = -\phi e.$$

We obtain [2]

[1] Here measurable is to be taken in the physical sense, not in the mathematical one.

[2] Let $F : I\!R^n \to I\!R^p$. Then $\frac{\partial F}{\partial x} \in I\!R^{p \times n}$ whereas $\nabla F = \frac{\partial F}{\partial x}^T \in I\!R^{n \times p}$, are the Jacobian and the gradient of $F$, respectively.

$$\dot{V} = -\left(\frac{\partial V}{\partial \tilde{\theta}}\right)\left(\frac{\partial V}{\partial \tilde{\theta}}\right)^T < 0$$

Let $V = \frac{1}{2}\tilde{\theta}^T\tilde{\theta}$, then $\dot{V} = \tilde{\theta}^T \dot{\tilde{\theta}} = -\tilde{\theta}^T\phi e$. Integrating we obtain

$$\int_0^t (-\tilde{\theta}^T\phi)edt = V(\tilde{\theta}(t)) - V(\tilde{\theta}(0)) \geq -V(\tilde{\theta}(0))$$

The operator $H : e \rightarrow -\tilde{\theta}^T\phi$ is therefore passive.

**Example 4.1 (Adaptive control of a simple NL system)** Let

$$\dot{x} = f(x)^T\theta' + bu$$

where $u, x \in I\!R$. Define

$$\begin{cases} \theta = \frac{\theta'}{b} \\ \tilde{\theta} = \hat{\theta} - \theta \\ \dot{\hat{\theta}} = fx \\ u = -\hat{\theta}^T f - x + v \end{cases}$$

and

$$V = \frac{b}{2}\tilde{\theta}^T\tilde{\theta} + \frac{1}{2}x^2$$

Then

$$\begin{aligned} \dot{V} &= b\tilde{\theta}^T\dot{\tilde{\theta}} + x\dot{x} \\ &= b\tilde{\theta}^T fx + x\left(f(x)^T\theta' + bu\right) \\ &= bx[(\hat{\theta} - \theta)^T f + \theta^T f + u] \\ &= -bx^2 + bxv \end{aligned} \tag{4.6}$$

From the last equation it follows that for $v = 0$, $V$ is a non-increasing function and thus $V, x, \tilde{\theta} \in \mathcal{L}_\infty$. Integrating the last equation it follows that $x \in \mathcal{L}_2 \cap \mathcal{L}_\infty$. Assume that $x \in \mathcal{L}_\infty \Rightarrow f(x) \in \mathcal{L}_\infty$. Therefore $u \in \mathcal{L}_\infty$ and also $\dot{x} \in \mathcal{L}_\infty$. $x \in \mathcal{L}_2$ and $\dot{x} \in \mathcal{L}_\infty$ implies $x \rightarrow 0$. Otherwise the operator $H : v \rightarrow x$ is output strictly passive (OSP) as will be defined later. ♠♠

In order to present the Passivity theorem and the Small gain theorem we will require the notion of extended spaces. We will next present a brief introduction to extended spaces. For a more detailed presentation the reader is referred to [45].

### 4.3.2 Linear maps

**Definition 4.4 (Linear maps)** Let $E$ be a linear space over $K$ ($I\!R$ *or* $\mathbb{C}$). Let $\tilde{\mathcal{L}}(E,E)$ be the class of all linear maps from $E$ into $E$. $\tilde{\mathcal{L}}(E,E)$ is a linear space satisfying the following properties $\forall\, x \in E$, $\forall\, A, B \in \tilde{\mathcal{L}}(E,E)$, $\forall\, \alpha \in K$

$$\begin{aligned}(A+B)x &= Ax + Bx \\ (\alpha A)x &= \alpha(Ax) \\ (AB)x &= A(Bx)\end{aligned} \tag{4.7}$$

♠♠

### 4.3.3 Induced norms

**Definition 4.5 (Induced Norms)** Let $|.|$ be a norm on $E$ and $A \in \tilde{\mathcal{L}}(E,E)$. The induced norm of the linear map $A$ is defined as

$$\begin{aligned}||A|| &\triangleq \sup_{x\neq 0} \frac{|Ax|}{|x|} \\ &= \sup_{|z|=1} |Az|\end{aligned} \tag{4.8}$$

### 4.3.4 Properties of induced the norms

If $||A|| < \infty$ and $||B|| < \infty$ then the following properties hold for all $x \in E, \alpha \in k$.

1. $|Ax| \leq ||A|| |x|$
2. $||\alpha A|| = |\alpha| ||A||$
3. $||A + B|| \leq ||A|| + ||B||$
4. $||AB|| \leq ||A|| ||B||$

**Example 4.2** Let $H$ be a linear map defined on $E$ in terms of an integrable function $h : I\!R \to I\!R$.

$$H : u \to Hu \triangleq h * u,\ \forall\, u \in \mathcal{L}^{\infty}$$

i.e.

$$(Hu)(t) = \int_0^t h(t-\tau)u(\tau)d\tau, \ \forall\, t \in I\!R^+$$

Assume that $||h||_1 = \int_0^\infty |h(t)|dt < \infty$.

**Theorem 4.2** Under those conditions the following properties hold:

a) $H : \mathcal{L}^\infty \to \mathcal{L}^\infty$

b) $||H||_\infty = ||h||_1;\ i.e. ||h * u||_\infty \leq ||h||_1 ||u||_\infty,\ \forall u \in \mathcal{L}^\infty$

and the right-hand side can be made arbitrarily close to the left-hand side of the inequality by appropriate choice of $u$. ♠♠

**Proof**

By definition

$$\begin{aligned} ||H||_\infty &= \sup_{||u||_\infty = 1} ||h * u||_\infty \\ &= \sup_{||u||_\infty = 1} \sup_{t \geq 0} |(h * u)(t)| \\ &= \sup_{||u||_\infty = 1} \left[ \sup_{t \geq 0} \left| \int_0^t h(t-\tau)u(\tau)d\tau \right| \right] \\ &\leq \sup_{||u||_\infty = 1} \left[ \sup_{t \geq 0} \int_0^t |h(t-\tau)|\, |u(\tau)|\, d\tau \right] \end{aligned}$$

Since $||u||_\infty = 1$ we have

$$\begin{aligned} ||H||_\infty &\leq \sup_{t \geq 0} \int_0^t |h(t-\tau)|\, d\tau \\ &= \sup_{t \geq 0} \int_0^t |h(t-\tau)|\, d\tau \\ &= \sup_{t \geq 0} \int_0^t |h(t')|\, dt' \\ &\leq \int_0^\infty |h(t')|\, dt' = ||h||_1 \end{aligned}$$

We can choose $u_t(\tau) = \text{sgn}[h(t-\tau)], t \in I\!N$. Thus

$$(h * u_t)(t) = \int_0^t |h(t-\tau)|d\tau \leq ||h * u_t||_\infty$$

therefore

$$\begin{aligned}
\int_0^t |h(\tau')|d\tau' &= \int_0^t |h(t-\tau)|d\tau \\
&\leq \|h * u_t\|_\infty \\
&\leq \|H\|_\infty \\
&\leq \|h\|_1 \\
&= \int_0^\infty |h(t')|dt'
\end{aligned} \tag{4.9}$$

Letting $t \to \infty$ it follows that $\|H\|_\infty = \|h\|_1$. ♠♠

### 4.3.5 Extended spaces

Consider a function $f(t)$ for $t \geq 0$. Define the truncated function

$$f_T(t) = \begin{cases} f(t) & t \leq T \\ 0 & t > T \end{cases} \tag{4.10}$$

The function $f_T$ is obtained by truncating $f$ at time $T$. Let us introduce the following definitions

$\mathcal{T}$: Subset of $\mathbb{R}^+$ (typically, $\mathcal{T} = \mathbb{R}^+$ or $\mathbb{Z}^+$)

$V$: Normed space with norm $\|.\|$ (typically $V = \mathbb{R}, \mathbb{R}^n, \mathbb{C}, \mathbb{C}^n$)

$\mathcal{F}$: $\{ f : \mathcal{T} \to \mathcal{V} \}$ the set of all functions mapping $\mathcal{T}$ into $V$

The normed linear subspace $\mathcal{L}$ is given by

$$\mathcal{L} \triangleq \{f : \mathcal{T} \to V | \|f\| < \infty\}$$

Associated with $\mathcal{L}$ is the extended space $\mathcal{L}_e$ defined by

$$\mathcal{L}_e \triangleq \{f : \mathcal{T} \to V | \forall T \in \mathcal{T}, \|f_T\| < \infty\}$$

The following properties hold for all $f \in \mathcal{L}$:

1. The map $T \to \|f_T\|$ is monotonically increasing.
2. $\|f_T\| \to \|f\|$ as $T \to \infty$

We can now introduce the notion of gain of an operator which will be used in the small gain theorem and the passivity theorem.

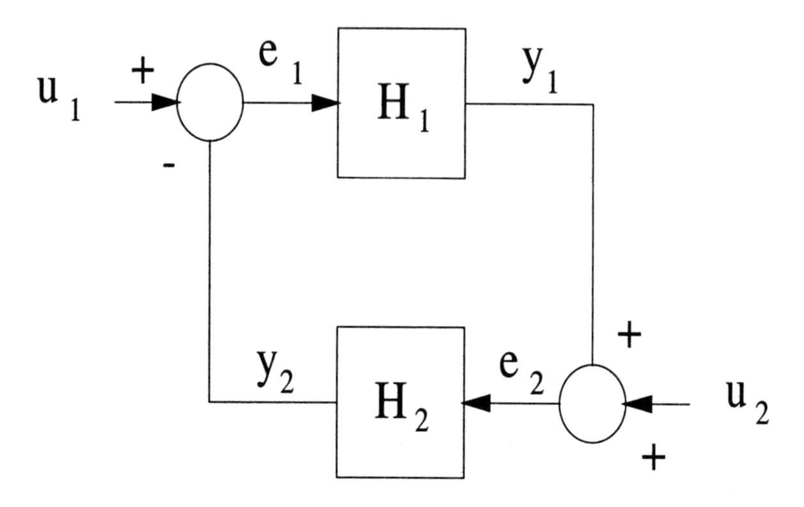


Figure 4.4: Closed-loop system with two external inputs.

### 4.3.6 Gain of an operator

**Definition 4.6 ([65])** Consider an operator $H : \mathcal{L}_e \to \mathcal{L}_e$. $H$ is weakly finite-gain stable (WFGS) if there exists a function $\beta$ and a constant $k$ such that

$$\|(Hu)_T\| \leq k\|u_T\| + \beta(x_0)$$

for all $u, x(0)$. If $\beta(x_0) = 0$, we call $H$ finite-gain stable (FGS). ♠♠

### 4.3.7 Small gain theorem

This theorem gives sufficient conditions under which a bounded input produces a bounded output (BIBO).
We call gain of $H$ the number $k$ (or $k(H)$) defined by

$$k(H) = \inf\{\bar{k} \in I\!R^+ / \; \exists \bar{\beta} : \|(Hu)_T\| \leq \bar{k}\|u_T\| + \bar{\beta}, \; \forall u \in \mathcal{L}_e, \forall T \in I\!R^+\}$$

**Theorem 4.3 (Small gain)** Consider $H_1 : \mathcal{L}_e \to \mathcal{L}_e$ and $H_2 : \mathcal{L}_e \to \mathcal{L}_e$. Let $e_1, e_2 \in \mathcal{L}_e$ and define (see figure 4.4 )

$$\begin{array}{rcl} u_1 & = & e_1 + H_2 e_2 \\ u_2 & = & e_2 - H_1 e_1 \end{array} \tag{4.11}$$

Suppose there are constants $\beta_1, \beta_2, \gamma_1, \gamma_2 \geq 0$ such that for all $T \in I\!R^+$:

$$\begin{array}{rcl} \|(H_1 e_1)_T\| & \leq & \gamma_1 \|e_{1T}\| + \beta_1 \\ \|(H_2 e_2)_T\| & \leq & \gamma_2 \|e_{2T}\| + \beta_2 \end{array} \tag{4.12}$$

under those conditions, if $\gamma_1 \gamma_2 < 1$, then

i)

$$\begin{aligned}
\|e_{1T}\| &\leq (1-\gamma_1\gamma_2)^{-1}(\|u_{1T}\| + \gamma_2\|u_{2T}\| + \beta_2 + \gamma_2\beta_1) \\
\|e_{2T}\| &\leq (1-\gamma_1\gamma_2)^{-1}(\|u_{2T}\| + \gamma_1\|u_{1T}\| + \beta_1 + \gamma_1\beta_2)
\end{aligned}$$

ii) If in addition, $\|u_1\|, \|u_2\| < \infty$ then $e_1, e_2, y_1, y_2$ have finite norms. ♠♠

**Proof**

From (4.11) we have

$$\begin{aligned}
e_{1T} &= u_{1T} - (H_2 e_2)_T \\
e_{2T} &= u_{2T} + (H_1 e_1)_T
\end{aligned} \tag{4.13}$$

then

$$\|e_{1T}\| \leq \|u_{1T}\| + \|(H_2e_2)_T\| \leq \|u_{1T}\| + \gamma_2\|e_{2T}\| + \beta_2$$

$$\|e_{2T}\| \leq \|u_{2T}\| + \|(H_1e_1)_T\| \leq \|u_{2T}\| + \gamma_1\|e_{1T}\| + \beta_1$$

Combining these two inequalities we get

$$\|e_{1T}\| \leq \|u_{1T}\| + \beta_2 + \gamma_2(\|u_{2T}\| + \gamma_1\|e_{1T}\| + \beta_1)$$

$$\|e_{2T}\| \leq \|u_{2T}\| + \beta_1 + \gamma_1(\|u_{1T}\| + \gamma_2\|e_{2T}\| + \beta_2)$$

Finally

$$\|e_{1T}\| \leq (1-\gamma_1\gamma_2)^{-1}\left[\|u_{1T}\| + \gamma_2\|u_{2T}\| + \beta_2 + \gamma_2\beta_1\right]$$

$$\|e_{2T}\| \leq (1-\gamma_1\gamma_2)^{-1}\left[\|u_{2T}\| + \gamma_1\|u_{1T}\| + \beta_1 + \gamma_1\beta_2\right]$$

The remainder of the proof follows immediately. ♠♠

## 4.4 Dissipative systems

We will now review the definitions and properties of dissipative systems. Most of the mathematical foundations on this subject are due to Willems [214], and Hill and Moylan [65], [66].

Consider a causal nonlinear system $\Sigma : u(t) \to y(t);\ u(t) \in \mathcal{L}_{pe}, y(t) \in \mathcal{L}_{pe}$ represented by the following state-space representation affine in the input:

$$(\Sigma)\begin{cases} \dot{x} &= f(x) + g(x)u \\ y &= h(x) + j(x)u \end{cases} \tag{4.14}$$

where $x \in \mathbb{R}^n$, $u, y \in \mathbb{R}^m$, $f, g, h$ and $j$ are smooth and $f(0) = h(0) = 0$.

Let us call $w(t) = w(u(t), y(t))$ the <u>supply rate</u> and be such that for all $u$ and $x(0)$ and for all $t \in \mathbb{R}^+$

$$\int_0^t |w(s)|ds < \infty \tag{4.15}$$

i.e. we are assuming $w$ to be locally Lebesgue integrable independently of the input and the initial conditions. In an electric circuit $\int_0^t w(s)ds$ can be associated with the energy supplied to the circuit in the interval $(0, t)$, i.e. $\int_0^t vi\; dt$ where $v$ is the voltage at the terminals and $i$ the current entering the circuit, see the example in chapter 1.

**Definition 4.7 (Dissipative system)** The system $\Sigma$ is said to be dissipative if there exists a so-called storage function $V(x) \geq 0$ such that the following dissipation inequality holds

$$V(x) \leq V(x(0)) + \int_0^t w(s)ds; \; \forall\; u, \forall\; x(0), \forall\; t \geq 0 \tag{4.16}$$

♠♠

**Remark 4.3** A local definition of dissipative systems is possible. Roughly, the dissipativity inequality should be satisfied as long as the system's state remains inside a closed domain of the state space [165].

The next natural question is, given a system, how can we find $V(x)$? This question is closely related to the problem of finding a suitable Lyapunov function in the Lyapunov second method. As will be seen next, a storage function can be found by computing the maximum amount of energy that can be extracted from the system as defined next.

**Definition 4.8 (Available storage)** The available storage $V_a(x)$ of the system $\sum$ is given by:

$$0 \leq V_a(x) = \sup_{x=x(0),u,t\geq 0} -\left\{\int_0^t w(s)ds\right\} \tag{4.17}$$

where $V_a(x)$ is the maximum amount of energy which can be extracted from the system. ♠♠

Another storage function plays an important role in dissipative systems:

**Definition 4.9 (Required supply)** The required supply $V_r(x)$ of the system $\sum$ is given by:

$$0 \leq V_r(x) = \inf_{u,t\geq 0} \left\{\int_{-t}^0 w(s)ds\right\} \tag{4.18}$$

with $x(-t) = x^\star$, $x(0) = x$, and the system is reachable from $x^\star$. The function $V_r(x)$ is the required amount of energy to be injected in the system to go from $x(-t)$ to $x(0)$. ♠♠

**Remark 4.4** The required supply is not necessarily positive. It is positive if $x^\star = 0$ [55, pp.142-144].

The following results relate the boundedness of the storage functions in definitions 4.8 and 4.9 to the dissipativeness of the system. As an example, consider again an electric circuit. If there is an ideal battery in the circuit, the energy that can be extracted is not finite. Such this circuit is not dissipative. The following results are due to Willems [212] [213].

**Theorem 4.4 (Willems, 1972)** The available storage $V_a$ in (4.17), is finite for all $x \in X$ if and only if $\Sigma$ (4.14) is dissipative. Moreover, $0 \leq V_a \leq V$ for dissipative systems and $V$ is itself a possible storage function. ♠♠

**Proof**

(⇒) In order to show that $V_a(x) < \infty \Rightarrow$ the system $\Sigma$ in (4.14) is dissipative, it suffices to show that the available storage $V_a$ in (4.17) is a storage function *i.e.* it satisfies the dissipation inequality

$$V_a(x(t)) \leq V_a(x_0) + \int_0^t w(t)dt$$

But this is certainly the case because the available storage $V_a(x(t))$ at time $t$ is not larger than the available storage $V_a(x_0)$ at time 0 plus the energy introduced into the system in the interval $[0, t]$.

(⇐) Let us now prove that if the system $\Sigma$ is dissipative then $V_a(x) < \infty$. If $\Sigma$ is dissipative then there exists $V(x) \geq 0$

$$V(x_0) + \int_0^t w(t)dt \geq V(x(t)) \geq 0$$

From the above and (4.17) it follows that

$$V(x_0) \geq \sup_{x=x(0),t\geq 0,u} \left\{ -\int_0^t w(t)dt \right\} = V_a(x)$$

Since the initial storage function $V(x_0)$ is finite it follows that $V_a(x) < \infty$.♠♠
Similarly we have (assume that $x^* = 0$ in definition 4.9):

**Theorem 4.5 (Willems, 1972)** The system $\Sigma$ (4.14) is dissipative if and only if the required supply satisfies $V_r(x) \geq -K > -\infty$ for all $x \in X$ and some $K \in I\!R$. Moreover, $0 \leq V_a \leq V \leq V_r$ for dissipative systems. ♠♠

The following is a consequence of theorem 2.1.

**Theorem 4.6 (Passive systems)** Suppose that the system $\Sigma$ in (4.14) is dissipative with supply rate $w = u^T y$ and storage function $V$ with $V(0) = 0$, i.e. for all $t \geq 0$:

$$V(x) \leq V(x(0)) + \int_0^t u^T(s)y(s)ds \tag{4.19}$$

then the system is passive. ♠♠

**Definition 4.10 (Strictly state passive systems)** A system $\Sigma$ in (4.14) is said to be **strictly state passive** if it dissipative with supply rate $w = u^T y$ and the storage function $V$ with $V(0) = 0$ and there exists a positive definite function $\mathcal{S}(x)$ such that for all $t \geq 0$:

$$V(x(t)) \leq V(x(0)) + \int_0^t u^T(s)y(s)ds + \int_0^t \mathcal{S}(x(t))dt \tag{4.20}$$

If the equality holds in the above and $\mathcal{S}(x) \equiv 0$, then the system is said to be **lossless** [38]. ♠♠

**Remark 4.5** If the system $\Sigma$ in (4.14) is dissipative with supply rate $w = u^T y$ and the storage function $V$ satisfies $V(0) = 0$ with $V(x)$ positive definite, then the system and its zero dynamics are Lyapunov stable. This can be seen from the dissipativity inequality (4.16) by taking $u$ or $y$ equal to zero.

A general supply rate has been introduced by [66] which is useful to distinguish different types of strictly passive systems and will be useful in the Passivity theorem presented in the next section. Let us reformulate some notions introduced in definition 2.1 in terms of supply rate.

**Definition 4.11 (General Supply Rate)** Let us consider a dissipative system, with supply rate

$$w(u, y) = y^T Qy + u^T Ru + 2y^T Su \tag{4.21}$$

with $Q = Q^T$, $R = R^T$.
If $Q = 0$, $R = -\varepsilon I_m$, $\varepsilon > 0$, $S = \frac{1}{2}I_m$, the system is said to be input strictly passive (ISP), i.e.

$$\int_0^t y^T udt \geq \beta + \epsilon \int_0^t u^T udt$$

If $R = 0$, $Q = -\delta I_m$, $\delta > 0$, $S = \frac{1}{2}I_m$, the system is said to be output strictly passive (OSP), i.e.

$$\int_0^t y^T u dt \geq \beta + \delta \int_0^t y^T y dt$$

If $Q = -\delta I_m$, $\delta > 0$, $R = -\varepsilon I_m$, $\varepsilon > 0$, $S = \frac{1}{2} I_m$, the system is said to be very-strictly passive (VSP), *i.e.*

$$\int_0^t y^T u dt + \beta \geq \delta \int_0^t y^T y dt + \epsilon \int_0^t u^T u dt$$

♠♠

Note that definitions 4.10 and 4.11 do not imply in general asymptotic stability of the considered system. For instance $\frac{s+a^2}{s}$ is ISP as stated in definition 4.11, see also theorem 2.3.

**Example 4.3** It is noteworthy that the function $\beta(.)$ in the definition 2.1 plays an important role in the Lyapunov stability of dissipative systems. For instance let us consider the following example, brought to our attention by David Hill, where the open-loop system is unstable:

$$\begin{aligned} \dot{x} &= x + u \\ y &= -\tfrac{\alpha x}{1+x^4} \end{aligned} \tag{4.22}$$

Let us note that

$$\begin{aligned} \textstyle\int_{t_0}^{t_1} u(t)y(t)dt &= -\textstyle\int_{t_0}^{t_1} (\dot{x} - x)\tfrac{\alpha x}{1+x^4} dt \\ &\geq -\tfrac{\alpha}{2}[\arctan(x^2(t_1)) - \arctan(x^2(t_0))] \end{aligned} \tag{4.23}$$

Thus the system is passive with respect to the storage function $V(x) = \frac{\alpha}{2}(\frac{\pi}{2} - \arctan(x^2))$. Hence the system is externally dissipative despite the fact that the open-loop is unstable. Note however that $-V(0) = \beta(0) < 0$ and that the system loses its observability at $x = \infty$. This system is even not zero state detectable, since $u \equiv 0$ and $y \equiv 0$ do not imply $x \to 0$ as $t \to +\infty$.

**Example 4.4** If a system $(A, B, C, D)$ is SPR and the relative degree $n^\star = 1$ (i.e. $D = 0$), then the system is OSP. Indeed from the KYP lemma, defining $V = x^T P x$ one gets $\dot{V} = -x^T(LL^T + Q)x + 2y^T u$. Integrating and taking into account that $Q$ is full rank, the result follows.

**Example 4.5** Consider the non-proper system $y = \dot{u} + au$, $a > 0$, with relative degree $n^\star = -1$. This system is SSPR and ISP since $\mathbf{Re}\,[j\omega + a] = a$ and

$$\int_0^t uy dt = \frac{u^2(t)}{2} + a \int_0^t u^2 dt$$

This plant belongs to the descriptor-variable systems, with state space representation:

$$\begin{array}{rcl} \dot{x}_1 & = & x_2 \\ 0 & = & -x_1 + u \\ y & = & x_2 + au \end{array}$$

**Example 4.6** If a system $(A, B, C, D)$ is SPR and if the matrix

$$\bar{Q} \triangleq \begin{pmatrix} Q + LL^T & -LW \\ -W^T L^T & D + D^T \end{pmatrix}$$

is positive definite, then the system is VSP. This can be proved by using again $V = x^T P x$. Let us denote $\bar{x} = \begin{pmatrix} x \\ u \end{pmatrix}$. Differentiating and using the KYP lemma, one gets $\dot{V} = -\bar{x}^T \bar{Q} \bar{x} + 2y^T u$. One deduces that $\int_{t_0}^{t_1} u^T(t) y(t) dt \geq \delta u^T u + \alpha y^T y$ for some $\delta > 0$ and $\alpha > 0$ small enough. Note that the condition $\bar{Q} > 0$ implies that $n^\star = 0$.

**Example 4.7** If a system $(A, B, C, D)$ is SPR, then it is strictly passive with $\mathcal{S}(x) = x^T Q x$. This can be proved using the KYP lemma.

**Example 4.8** Consider the system $H(s) = \frac{1}{s+a}, a > 0$ . We will now prove that the system is $H(s)$ is OSP.

The system is described by

$$\dot{y} = ay + u$$

Let us consider the positive definite functions

$$V = \frac{1}{2} y^2$$

Thus

$$\dot{V} = y\dot{y} = -a^2 + uy$$

Integrating we obtain:

$$-V(0) \leq V - V(0) = -a \int_0^t y^2 dt + \int_0^t uy dt$$

$$\Rightarrow \int_0^t uy dt + V(0) \geq a \int_0^t y^2 dt$$

Thus the system is OSP. Taking $a = 0$, we can see that $\frac{1}{s}$ is a passive system.

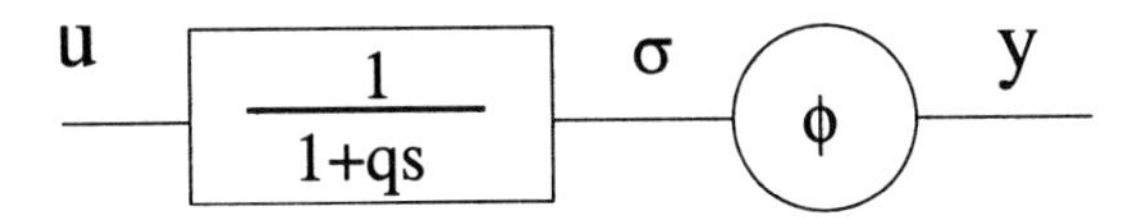


Figure 4.5: A linear system and a static nonlinearity in cascade.

**Remark 4.6** As we saw in section 2.8 for linear systems, there exists a relationship between passive systems and $\mathcal{L}_2$−gain [45]. Let $\Sigma : u \to y$ be a passive system. Define the variable transformation $u = \gamma w + z; y = \gamma w - z$, then

$$\beta \leq \int_0^t u^T y dt = \int_0^t (\gamma^2 w^T w - z^T z) dt$$

which is equivalent to

$$\int_0^t z^T z dt \leq \int_0^t \gamma^2 w^T w dt - \beta$$

which means that the system $\Sigma' : w \to z$ has a finite $\mathcal{L}_2$−gain.

**Proposition 4.2** Consider the system represented in figure 4.5, where $\phi$ is a static nonlinearity, $q \geq 0$ and $\phi \in [0, \infty)$. Then $H : u \to y$ is externally passive (internally for some $\phi$). ♠♠

**Proof**
Define $\langle u|y \rangle_t \triangleq \int_0^t uydt$, then

$$\begin{aligned}
\langle y|u \rangle_t &= \langle \phi(\sigma)|u \rangle_t \\
&= \langle \phi(\sigma)|q\dot{\sigma} + \sigma \rangle \\
&= q \int_0^t \phi[\sigma(t)]\dot{\sigma} dt + \int_0^t \sigma(t)\phi[\sigma(t)]dt \qquad (4.24) \\
&= q \int_{\sigma(0)}^{\sigma(t)} \phi[\sigma]d\sigma + \int_0^t \sigma(t)\phi[\sigma(t)]dt \\
&\geq q \int_0^{\sigma(t)} \phi[\sigma]d\sigma - q \int_0^{\sigma(0)} \phi[\sigma]d\sigma
\end{aligned}$$

where we have used the fact that $\sigma\phi(\sigma) \geq 0$. Note that $V(\sigma) = \int_0^\sigma \sigma[\xi]d\xi \geq 0$.

**Example 4.9** If a system is output-strictly passive, then it is also weakly finite gain stable, i.e. OSP $\Rightarrow$ WFGS.

**Proof**

$$\begin{aligned} \delta \int_0^t y^2 dt & \leq \beta + \int_0^t uydt \\ & \leq \beta + \int_0^t uydt + \tfrac{1}{2}\int_0^t (\sqrt{\lambda}u - \tfrac{y}{\sqrt{\lambda}})^2 dt \\ & = \beta + \tfrac{\lambda}{2}\int_0^t u^2 dt + \tfrac{1}{2\lambda}\int_0^t y^2 dt \end{aligned} \tag{4.25}$$

Choosing $\lambda = \frac{1}{\delta}$

$$\frac{\delta}{2}\int_0^t y^2 dt \leq \beta + \frac{1}{2\delta}\int_0^t u^2 dt$$

♠♠

**Example 4.10** It is known that OSP implies $\mathcal{L}_2$-gain. In [214] it is shown that ISP and finite $\mathcal{L}_2$-gain yield asymptotic stability.

**Example 4.11** Let us consider two linear systems in parallel, i.e.:

$$\begin{aligned} y_1 &= k_1 u \\ \dot{y}_2 &= -ay_2 + k_2 u \\ y &= y_1 + y_2 \end{aligned}$$

where $a > 0$. Thus, for some constants $\beta$ and $k_3$

$$\begin{aligned} \int_0^t uydt & = \int_0^t uy_1 dt + \int_0^t uy_2 dt \\ & \geq k_1 \int_0^t u^2 dt + \beta + k_3 \int_0^2 y_2^2 dt \\ & \geq \tfrac{k_1}{2}\int_0^t u^2 dt + \beta + k' \int_0^t (y_1^2 + y_2^2) dt \\ & \geq \tfrac{k_1}{2}\int_0^t u^2 dt + \beta + \tfrac{k'}{2}\int_0^t (y_1 + y_2)^2 dt \end{aligned} \tag{4.26}$$

where $k' \leq k_3$ and $k' \leq \frac{1}{2k_1}$. So the system $\Sigma : u \to y$ is VSP.

## 4.5 Passivity theorems

In this section we will study the stability of the interconnection in negative feedback of different type of passive systems. We will first study closed-loop interconnections with one external input (One-channel results) and then interconnections with two external inputs (Two-channel results). The implicit assumption in the passivity theorems is that the problem is well-possed, i.e. that all the signals belong to $\mathcal{L}_{2e}$. In others words we assume that there exists a time interval $(0, t')$ within which the closed-loop signals remain bounded.

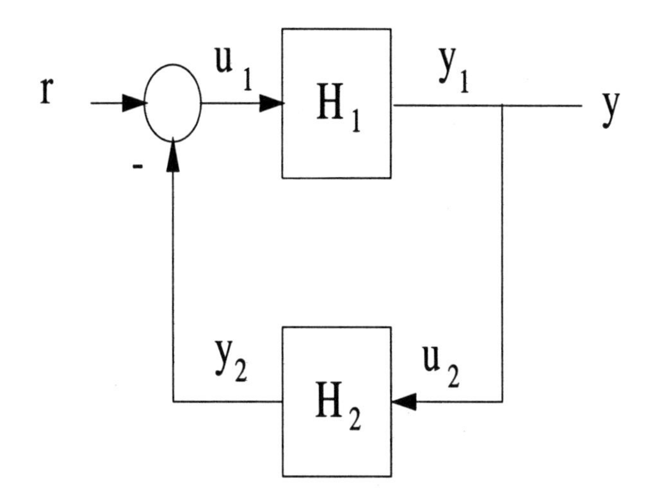


Figure 4.6: Closed-loop system with one external input.

**Remark 4.7** Different version of passivity theorems can be obtained depending on the properties of the subsystems in the interconnections. We will only consider here the most classical versions.

### 4.5.1 One channel results

**Theorem 4.7 (Passivity (one channel) [66])** Assume that both $H_1, H_2$ are pseudo-VSP, i.e.

$$\int_0^t y_1^T u_1 dt + \beta_1 \geq \delta_1 \int_0^t y_1^T y_1 dt + \epsilon_1 \int_0^t u_1^T u_1 dt$$

$$\int_0^t y_2^T u_2 dt + \beta_2 \geq \delta_2 \int_0^t y_2^T y_2 dt + \epsilon_2 \int_0^t u_2^T u_2 dt$$

with

$$\delta_1 + \epsilon_1 > 0; \quad \delta_2 + \epsilon_2 > 0$$

The feedback closed-loop system (see figure 4.6) is finite gain stable if:

$$\delta_2 \geq 0, \epsilon_1 \geq 0, \epsilon_2 + \delta_1 > 0$$

where $\epsilon_2$ or $\delta_1$ may be negative. ♠♠

**Corollary 4.1** The feedback system in figure 4.6 is finite gain stable if:

1. $H_1$ is passive and $H_2$ is ISP i.e. $\epsilon_1 \geq 0, \epsilon_2 > 0, \delta_1 \geq 0, \delta_2 \geq 0$.
2. $H_1$ is OSP and $H_2$ is passive i.e. $\epsilon_1 \geq 0, \epsilon_2 \geq 0, \delta_1 > 0, \delta_2 \geq 0$.

♠♠

**Proof**

$$\begin{aligned} \langle r|y\rangle_T &= \langle u_1 + y_2|y\rangle_T \\ &= \langle u_1|y_1\rangle_T + \langle y_2|u_2\rangle_T \\ &\geq \beta_1 + \epsilon_1||u_1||^2 + \delta_1||y_1||_T^2 + \beta_2 + \epsilon_2||u_2||^2 + \delta_2||y_2||_T^2 \\ &\geq \beta_1 + \beta_2 + (\delta_1 + \epsilon_2)||y||_T^2 \end{aligned} \tag{4.27}$$

where $||y||_T^2 = \langle y|y\rangle_T$. Using the Schwartz inequality we have

$$\langle r|y\rangle_T = \int_0^T rydt \leq \left[\int_0^T r^2dt\right]^{\frac{1}{2}} \left[\int_0^T y^2dt\right]^{\frac{1}{2}} = ||r||_T||y||_T$$

then

$$||r||_T||y||_T \geq \langle r|y\rangle_T \geq \beta_1 + \beta_2 + (\delta_1 + \epsilon_2)||y||_T^2$$

for any $\lambda \in I\!R$ the following holds

$$\begin{aligned} \tfrac{1}{2\lambda}||r||_T^2 + \tfrac{\lambda}{2}||y||_T^2 &= \tfrac{1}{2}\left(\tfrac{1}{\sqrt{\lambda}}||r||_T - \sqrt{\lambda}||y||_T\right)^2 + ||r||_T||y||_T \\ &\geq \beta_1 + \beta_2 + (\delta_1 + \epsilon_2)||y||_T^2 \end{aligned} \tag{4.28}$$

choosing $\lambda = \delta_1 + \epsilon_2$ we get

$$\frac{||r||_T^2}{2(\delta_1 + \epsilon_2)} \geq \beta_1 + \beta_2 + \frac{(\delta_1 + \epsilon_2)}{2}||y||_T^2$$

which concludes the proof. ♠♠

### 4.5.2 Two-channel results

Consider now the system depicted in figure 4.7 where $r_1, r_2$ can represent disturbances, initial condition responses or commands. Assume well-posedness.

**Theorem 4.8 (passivity (two channel) [206])**
Assume $H_1, H_2$ are pseudo VSP. The feedback system is finite -gain stable if

$$\epsilon_1 + \delta_2 > 0$$

$$\epsilon_2 + \delta_1 > 0$$

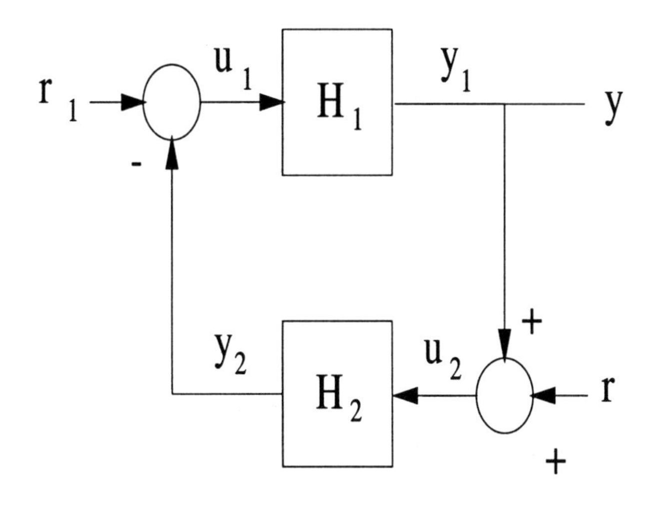


Figure 4.7: Closed-loop system with two external inputs.

where $\epsilon_i, \delta_i$ may be negative. ♠♠

**Corollary 4.2** The feedback system is finite-gain stable if

1. $H_1, H_2$ are ISP ($\epsilon_1 > 0, \epsilon_2 > 0, \delta_1 = \delta_2 = 0$).
2. $H_1, H_2$ are OSP ($\delta_1 > 0, \delta_2 > 0, \epsilon_1 = \epsilon_2 = 0$).
3. $H_1$ is VSP, $H_2$ is passive ($\epsilon_1 > 0, \delta_1 > 0, \delta_2 = \epsilon_2 = 0$).
4. $H_1$ is passive, $H_2$ is VSP ($\epsilon_2 > 0, \delta_2 > 0, \delta_1 = \epsilon_1 = 0$).

♠♠

**Proof**

$$\begin{aligned}
\langle u_1|y_1\rangle_T + \langle y_2|u_2\rangle_T &= \langle r_1 - y_2|y_1\rangle_T + \langle y_2|y_1 + r_2\rangle_T \\
&= \langle r_1|y_1\rangle_T + \langle y_2|r_2\rangle_T \\
&\geq \beta_1 + \epsilon_1||u_1||^2 + \delta_1||y_1||_T^2 + \beta_2 \\
&+\epsilon_2||u_2||_T^2 + \delta_2||y_2||_T^2
\end{aligned} \tag{4.29}$$

Note that

$$\begin{aligned}
||u_1||_T^2 &= \int_0^T u_1^T u_1 dt \\
&= \int_0^T (r_1 - y_2)^T (r_1 - y_2) dt \\
&\geq -2\langle r_1|y_2\rangle_T + ||y_2||_T^2
\end{aligned} \tag{4.30}$$

and similarly

$$||u_2||_T^2 \geq 2\langle r_2|y_1\rangle_T + ||y_1||_T^2$$

Then

$$\begin{aligned}\langle r_1|y_1\rangle_T + \langle y_2|r_2\rangle_T & +2\epsilon_1\langle r_1|y_2\rangle_T - 2\epsilon_2\langle r_2|y_1\rangle_T \geq \\ & \beta_1+\beta_2+(\epsilon_1+\delta_2)||y_2||_T^2+(\epsilon_2+\delta_1)||y_1||_T^2\end{aligned} \tag{4.31}$$

Note that for any $\lambda \in I\!R$, for $i = 1,2$ we have

$$\begin{aligned}\langle r_i|y_i\rangle_T & \leq ||y_i||_T||r_i||_T + \frac{1}{2}(\frac{1}{\sqrt{\lambda_i}}||r_i||_T - \sqrt{\lambda_i}||y_i||_T)^2 \\ & \leq \frac{1}{2\lambda_i}||r_i||_T^2 + \frac{2\lambda_i}{2}||y_i||_T^2\end{aligned} \tag{4.32}$$

We choose $\lambda_1 = \frac{\epsilon_2+\lambda_1}{2}$ and $\lambda_2 = \frac{\epsilon_1+\epsilon_2}{2}$

- if $\epsilon_1 = 0$ then $2\epsilon_1\langle r_1|y_2\rangle_T \leq 0$
- if $\epsilon_1 > 0$ then for any $\lambda' \in I\!R$

$$2\epsilon_1\langle r_1|y_2\rangle_T \leq \frac{\epsilon_1}{\lambda_1'}||r_1||_T^2 + \epsilon_1\lambda_1'||y_2||_T^2$$

Let us choose $\lambda_1' = \frac{\lambda_1"}{\epsilon_1}$ and $\lambda_1" = \frac{\epsilon_1+\delta_2}{4}$.
Therefore

$$\begin{aligned}& \beta_1+\beta_2+\frac{(\epsilon_1+\delta_2)}{4}||y_2||_T^2+\frac{(\epsilon_2+\delta_1)}{4}||y_1||_T^2 \\ & \leq ||r_1||_T^2(\frac{1}{\epsilon_2+\delta_1}+\frac{4\epsilon_1^2}{\epsilon_1+\delta_2})+||r_2||_T^2(\frac{1}{\epsilon_1+\delta_2}+\frac{4\epsilon_1^2}{\epsilon_1+\delta_2})\end{aligned} \tag{4.33}$$

which concludes the proof. ♠♠

Boundedness of the closed-loop signals can be ensured if $H_1$ and $H_2$ have finite gain as can be seen from the following lemma.

**Lemma 4.2** Consider again the negative feedback interconnection of $H_1$ and $H_2$ as in figure 4.7. Assume that $H_1, H_2$ are pseudo VSP i.e.

$$\int_0^T u_i^T y_i dt = V_i(x_i) - V_i(x_i(0)) + \epsilon_i \int_0^t u_i^T u_i dt + \delta_i \int_0^t y_i^T y_i dt$$

then the origin is an asymptotically stable equilibrium point if:

$$\epsilon_1 + \delta_2 > 0$$

and

$$\epsilon_2 + \delta_1 > 0$$

and $H_1$ and $H_2$ are zero-state detectable (i.e. $u_i \equiv 0, y_i \equiv 0 \Rightarrow x_i = 0$). ♠♠

**Proof**

Consider the positive definite function

$$V(x) = V_1(x_1) + V_2(x_2) \geq 0$$

Then using the dissipativity inequalities we get

$$\begin{aligned} \dot{V}(x) &= \sum_{i=1}^{2} \left[u_i^T y_i - \epsilon_i u_i^T u_i - \delta_i y_i^T y_i\right] \\ &= -(\epsilon_1 + \delta_2) u_1^T u_1 - (\epsilon_2 + \delta_1) y_1^T y_1 \end{aligned} \tag{4.34}$$

The result follows from Krasovskii-LaSalle theorem and the assumption guaranteeing that $y_i \equiv 0, u_i \equiv 0 \Rightarrow x_i = 0$. ♠♠

It is known that the feedback interconnection of a PR and a SPR blocks yields an asymptotically stable system, see lemma 3.5. In the case of nonlinear systems, the passivity theorem provides $\mathcal{L}_2$-stability results for the interconnection of an externally-passive block with an externally-ISP, -OSP or a -VSP block (see e.g. [206]). Lyapunov stability can be obtained when the blocks are internally-passive. The goal of the following section is to present stability results with slightly relaxed requirements on the feedback block. More precisely, we will deal with the interconnection of lossless blocks with WSPR blocks. The results presented in this section relax the conditions of the passivity theorem as was conjectured in [105].

### 4.5.3 Lossless and WSPR blocks interconnection

We now consider the negative feedback interconnection of a lossless (possibly nonlinear) system with a linear WSPR system and prove the stability of the closed-loop system.

**Lemma 4.3** Assume that $\mathrm{H}_1$ in figure 4.6 is lossless with a positive definite storage function, whereas $\mathrm{H}_2$ is WSPR. Then the feedback interconnection of $\mathrm{H}_1$ and $\mathrm{H}_2$ is Lyapunov asymptotically stable. ♠♠

**Proof**

Consider $V = x_2^T P_2 x_2 + 2V_1(x_1)$, where $V_1$ is a positive definite storage function for $H_1$. In view of the assumptions and of the KYP lemma, there exists matrices $P_2$, $L_2$, $W_2$ such that (3.2) are satisfied for $H_2$. Moreover the transfer function $\mathcal{H}_2(s) = W_2 + L_2(sI - A_2)^{-1}B_2$ has no zeros on the imaginary axis.

Then:

$$
\begin{aligned}
\dot{V} &= -x_2^T L_2^T L_2 x_2 + 2x_2^T P_2 B_2 u_2 + 2u_1^T y_1 \\
&= -x_2 L_2^T L_2 x_2 + 2u_2^T (C_2 - W_2^T L_2^T) x_2 + 2u_1^T y_1 \\
&= -x_2^T L_2^T L_2 x_2 - 2u_2^T (W_2^T L_2^T x_2 + D_2 u_2) \\
&= -x_2^T L_2^T L_2 x_2 - 2u_2^T W_2^T L_2^T x_2 - u_2^T (D_2 + D_2^T) u_2 \\
&= -(u_2^T W_2^T + x_2^T L_2^T)(W_2 u_2 + L_2 x_2) \\
&= -\bar{y}_2^T \bar{y}_2
\end{aligned}
\tag{4.35}
$$

Note that the set $\bar{y}_2 \equiv 0$ corresponds to the zero-dynamics of $\mathcal{H}_2$, which either exponentially diverges or converges. However unstable modes cannot exist in the overall system since the state is bounded and the state space representations are minimal. Hence from the Krasovskii-La Salle invariance set theorem, the largest invariant set $S$ inside the set $\bar{y}_2 \equiv 0$ is reduced to $x_1 = x_2 = 0$ plus all the trajectories in the zero-dynamics of $\mathcal{H}_2$ that converge to zero, which ends the proof. ♠♠

## 4.6 Nonlinear KYP lemma

The KYP lemma for linear systems can be extended for nonlinear systems having state-space representations affine in the input. In this section we will consider the case when the plant output $y$ is not a function of the input $u$. A more general case will be studied in the next section. Consider the following nonlinear system

$$
(\Sigma) \left\{ \begin{array}{rcl} \dot{x} &=& f(x) + g(x)u \\ y &=& h(x) \end{array} \right. \tag{4.36}
$$

where $x \in I\!R^n, u \in I\!R^m, y \in I\!R^m$, and $f(\cdot) \in I\!R^{n\times 1}$ with $f(0) = 0$, $g(\cdot) \in I\!R^{n\times m}, h(\cdot) \in I\!R^{m\times 1}$ are smooth functions of $x$. We then have the following result.

**Lemma 4.4 (KYP lemma for nonlinear systems)** Consider the nonlinear system (4.36). The following statements are equivalent.

**(1)** There exists a $\mathcal{C}^1$ storage function $V(x) \geq 0$, $V(0) = 0$ and a function $\mathcal{S}(x) \geq 0$ such that for all $t \geq 0$:

$$
V(x) - V(x(0)) = \int_0^t y^T(s)u(s)ds - \int_0^t \mathcal{S}(x(s))ds \tag{4.37}
$$

The system is Strictly Passive for $\mathcal{S}(x) > 0$, Passive for $\mathcal{S}(x) \geq 0$ and lossless for $\mathcal{S}(x) = 0$.

**(2)** There exists a $\mathcal{C}^1$ non-negative function $V : X \to I\!R$ with $V(0) = 0$, such that

$$\begin{array}{rcl} L_f V(x) & = & -\mathcal{S}(x) \\ L_g V(x) & = & h^T(x) \end{array} \tag{4.38}$$

where $L_g V(x) = \frac{\partial V(x)}{\partial x} g(x)$. ♠♠

**Remark 4.8** Note that if $V(x)$ is a positive definite function (i.e. $V(x) > 0$), then the system $\dot{x} = f(x)$ has a stable equilibrium point at $x = 0$. If in addition $\mathcal{S}(x) > 0$ then $x = 0$ is an asymptotically stable equilibrium point.

**Proof**

- (1)⇒ (2). By assumption we have:

$$V(x) - V(x(0)) = \int_0^t y^T(s) u(s) ds - \int_0^t \mathcal{S}(x(s)) ds \tag{4.39}$$

Taking the derivative with respect to $t$ and using (4.36)

$$\begin{array}{rcl} \frac{dV(x(t))}{dt} & = & \frac{\partial V(x)}{\partial x} \dot{x} \\ & = & \frac{\partial V(x)}{\partial x} (f(x) + g(x)u) \\ & \triangleq & L_f V(x(t)) + L_g V(x(t)) u \\ & = & y^T u - \mathcal{S}(x) \quad \text{(see (4.36))} \end{array} \tag{4.40}$$

Taking the partial derivative with respect to $u$, we get $L_f V(x) = -\mathcal{S}(x)$ and therefore $L_g V(x) = h^T(x)$.

- (2)⇒(1). From (4.36)(4.38) we obtain

$$\frac{dV(x(t))}{dt} = L_f V(x) + L_g V(x) u = -\mathcal{S}(x) + h^T(x) u$$

Integrating the above we obtain (4.36). ♠♠

**Remark 4.9** From these developments it is clear that the dissipativity equality in (4.37) is equivalent to its infinitesimal version $\dot{V} = L_f V + L_g V u = h^T(x) - \mathcal{S}(x) = \langle u, y \rangle - \mathcal{S}(x)$.

**Remark 4.10 (Inverse optimal control)** We have pointed out in remark 3.1 the relationship between optimal control and Positive Real systems. Consider now the control system in (4.36). An optimal control problem is to find the control input $u(\cdot)$ that minimizes the integral action $\int_0^{\infty}[q(x)+u^Tu]dt$ under the dynamics in (4.36), where $q(x)$ is continuously differentiable and positive definite. From standard dynamic programming arguments it is known that the optimal input $u^{\star}=-\frac{1}{2}g^T(x)\frac{\partial V^{\star}}{\partial x}^T$, where $V^{\star}$ is the solution of the partial differential equation (called a Hamilton-Jacobi-Bellman equation):

$$\frac{\partial V^{\star}}{\partial x}f(x)-\frac{1}{4}\left(\frac{\partial V^{\star}}{\partial x}g(x)g^T(x)\frac{\partial V^{\star}}{\partial x}^T\right)+q(x)=0 \tag{4.41}$$

Moreover $V^{\star}(x(t))=\inf_{u(\cdot)}\int_t^{\infty}[q(x(\tau))+u^T(\tau)u(\tau)]d\tau$, $V^{\star}(0)=0$. One recognizes that $u^{\star}$ is nothing else but a static feedback of the passive output of the system (4.36) with storage function $V^{\star}$. Applying some of the results in this section and in section 4.8 one may additionally study the stability of the closed-loop system with the optimal input (see in particular theorem 4.9). There is therefore a close connection between dissipative systems and optimal control. One may start from the feedback of the passive output (given by the KYP lemma) and associate it to an optimal control problem. These problems have been studied in [81] [140]. See also the book [180] for more informations.

### 4.6.1 Nonlinear KYP lemma in the general case

We will now consider the more general case in which the system is described by the following state-space representation affine in the input:

$$(\Sigma)\left\{\begin{array}{rcl}\dot{x} & = & f(x)+g(x)u\\ \\ y & = & h(x)+j(x)u\end{array}\right. \tag{4.42}$$

where $x\in\mathbb{R}^n$, $u\in\mathbb{R}^m$, $y\in\mathbb{R}^m$, and $f(\cdot)\in\mathbb{R}^{n\times 1}$, $g(\cdot)\in\mathbb{R}^{n\times m}$, $h(\cdot)\in\mathbb{R}^{m\times 1}$, $j(\cdot)\in\mathbb{R}^{m\times m}$ are smooth functions of $x$ with $f(0)=0, h(0)=0$. We assume that the system is reachable from the origin. Consider the general supply rate:

$$\begin{array}{rcl}w(u,y) & = & y^TQy+2y^TSu+u^TRu\\ \\ & = & [y^Tu^T]\left[\begin{array}{cc}Q & S\\ S^T & R\end{array}\right]\left[\begin{array}{c}y\\ u\end{array}\right]\end{array} \tag{4.43}$$

with $Q=Q^T, R=R^T$. We then have the following theorem which is due to Hill and Moylan [66].

**Lemma 4.5 (NL KYP lemma: general case)** The nonlinear system (4.42) is dissipative with respect to the supply rate $w(u,y)$ in (4.43) if and only if there exists functions $V \in I\!R, L \in I\!R^q$ and $W \in I\!R^{q\times m}$ (for some integer $q$) such that:

$$\begin{aligned} V(x) &\geq 0 \\ V(0) &= 0 \\ \nabla^T V(x) f(x) &= h^T(x) Q h(x) - L^T(x) L(x) \\ \tfrac{1}{2} g^T(x) \nabla V(x) &= \hat{S}^T(x) h(x) - W^T(x) L(x) \\ \hat{R}(x) &= W^T(x) W(x) \end{aligned} \tag{4.44}$$

where

$$\begin{aligned} \hat{S}(x) &\triangleq Qj(x) + S \\ \hat{R}(x) &\triangleq R + j^T(x) S + S^T j(x) + j^T(x) Q j(x) \end{aligned} \tag{4.45}$$

♠♠

**Proof**
**Sufficiency**. From (4.43), (4.42), (4.44) and (4.45) we obtain

$$\begin{aligned} w(u,y) &= y^T Q y + 2 y^T S u + u^T R u \\ &= (h + ju)^T Q (h + ju) + 2(h + ju)^T S u + u^T R u \\ &= h^T Q h + 2u^T j^T Q h + u^T j^T Q j u + u^T R u + 2 u^T j^T S u \\ &\quad + 2 h^T S u \\ &= h^T Q h + 2 u^T j^T Q h + u^T \hat{R} u + 2 h^T S u \\ &= \nabla^T V f + L^T L + u^T \hat{R} u + 2 u^T [S^T + j^T Q] h \\ &= \nabla^T V f + L^T L + u^T \hat{R} u + 2 u^T \hat{S}^T h \\ &= \nabla^T V f + L^T L + u^T W^T W u + u^T g^T \nabla V + 2 u^T W^T L \\ &= \nabla^T V \dot{x} + (L + Wu)^T (L + Wu) \\ &\geq \nabla^T V \dot{x} = \dot{V} \end{aligned} \tag{4.46}$$

Integrating the above

$$\int_0^t w(t)dt \geq V(x(t)) - V(x(0)) \tag{4.47}$$

**Necessity.** We will show that the available storage function $V_a(x)$ is a solution to the set of equations (4.44) for some $L$ and $W$. Since the system is reachable from the origin, there exists $u(.)$ defined on $[t_{-1}, 0]$ such that $x(t_{-1}) = 0$ and $x(0) = x_0$. Since the system (4.42) is dissipative, then there exists $V(x) \geq 0, V(0) = 0$ such that:

$$\begin{aligned} \int_{t_{-1}}^{T} w(t)dt &= \int_{t_{-1}}^{0} w(t)dt + \int_{0}^{T} w(t)dt \\ &\geq V(x(T)) - V(x(t_{-1})) \\ &\geq 0 \end{aligned}$$

Remember that $\int_{t_{-1}}^{T} w(t)dt$ is the energy introduced into the system. From the above we have

$$\int_{0}^{T} w(t)dt \geq -\int_{t_{-1}}^{0} w(t)dt$$

The right-hand side of the above depends only on $x_0$. Hence, there exists a bounded function $C(\cdot) \in I\!R$ such that

$$\int_{0}^{T} w(t)dt \geq C(x_0) > -\infty$$

Therefore the available storage is bounded:

$$0 \leq V_a(x) = \sup_{x=x(0),t_1\geq 0,u} \left\{ -\int_0^t w(s)ds \right\} \leq \infty$$

Dissipativeness implies (see theorem 4.4) that $V_a(0) = 0$ (since $0 \leq V_a \leq V$) and the available storage $V_a(x)$ is itself a storage function, i.e.

$$V_a(x(t)) - V_a(x(0)) \leq \int_0^t w(s)ds \;\; \forall \; t \geq 0$$

or

$$0 \leq \int_0^t (w(s) - \frac{dV_a}{dt}dt)ds \;\; \forall \; t \geq 0$$

Since the above inequality holds for all $t \geq 0$, taking the derivative in the above it follows that

$$0 \leq w(u, y) - \frac{dV_a(x)}{dt} \triangleq d(x, u)$$

Introducing (4.42)

$$\begin{aligned} d(x, u) &= w(u, y) - \frac{dV_a(x)}{dt} \\ &= w[u, h(x) + j(x)u] - \frac{\partial V_a(x)}{\partial x}[f(x) + g(x)u] \qquad (4.48) \\ &\geq 0 \end{aligned}$$

Since $d(x,u) \geq 0$ and since $w = y^T Qy + 2y^T Su + u^T Ru$, it follows that $d(x,u)$ is quadratic in $u$ and may be factored as

$$d(x,u) = [L(x) + W(x)u]^T [L(x) + W(x)u]$$

for some $L(x) \in I\!R^q, W(x) \in I\!R^{q\times m}$ and some integer $q$. Therefore from the two previous equations and the system (4.42) and the definitions (4.45) we obtain

$$\begin{aligned} d(x,u) &= -\frac{dV_a(x)}{dt}^T [f + gu] + (h + ju)^T Q(h + ju) \\ &\quad +2(h + ju)^T Su + u^T Ru \\ &= -\nabla V_a^T f - \nabla V_a^T gu + h^T Qh + 2h^T [Qj + S] u \\ &\quad +u^T \left[R + j^T S + S^T j + j^T Qj\right] u \\ &= -\nabla V_a^T f - \nabla V_a^T gu + h^T Qh + 2h^T \hat{S}u + u^T \hat{R}u \\ &= L^T L + 2L^T Wu + u^T W^T Wu \end{aligned} \tag{4.49}$$

which holds for all $x, u$. Equating coefficients of like powers of $u$ we get:

$$\begin{aligned} \nabla V_a^T f &= h^T Qh - L^T L \\ \tfrac{1}{2} g^T \nabla V_a &= \hat{S}^T h - W^T L \\ \hat{R} &= W^T W \end{aligned} \tag{4.50}$$

which concludes the proof. ♠♠

Using the sufficiency part of the proof of the above theorem we have the following Corollary:

**Corollary 4.3** If system (4.42) is dissipative with respect to the supply rate $w$ in (4.43), then there exists $V(x) \geq 0, V(0) = 0$ and some $L(x) \in I\!R^q, W(x) \in I\!R^{q\times m}$ such that

$$\frac{dV(x)}{dt} = -[L(x) + W(x)u]^T [L(x) + W(x)u] + w(u,y)$$

♠♠

**Remark 4.11** The lemma (4.4) is a special case of lemma 4.5 for

$$Q = 0,\ R = 0,\ S = \frac{1}{2}I,\ j = 0$$

In that case (4.44) reduces to

$$\begin{aligned} \nabla V^T(x) f(x) &= -L^T(x)L(x) = -S(x) \\ g^T(x)\nabla V(x) &= h(x) \end{aligned} \tag{4.51}$$

**Remark 4.12** Storage functions that satisfy (4.44) can also be shown to be the solutions of the following partial differential inequation where we have omitted the argument $x$:

$$\nabla V^T f + (h^T - \frac{1}{2}\nabla V^T g)\hat{R}^{-1}(h - \frac{1}{2}g^T \nabla V) \leq 0 \tag{4.52}$$

when $\hat{R} = j(x) + j^T(x)$ is full-rank, $R = 0$, $Q = 0$, $S = \frac{1}{2}I$. The available storage and the required supply satisfy this formula (that is similar to a Riccati equation) as an equality [55].

**Remark 4.13** If $j(x) \equiv 0$, then the system in (4.42) cannot be ISP (that corresponds to having $R = -\epsilon I$ in (4.43) for some $\epsilon > 0$). Indeed if (4.42) is dissipative with respect to (4.43) we obtain:

$$\begin{aligned}
\dot{V} &= w(u, y) \\
&= h^T Q h - LLT + 2h^T \hat{S} u - L^T W u \\
&= (y - ju)^T Q (y - ju) - LL^T + 2(y - ju)^T [Qj + S] u - L^T W \\
&= y^T Q y - 2y^T Q j u + u^T j^T Q j u - LL^T \\
&\quad + 2y^T Q j u + 2y^T S u - 2u^T j^T Q j u - 2u^T j^T S u \\
&= y^T Q y + 2y^T S u - \epsilon u^T u
\end{aligned} \tag{4.53}$$

If $j(x) = 0$ we get $-LL^T(x) = -\epsilon u^T u$ which obviously cannot be satisfied with $x$ and $u$ considered as independent variables (except if both sides are identically zero). This result is consistent with the linear case (a PR or SPR function has to have relative degree 0 to be ISP).

## 4.7 Stability of dissipative systems

In this section we will study the relationship between dissipativeness and stability of dynamical systems. Let us first recall that in the case of linear systems, the plant is required to be asymptotically stable to be WSPR, SPR or SSPR. For a PR system it is required that its poles be in the left-half plane and the poles in the $j\omega$−axis be simple and have non-negative associated residues. Consider a dissipative system as in definition 4.7. It can be seen that if $u = 0$ or $y = 0$, then $V(x(t)) \leq V(x(0))$. If in addition the storage function is positive definite i.e. $V(x(t)) > 0$, then we can conclude that the system $\dot{x}= f(x)$, and the system's zero dynamics are stable. Furthermore, if the system is strictly passive (i.e. $\mathcal{S}(x) > 0$ in (4.10)) then the system $\dot{x}= f(x)$, and the system's zero dynamics are both asymptotically stable (see theorem 4.1).

Let us now consider passive systems as given by definition 2.1. The two following lemmae will be used to establish the conditions under which a passive system is asymptotically stable.

**Definition 4.12 (locally ZSD)** A nonlinear system (4.55) is locally zero-state detectable
(ZSD)[ZSO (Zero state sbservable)] if there exists a neighborhood $N$ of 0 such that for all $x \in N$

$$h(x(t)) = 0 \ \forall \ t \Rightarrow x(t) \to 0 \ (x(t) = 0)$$

If $N = \mathbb{R}^n$ the system is ZSD (ZSO). ♠♠

**Lemma 4.6 ([66])** Consider a dissipative system with supply rate $w(u, y)$. Assume that:

1. The system is zero-state detectable
2. There exists $u(t)$ such that $w(u, y) < 0$ for all $y(t)$

then all the solutions to the NL-KYP set of equations (4.44) are positive definite.

♠♠

**Proof**
We have already seen that the available storage

$$V_a(x) = \sup_{x=x(0), t \geq 0, u} \left\{ - \int_0^t w(s) ds \right\}$$

is a (minimum) solution of the KYP-NL set of equations (4.44), see the necessity part of the proof of lemma 4.5 and theorem 4.4. Recall that $0 \leq V_a(x) \leq V(x)$. If we choose $u$ such that $w(u, y) \leq 0$ on $[t_0, \infty)$, with strict inequality on a subset of positive measure, then $V_a(x) > 0$, $\forall y \neq 0$. Note from the equation above that the available storage $V_a(x)$ does not depend on $u(t)$ for $t \in [t_0, \infty)$. When $y = 0$ we can choose $u = 0$ and therefore $x = 0$ in view of the zero-state detectability assumption. We conclude that $V_a(x)$ is positive definite and that $V(x)$ is also positive definite. ♠♠

**Lemma 4.7** Under the same conditions of the previous lemma, the free system $\dot{x} = f(x)$ is (Lyapunov) stable if $Q \leq 0$ and asymptotically stable if $Q < 0$, where Q is the weighting matrix in the general supply rate (4.43). ♠♠

**Proof**
From corollary 4.3 and lemma 4.5 there exists $V(x) > 0$, for all $x \neq 0$, such that (using (4.44) and (4.45))

$$\begin{aligned} \frac{dV(x)}{dt} &= -[L+Wu]^T[L+Wu] + y^TQy + 2y^TSu + u^TRu \\ &= -L^TL - 2L^TWu - u^TW^TWu + (h+ju)^TQ(h+ju) \\ &\quad +2(h+ju)^TSu + u^TRu \\ &= -L^TL - u^TW^TWu + u^T[R + j^TQj + j^TS + S^Tj]u \\ &\quad +2[-L^TW + h^T(Qj+S)]u + h^TQh \\ &= -L^TL - u^T\hat{R}u + u^T\hat{R}u + 2[-L^TW + h^T\hat{S}]u + h^TQh \\ &= -L^TL + \nabla V^T(x)g(x)u + h^TQh \end{aligned} \tag{4.54}$$

For the free system $\dot{x} = f(x)$ we have

$$\frac{dV(x)}{dt} = -L^T(x)L(x) + h^T(x)Qh(x) \leq h^T(x)Qh(x) \leq 0$$

If $Q \leq 0$ then $\frac{dV(x)}{dt} \leq 0$ which implies stability of the system. If $Q \leq 0$ we use Krasovskii-LaSalle Invariance Principle. The invariance set is given by $\Omega : \{\xi | h(\xi) = y = 0\}$ and therefore $x$ converges to the set $\Omega$. In view of the zero-state detectability we conclude that $x \to 0$ asymptotically. ♠♠

**Corollary 4.4 ([66])** Consider a dissipative system with a general supply rate $w(u, y)$. Assume that:

1. The system is zero-state detectable(i.e. $u(t) \equiv 0$ and $y(t) \equiv 0 \Rightarrow x(t) = 0$)
2. There exists $u(t)$ such that $w(u, y) < 0$ for all $y(t)$,

then Passive systems (i.e.$Q = R = 0, S = I$) and input strictly passive systems (ISP) (i.e.$Q = 0, S = I, R = -\epsilon$) are stable, while output passive systems (OSP) (i.e.$Q = -\delta, S = I, R = 0$) and very strictly passive systems (VSP) (i.e. $Q = -\delta, S = I, R = -\epsilon$) are asymptotically stable. ♠♠

## 4.8 Stabilization by output feedback

Consider a causal nonlinear system $\Sigma : u(t) \to y(t)$; $u(t) \in \mathcal{L}_{pe}, y(t) \in \mathcal{L}_{pe}$ represented by the following state-space representation affine in the input:

$$(\Sigma)\left\{\begin{array}{rcl}\dot{x} & = & f(x)+g(x)u \\ y & = & h(x)+j(x)u\end{array}\right. \tag{4.55}$$

where $x \in I\!R^n$, $u, y \in I\!R^m$, $f, g, h$,and $J$ are smooth and $f(0) = h(0) = 0$. Before presenting a result on stabilization of passive systems, let us introduce the following definition:

**Definition 4.13 (Proper function)** A function $V : x \to I\!R$ is said to be proper if for each $a > 0$, the set $V^{-1}[0, a] = \{x : 0 \leq V(x) \leq a\}$ is compact (closed [3] and bounded). ♠♠

We can now state the following result:

**Theorem 4.9 (Asymptotic stabilization [38])** Suppose (4.55) is passive and locally ZSD. Let $\phi(y)$ be any smooth function such that $\phi(0) = 0$ and $y^T\phi(y) > 0$, $\forall y \neq 0$. Assume that the storage function $V(x) > 0$ is proper. Then, the control law $u = -\phi(y)$ asymptotically stabilizes the equilibrium point $x = 0$. If in addition (4.55) is ZSD then $x = 0$ is globally asymptotically stable. ♠♠

**Proof**
By assumption, $V(x) > 0$. Replacing $u = -\phi(y)$ in (4.19) we obtain

$$V(x(t)) - V(x(0)) \leq -\int_0^t y^T(s)\phi(y(s))ds \leq 0$$

It follows that $V(x) \leq V(x(0)) < \infty$, which implies that $||x|| < \infty$, and thus $||y|| < \infty$. Therefore $V(x)$ is non-increasing and thus converges. In the limit, the left hand side of the inequality is 0 , i.e. $\int_0^t y^T(s)\phi(y(s))ds \to 0$ as $t \to \infty$. Thus $y \to 0$ as $t \to \infty$ and $u$ also converges to 0. Since the system is locally ZSD, then $x(t) \to 0$. If in addition the system is globally ZSD, then $x = 0$ is globally asymptotically stable. ♠♠

**Lemma 4.8** Suppose the system (4.55) is passive and ZSO (see definition 4.12) with feedback control law $u = -\phi(y)$. Then the storage function is positive definite, i.e. $V(x) > 0$, for all $x \neq 0$. ♠♠

**Proof**
Recall that the available storage satisfies $0 \leq V_a(x) \leq V(x)$ and

$$\begin{array}{rcl} V_a(x) & = & \sup\limits_{x=x(0),t\geq 0,u} \left\{-\int_0^t y^T(s)u(s)ds\right\} \\ & = & \sup\limits_{x=x(0),t\geq 0,u} \left\{\int_0^t y^T(s)\phi(y(s))ds\right\} \end{array} \tag{4.56}$$

If $V_a(x) = 0$, then necessarily $y(t) = 0$. In view of zero state observability, $y = 0 \Rightarrow x = 0$. Thus $V_a(x)$ vanishes only at $x = 0$ and so does $V(x)$. ♠♠

[3] A set is closed if it contains its limit points.

### 4.8.1 WSPR does not imply OSP

In this subsection we prove that if a system is WSPR (Weakly Strictly Positive Real), it does not necessarily imply that the system is OSP (Output Strictly Passive). The proof is established by presenting a counterexample. The passivity theorems concern interconnections of two blocks, where the feedback block must be either ISP, OSP or VSP. The interest of the results in section 4.5.3 is that the conditions on the feedback block are relaxed to WSPR. We prove now that the following transfer function (which is WSPR, see example 2.17)

$$H(s) = \frac{s+a+b}{(s+a)(s+b)} \tag{4.57}$$

is not OSP. This proves that in general WSPR $\not\Rightarrow$ OSP. A minimal state space representation $(A, B, C)$ for $H(s)$ is given by $A = \begin{pmatrix} 0 & 1 \\ -ab & -a-b \end{pmatrix}$, $B = \begin{pmatrix} 1 \\ 0 \end{pmatrix}$, $C = (1, 0)$. Let us choose $a = 1$, $b = 2$, $x(0) = 0$, $u = \sin(\omega t)$. Then

$$y(t) = \int_0^t [2\exp(\tau - t) - \exp(2\tau - 2t)] \sin(\omega\tau) d\tau \tag{4.58}$$

It can be shown that

$$y(t) = f_1(\omega)\cos(\omega t) + f_2(\omega)\sin(\omega t) \tag{4.59}$$

with $f_1(\omega) = -\frac{\omega^3 - 7\omega}{(1+\omega^2)(4\omega^2)}$, and $f_2(\omega) = \frac{6}{(1+\omega^2)(4\omega^2)}$. It can also be proved that

$$\int_0^t u(\tau)y(\tau)d\tau = -\frac{f_1(\omega)}{4\omega}[\cos(2\omega t) - 1] + \frac{f_2(\omega)}{2}[t - \frac{\sin(2\omega t)}{2\omega}] \tag{4.60}$$

and that

$$\begin{aligned} \int_0^t y^2(\tau)d\tau &= f_1^2(\omega)\left[\frac{t}{2} + \frac{\sin(2\omega t)}{4\omega}\right] + f_2^2(\omega)\left[\frac{t}{2} - \frac{\sin(2\omega t)}{\omega}\right] \\ &\quad - f_1(\omega)f_2(\omega)\left[\frac{\cos(2\omega t)}{2\omega} - 1\right] \end{aligned} \tag{4.61}$$

Let us choose $t_n = \frac{2n\pi}{\omega}$ for some integer $n > 0$. When $\omega \to +\infty$, then $\int_0^{t_n} u(\tau)y(\tau)d\tau = \frac{f_2(\omega)2n\pi}{4\omega}$, whereas $\int_0^{t_n} y^2(\tau)d\tau = \frac{2n\pi(f_1^2(\omega)+f_2^2(\omega))}{4\omega} + f_1(\omega)f_2(\omega)\left(1 - \frac{1}{2\omega}\right)$. It follows that $\int_0^{t_n} u(\tau)y(\tau)d\tau \underset{\omega \to \infty}{\simeq} \frac{\alpha}{\omega^5}$ while $\int_0^{t_n} y^2(\tau)d\tau \underset{\omega \to \infty}{\simeq} \frac{\gamma}{\omega^3}$ for some positive real $\alpha$ and $\gamma$. Therefore we have found an input $u(t) = \sin(\omega t)$ and a time $t$ such that the inequality $\int_0^t u(\tau)y(\tau)d\tau \geq \delta \int_0^t y^2(\tau)d\tau$ cannot be satisfied for any $\delta > 0$, as $\omega \to +\infty$. ♠♠

## 4.9 Equivalence to a passive system

Byrnes, Isidori and Willems [38] have found sufficient conditions for a nonlinear system to be feedback equivalent to a passive system with positive definite storage function. See chapter 9 for a short review on nonlinear systems. Consider a nonlinear system described by

$$\begin{aligned} \dot{x} &= f(x) + g(x)u \\ y &= h(x) \end{aligned} \tag{4.62}$$

The system has relative degree $\{1, \ldots, 1\}$ at $x = 0$ if $L_g h(0) = \frac{\partial h(x)}{\partial x} g(x)|_{x=0}$ is a non singular $(m \times m)$ matrix. If in addition the vector field $g_1(x), \ldots, g_m(x)$ is involutive then the system can be written in the normal form

$$\begin{aligned} \dot{z} &= q(z, y) \\ \dot{y} &= b(z, y) + a(z, y)u \end{aligned} \tag{4.63}$$

where

$$b(z, y) = L_f h(x)$$

$$a(z, y) = L_g h(x)$$

The normal form is globally defined if and only if

H1: $L_g h(x)$ is non singular for all $x$.

H2: The columns of $g(x)[L_g h(x)]^{-1}$ form a complete vector field.

H3: The vector field formed by the columns of $g(x)[L_g h(x)]^{-1}$ commute.

The zero dynamics describe the internal dynamics of the system when $y \equiv 0$ and is characterized by:

$$\dot{z} = q(z, 0)$$

Define the manifold

$$Z^* = \{x : h(x) = 0\}$$

and

$$\tilde{f}(x) = f(x) + g(x)u^*(x) \tag{4.64}$$

with

$$u^*(x) = -[L_g h(x)]^{-1} L_f h(x) \tag{4.65}$$

Let $f^*(x)$ be the restriction to $Z^*$ of $\tilde{f}(x)$. Then the zero dynamics is also described by

$$\dot{x} = f^*(x) \text{ for all } x \in Z^*. \tag{4.66}$$

**Definition 4.14** Assume that the matrix $L_g h(0)$ is nonsingular. Then the system (4.62) is said to be:

1. minimum phase if $z = 0$ is an asymptotic stable equilibrium of (4.66)
2. weakly minimum phase if $\exists W^*(z) \in C^r, r \geq 2$, with $W^*(z)$ positive definite, proper and such that $L_{f^*} W^*(z) \leq 0$ locally around $z = 0$ ♠♠

These definitions become global if they hold for all $z$ and H.1-H.3 hold.

**Definition 4.15** $x^0$ is a regular point of (4.62) if rank$\{L_g h(0)\}$ is constant in a neighborhood of $x^0$. ♠♠

Recall that a necessary condition for a strictly proper transfer to be PR is to have relative degree equal to 1. The next theorem extends this fact for multivariable nonlinear systems. We will assume in the sequel that rank $g(0) =$ rank $dh(0) = m$.

**Theorem 4.10** [38] Assume that the system (4.62) is passive with a $C^2$ positive definite storage function $V(x)$. Suppose $x = 0$ is a regular point. Then $L_g h(0)$ is nonsingular and the system has a relative degree $\{1, \ldots, 1\}$ at $x = 0$. ♠♠

**Proof**
If $L_g h(0)$ is singular, there exists $u(x) \neq 0$ for $x$ in the neighborhood of $x = 0$ ($N(0)$) such that

$$L_g h(x) u(x) = 0$$

Since rank$\{dh(x)\} = m$, for all $x \in N(0)$, we have

$$\gamma(x) = g(x) u(x) \neq 0$$

for all $x \in N(0)$. Given that the system (4.62) is passive it follows that $L_g V(x) = h^T(x)$ so that

$$L_\gamma^2 V(x) = L_\gamma[L_g V(x) u(x)] = L_\gamma[u^T(x) h(x)]$$

where

$$L_\gamma[u^T h] = \frac{\partial(u^T h)}{\partial x}\gamma$$

and

$$\begin{aligned}
\frac{\partial(u^T h)}{\partial x} &= \left[\frac{\partial(u^T h)}{\partial x_1}, \ldots, \frac{\partial(u^T h)}{\partial x_n}\right] \\
&= [\frac{\partial u^T}{\partial x_1} h + u^T \frac{\partial h}{\partial x_1}; \ldots ; \frac{\partial u^T}{\partial x_n} h + u^T \frac{\partial h}{\partial x_n}] \\
&= h^T [\frac{\partial u}{\partial x_1}, \ldots, \frac{\partial u}{\partial x_2}] + u^T [\frac{\partial h}{\partial x_1}, \ldots, \frac{\partial h}{\partial x_n}] \\
&= h^T \frac{\partial u}{\partial x} + u^T \frac{\partial h}{\partial x}
\end{aligned} \tag{4.67}$$

Then

$$\begin{aligned} L_\gamma^2 V &= L_\gamma[u^T h] \\ &= h^T L_\gamma u + u^T L\gamma h \\ &= (L_\gamma u)^T h + u^T L_\gamma h \\ &= v^T(x)h(x) \end{aligned} \tag{4.68}$$

with

$$v^T(x) = (L_\gamma u)^T + u^T L_\gamma$$

Let $\phi_t^\gamma(x_{t_0})$ denote the flow of the vector field $\gamma$, i.e. the solution of $\dot{\xi}(t) = \gamma(\xi(t))$ for $\xi_0 = x(t_0)$.
Define $f(t) = V(\phi_t^\gamma(0))$. Using Taylor's theorem for $n = 2$ we have

$$f(t) = f(0) + f^{(1)}(0)t + f^{(2)}(s)\frac{1}{2}t^2$$

where $0 \leq s \leq t$. Note that

$$\begin{aligned} f(t) &= V(\phi_t^\gamma(0)) \\ f^{(1)}(t) &= \tfrac{\partial V(\phi_t^\gamma(0))}{\partial \xi}\dot{\xi} = \tfrac{\partial V(\phi_t^\gamma(0))}{\partial \xi}\gamma(\xi(t)) = L_\gamma V(\phi_t^\gamma(0)) \\ f^{(2)}(t) &= \tfrac{\partial f^{(1)}(t)}{\partial \xi}\dot{\xi} = L_\gamma f^{(1)}(t) = L_\gamma^2 V(\phi_t^\gamma(0)) \end{aligned} \tag{4.69}$$

Therefore

$$V(\phi_t^\gamma(0)) = V(0) + L_\gamma V(0)t + L_\gamma^2 V(\phi_s^\gamma(0))\frac{1}{2}t^2$$

Given that $V(0) = 0$ we have

$$\begin{aligned} L_\gamma V(0) &= \tfrac{\partial V(x)}{\partial x} g(x)u(x)|_{x=0} \\ &= L_g V(x)u(x)|_{x=0} \\ &= h^T(0)u(0) \\ &= 0 \end{aligned} \tag{4.70}$$

Thus

$$V(\phi_t^\gamma(0)) = v^T(\phi_s^\gamma(0))h(\phi_s^\gamma(0))\frac{1}{2}t^2$$

Recall that $\frac{\partial h(x)}{\partial x}g(x)u(x) = 0$, for all $x$ and in particular we have $\frac{\partial h(\xi)}{\partial \xi}g(\xi)u(\xi) = 0$ which implies that $\frac{\partial h(\xi)}{\partial \xi}\dot{\xi} = 0 \Rightarrow \dot{h}(\xi) = 0 \Rightarrow h(\xi) = cte.$

Thus $h(\phi_t^\gamma(0)) = h(0) = 0$ and then $V(\phi_t^\gamma(0)) = 0 \Rightarrow \phi_t^\gamma(0) = 0 \Rightarrow \gamma(0)) = 0$ which is a contradiction. Therefore $L_g h(0)$ must be nonsingular. ♠♠
Recall that a necessary condition for a strictly proper transfer to be PR is to be minimum phase. The next theorem extends this fact for general nonlinear systems.

**Theorem 4.11** [38] Assume that the system (4.62) is passive with a $C^2$ positive definite storage function $V$. Suppose that either $x = 0$ is a regular point or that $V$ is non degenerate. Then the system zero-dynamics locally exists at $x = 0$ and the system is weakly minimum phase. ♠♠

**Proof**
In view of theorem 4.10 the system has relative degree $\{1 \dots 1\}$ at $x = 0$ and therefore its zero-dynamics locally exists at $x = 0$. Define the positive definite function $W^* = V|_{Z^*}$ with $Z^* = \{x : h(x) = 0\}$. Since the system is passive we have $0 \geq L_f V(x)$ and $L_g V(x) = h^T(x)$. Define $f^*(x) = f(x) + g(x)u^*(x)$ and $u^*(x) = -[L_g h(x)]^{-1} L_f h(x)$. Thus:

$$\begin{array}{rcl} 0 & \geq & L_f V(x) \\ & = & L_{f^*} V(x) - L_g V(x) u^*(x) \\ & = & L_{f^*} V(x) - h^T(x) u^*(x) \\ & = & L_{f^*} V(x) \end{array} \tag{4.71}$$

along any trajectory of the zero dynamics ($h(x) = 0$). ♠♠
The two theorems above show essentially that any passive system with a positive definite storage function, under mild regularity assumptions, necessarily has relative degree $\{1 \dots 1\}$ at $x = 0$ and is weakly minimum phase. These two conditions are shown to be sufficient for a system to be feedback equivalent to a passive system as stated in the following theorem.

**Theorem 4.12** [38] Suppose $x = 0$ is a regular point. Then the system (4.62) is locally feedback equivalent to a passive system with $C^2$ storage function $V$ which is positive definite, if and only if (4.62) has relative degree $\{1 \dots 1\}$ at $x = 0$ and is weakly minimum phase.

## 4.10 Cascaded systems

Cascaded systems are important systems that appear in many different practical cases. We will state here some results concerning this type of systems which will be used later in the book.
Consider a cascaded system of the following form

$$\begin{array}{rcl} \dot{\zeta} & = & f_0(\zeta) + f_1(\zeta, y) y \\ \dot{x} & = & f(x) + g(x) u \\ y & = & h(x) \end{array} \tag{4.72}$$

The 1st equation above is called the *driven system* while the 2nd and 3rd equation are called the *driving system.*

**Theorem 4.13** [38] Consider the cascaded system (4.72). Suppose that the *driven system* is globally asymptotically stable and the *driven system* is (strictly) passive with a $C^r$ , $r \geq 2$, storage function $V$ which is positive definite. The system (4.72) is feedback equivalent to a (strictly) passive system with a $C^r$ storage function which is positive definite. ♠♠

The cascaded system in (4.72) can also be globally stabilized using a smooth control law as is stated in the following theorem for which we need the following definitions. Concerning the *driving system* in (4.72) we define the the associate distribution [72], [148].

$$\mathcal{D} = \text{span}\{ad_f^k g_i : \ 0 \leq k \leq n-1, \ 1 \leq i \leq m\} \tag{4.73}$$

and the following set

$$\mathcal{S} = \{x \in X : \ L_f^m L_\tau \ V(x) = 0, \text{ for all } \ \tau \in \mathcal{D}, \ \ 0 \leq m < r\} \tag{4.74}$$

**Theorem 4.14** [38] Consider the cascaded system (4.72). Suppose that the *driven system* is globally asymptotically stable and the *driven system* is passive with a $C^r$ , $r \geq 1$, storage function $V$ which is positive definite and proper. Suppose that $\mathcal{S} = \{0\}$. Then the system (4.72) is globally asymptotically stabilizable by the smooth feedback

$$u^T(\zeta, x) = -L_{f_1(\zeta, h(x))} U(\zeta) \tag{4.75}$$

where $U$ is a Lyapunov function for the *driven system* part of the cascaded system (4.72). Some additional comments on the choice of $u$ in 4.75 are given in subsection 6.2.4. ♠♠

## 4.11 Passivity of linear delay systems

The above developments focus on particular classes of smooth finite dimensional dynamical systems. Let us investigate another type of systems that does not fit within these classes, namely time-delayed systems. Note that other type of infinite-dimensional systems have been studied in section 2.12.
Stability and control of linear systems with delayed state are problems of recurring interest since the existence of a delay in a system representation may induce instability, oscillations or bad performances for the closed-loop scheme. In this section we shall consider the *passivity* problem of a linear system described by differential equations with delayed state. The interconnection schemes with passive systems will be also treated. The proposed approach

is based on an appropriate Lyapunov-Krasovskii functional contruction. The material presented in this section follows the analysis given in [145], see also [147] and [99]. The corresponding results may include or not delay information and are expressed in terms of solutions of some algebraic Riccati equations. The results presented here can be extended to the multiple delays case by an appropriate choice of the Lyapunov functional.

### 4.11.1 Systems with state delay

Consider the following system

$$\begin{aligned} \dot{x} &= Ax + A_1x(t-\tau) + Bu \\ y &= Cx \end{aligned} \tag{4.76}$$

where $x \in I\!R^n, y \in I\!R^p, u \in I\!R^p$ are the state, the output and the input of the system and $\tau$ denotes the delay.
The main result of this section can be stated as follows:

**Lemma 4.9** If there exists positive definite matrices $P > 0$ and $S > 0$ and a scalar $\gamma \geq 0$ such that:

$$\begin{aligned} \Gamma &\triangleq A^TP + PA + PA_1S^{-1}A_1^TP + S \quad < \gamma C^TC \\ C &= B^TP \end{aligned} \tag{4.77}$$

then system (4.76) satisfies the following inequality:

$$\int_0^t u(s)^T y(s)ds \geq \tfrac{1}{2}\left[V(t) - V(0)\right] - \tfrac{1}{2}\gamma \int_0^t y(s)^T y(s)ds \tag{4.78}$$

where

$$V(t) = x(t)^T P x(t) + \int_{t-\tau}^t x(s)^T S x(s) ds \tag{4.79}$$

**Remark 4.14** Note that system (4.76) is passive only if $\gamma = 0$. Roughly speaking for $\gamma > 0$ we may say system (4.76) is less than output strictly passive. This gives us an extra degree of freedom for choosing $P$ and $S$ in (4.77) since inequality in (4.77) becomes more restrictive for $\gamma = 0$. We can expect to be able to stabilize system (4.76) using an appropriate passive controller as will be seen in the next section. Note that for $\gamma < 0$ the system is output strictly passive but this imposes stronger restrictions on the system (see (4.77)).

**Proof**
From (4.76) and the above conditions we have

$$
\begin{aligned}
2\int_0^t u^T y ds &= 2\int_0^t u(s)^T C x(s) ds \\
&= 2\int_0^t u(s)^T B^T P x(s) ds \\
&= \int_0^t u(s)^T B^T P x(s) ds + \int_0^t x(s)^T P B u(s) ds \\
&= \int_0^t \left\{ \left( \tfrac{dx}{ds} - Ax(s) - A_1 x(s-\tau) \right)^T P x(s) + \right. \\
&\quad \left. + x(s)^T P \left( \tfrac{dx}{ds} - Ax(s) - A_1 x(s-\tau) \right) \right\} ds \\
&= \int_0^t \left\{ \tfrac{d(x(s)^T P x(s))}{ds} - x(s)^T (A^T P + PA) x(s) - \right. \\
&\quad \left. - x(s-\tau)^T A_1^T P x(s) - x(s)^T P A_1 x(s-\tau) \right\} ds \\
&= \int_0^t \left\{ \tfrac{dV(s)}{ds} - x(s)^T \Gamma x(s) + I(x(s), x(s-\tau)) \right\} ds
\end{aligned} \tag{4.80}
$$

where $\Gamma$ is given by (4.77) and:

$$
\begin{aligned}
I(x(t), x(t-\tau)) &= \left[ S^{-1} A_1^T P x(t) - x(t-\tau) \right]^T S \\
&\quad \times \left[ S^{-1} A_1^T P x(t) - x(t-\tau) \right]
\end{aligned}
$$

Note that $V(t)$ is a positive definite function and $I(x(t), x(t-\tau)) \geq 0$ for all the trajectories of the system. Thus from (4.78) and (4.80) it follows that:

$$
\begin{aligned}
\int_0^t u(s)^T y(s) ds &\geq \tfrac{1}{2}[V(t) - V(0)] - \tfrac{1}{2}\int_0^t x(s)^T \Gamma x(s) ds \\
&\geq \tfrac{1}{2}[V(t) - V(0)] - \tfrac{1}{2}\gamma \int_0^t x(s)^T C^T C x(s) ds \\
&\geq -\tfrac{1}{2} V(0) - \tfrac{1}{2}\gamma \int_0^t y(s)^T y(s) ds \;\; \forall t > 0
\end{aligned} \tag{4.81}
$$

Therefore if $\gamma = 0$ then the system is passive. ♠♠

### 4.11.2 Interconnection of passive systems

Let us consider the block interconnection depicted in figure 4.6 where $H_1$ represents system (4.76) and $H_2$ is the controller which is input strictly passive as defined above i.e. for some $\varepsilon > 0$

$$
\int_0^t u_2(s)^T y_2(s) ds \geq -\beta_2^2 + \varepsilon \int_0^t u_2(s)^T u_2(s) ds \;\text{ for some }\; \beta \in I\!R, \forall t \tag{4.82}
$$

$H_2$ can be a finite dimensional linear system for example. For the sake of simplicity we will consider $H_2$ to be an asymptotically stable linear system. We will show next that the controller satisfying the above property will stabilize system (4.76).
From lemma 4.9, the interconnection scheme and (4.82) we have:

$$\begin{aligned} u_1 &= u \\ y_1 &= y \\ u_2 &= y_1 \\ y_2 &= -u_1 \end{aligned} \tag{4.83}$$

Therefore from (4.78) and (4.82) we have

$$\begin{aligned} 0 &= \int_0^t u_1(s)^T y_1(s)ds + \int_0^t u_2(s)^T y_2(s)ds \\ &\geq -\tfrac{1}{2}V(0) - \tfrac{1}{2}\gamma \int_0^t y_1(s)^T y_1(s)ds - \beta_2^2 + \varepsilon \int_0^t u_2(s)^T u_2(s)ds \\ &\geq -\beta^2 + (\varepsilon - \tfrac{1}{2}\gamma) \int_0^t y_1(s)^T y_1(s)ds \end{aligned}$$

where $\beta^2 = \frac{1}{2}V(0) + \beta_2^2$. If $\varepsilon - \frac{1}{2}\gamma > 0$ then $y_1$ is $\mathcal{L}_2$. Since $H_2$ is an asymptotically stable linear system with an $\mathcal{L}_2$ input, it follows that the corresponding output $y_2$ is also $\mathcal{L}_2$. Given that the closed-loop system is composed of two linear systems, the signals in the closed-loop can not have peaks. Therefore all the signals converge to zero which means the stability of the closed loop system.

### 4.11.3 Extension to a system with distributed state delay

Let us consider the following class of distributional convolution systems

$$\begin{aligned} \dot{x} &= \mathcal{A}{*}x + Bu \\ y &= Cx \end{aligned} \tag{4.84}$$

where $\mathcal{A}$ denotes a distribution of order 0 on some compact support $[-\tau, 0]$. Let us choose

$$\mathcal{A} = A\delta(\theta) + A_1\delta(\theta - \tau_1) + A_2(\theta) \tag{4.85}$$

where $\delta(\theta)$ represents the Dirac delta functional and $A_2(\theta)$ is a piece-wise continuous function. Due to the term $A_2(\theta)$ the system has a distributed delay. For the sake of simplicity we shall consider $A_2(\theta)$ constant. The system (4.84) becomes

$$\begin{aligned} \dot{x} &= Ax + A_1 x(t - \tau_1) + \int_{-\tau}^0 A_2 x(t + \theta)d\theta + Bu \\ y &= Cx \end{aligned} \tag{4.86}$$

**Lemma 4.10** If there exists positive definite matrices $P > 0$, $S_1 > 0$ and $S_2 > 0$ and a scalar $\gamma \geq 0$ such that:

$$\begin{aligned} \Gamma(\tau) &\triangleq A^TP + PA + PA_1S_1^{-1}A_1^TP + S_1 + \tau(PA_2S_2^{-1}A_2^TP + S_2) \\ &< \gamma C^TC \\ C &= B^TP \end{aligned} \tag{4.87}$$

then system (4.86) verifies the following inequality:

$$\int_0^t u(s)^Ty(s)ds \geq \tfrac{1}{2}[V(t) - V(0)] - \tfrac{1}{2}\gamma \int_0^t y(s)^Ty(s)ds \tag{4.88}$$

where

$$\begin{aligned} V(t) &= x(t)^TPx(t) + \int_{t-\tau_1}^t x(s)^TS_1x(s)ds \\ &+ \int_{-\tau}^0 (\int_{t+\theta}^t x(s)^TS_2x(s)ds)d\theta \end{aligned} \tag{4.89}$$

♠♠

**Proof**

We shall use the same steps as in the proof of lemma 4.9. Thus from (4.86) and the above conditions we have:

$$\begin{aligned} 2\int_0^t u^Tyds &= 2\int_0^t u(s)^TCx(s)ds \\ &= 2\int_0^t u(s)^TB^TPx(s)ds \\ &= \int_0^t u(s)^TB^TPx(s)ds + \int_0^t x(s)^TPBu(s)ds \\ &= \int_0^t \Big\{\tfrac{dx}{ds} - Ax(s) - A_1x(s-\tau_1) - \\ &\quad - \int_{-\tau}^0 A_2x(s+\theta)d\theta\Big\}^T Px(s)ds \\ &\quad + \int_0^t x(s)^TP\Big\{\tfrac{dx}{ds} - Ax(s) - A_1x(s-\tau_1) - \\ &\quad - \int_{-\tau}^0 A_2x(s+\theta)d\theta\Big\}ds \end{aligned} \tag{4.90}$$

we also have

$$
\begin{aligned}
2\int_0^t u^T y ds &= \int_0^t \left\{ \frac{d(x(s)^T P x(s))}{ds} - x(s)^T (A^T P + PA) x(s) - \right. \\
&\quad \left. - x(s-\tau_1)^T A_1^T P x(s) - x(s)^T P A_1 x(s-\tau_1) \right\} ds \\
&\quad - \int_0^t \left\{ x(s)^T P \int_{-\tau}^0 A_2 x(s+\theta) d\theta + \right. \\
&\quad \left. + \left[ \int_{-\tau}^0 x^T(s+\theta) A_2^T d\theta \right] P x(s) \right\} ds \\
&= \int_0^t \left\{ \frac{dV(s)}{ds} - x(s)^T \Gamma(\tau) x(s) + I_1(x(s), x(s-\tau_1)) + \right. \\
&\quad \left. + I_2(x(s), x(s+\theta)) \right\} ds
\end{aligned} \tag{4.91}
$$

where $\Gamma(\tau)$ is given by (4.87) and:

$$
\begin{aligned}
I_1(x(t), x(t-\tau_1)) &= \left[ S_1^{-1} A_1^T P x(t) - x(t-\tau_1) \right]^T S_1 \times \\
&\quad \times \left[ S_1^{-1} A_1^T P x(t) - x(t-\tau_1) \right]
\end{aligned} \tag{4.92}
$$

$$
\begin{aligned}
I_2(x(t), x(t+\theta)) &= \int_{-\tau}^0 \left[ S_2^{-1} A_2^T P x(t) - x(t+\theta) \right]^T S_2 \times \\
&\quad \times \left[ S_2^{-1} A_2^T P x(t) - x(t+\theta) \right] d\theta
\end{aligned} \tag{4.93}
$$

Note that $V(t)$ is a positive definite function and $I_1(x(t), x(t-\tau_1)) \geq 0$ and $I_2(x(t), x(t+\theta)) \geq 0$ for all the trajectories of the system. Thus from (4.87) and (4.89) it follows that:

$$
\begin{aligned}
\int_0^t u(s)^T y(s) ds &\geq \tfrac{1}{2}\left[ V(t) - V(0) \right] - \tfrac{1}{2} \int_0^t x(s)^T \Gamma(\tau) x(s) ds \\
&\geq \tfrac{1}{2}\left[ V(t) - V(0) \right] - \tfrac{1}{2}\gamma \int_0^t x(s)^T C^T C x(s) ds \\
&\geq -\tfrac{1}{2} V(0) - \tfrac{1}{2}\gamma \int_0^t y(s)^T y(s) ds \;\; \forall t > 0
\end{aligned} \tag{4.94}
$$

Therefore if $\gamma = 0$ then the system is passive. ♠♠

**Remark 4.15** The presence of a distributed delay term in the system (4.86) imposes extra constraints in the solution of inequality (4.87). Note that for $\tau = 0$ we recover the previous case having only a point state delay. Extensions of the result presented in this section can be found in [164].

**Remark 4.16** Note also that given that system (4.86) satisfies inequality (4.94), it can be stabilized by an input strictly passive system as described in the previous section. Furthermore due to the form of the Riccati equation the upper bound for the (sufficient) distributed delay $\tau$ (seen as a parameter) may be improved by feedback interconnection for the same Lyapunov-based construction. Such result does not contradict the theory since the derived condition is only sufficient, and not necessary and sufficient.

## 4.12 Passivity of a helicopter model

In the next chapter we shall investigate in detail the dissipativity properties of physical systems, including mechanical systems. However before this general presentation, it is useful to end this chapter on dissipative dynamical systems theory with an interesting example of a dissipative mechanical system. To this end let us present the model of a reduced scale helicopter and show that the rotational part of the model is passive. The material in this section is inspired from [121], [122] and [47].

Consider the figure 4.8. The airframe is denoted by the letter $\mathcal{A}$ and the helicopter as a whole by the letter $\mathcal{H}$. Let $\mathcal{I} = \{E_x, E_y, E_z\}$ define a right-hand inertial frame stationary with respect to the earth. Let the vector $\xi = (x, y, z)$ denote the position of the center of the mass of the helicopter relative to the air frame $\mathcal{I}$. Let $\mathcal{A} = \{E_1^a, E_2^a, E_3^a\}$ be a right-hand body fixed frame for $\mathcal{A}$.

The direction of the lift force is oriented perpendicular to the orientation of the main rotor disk. We will assume that the main rotor lift is $(0, 0, -u) \in \mathcal{A}$ oriented in the directions $\{E_1^a, E_2^a, E_3^a\}$. The three torque controls inputs are $\tau = (\tau_1, \tau_2, \tau_3) \in \mathcal{A}$ where $\tau_3$ is due to the tail rotor and $\tau_1, \tau_2$ are due to the cyclic pitch of the main rotor blades. It is assumed that the center of the mass of the helicopter and the hub of the main rotor are co-linear whit respect to the axis $E_3^a$.

The orientation of the helicopter is given by the Euler angles

$$\eta = (\phi, \theta, \psi) \tag{4.95}$$

which are the classical "yaw, pitch and roll" angles commonly used in aerodynamics applications [59]. Firstly, a rotation of angle $\phi$ around the axes $E_z$ is applied, corresponding to 'yaw'. Secondly, a rotation of angle $\theta$ around the rotated version of the $E_y$ axis is applied, corresponding to 'pitch' of the airframe. Lastly, a rotation of angle $\psi$ around the axes $E_1^a$ is applied. This corresponds to 'roll' of $\mathcal{A}$ around the natural axis $E_1^a$. The rotation matrix $R(\phi, \theta, \psi) \in SO(3)$ representing the orientation of the airframe $\mathcal{A}$ relative to a fixed inertial frame is

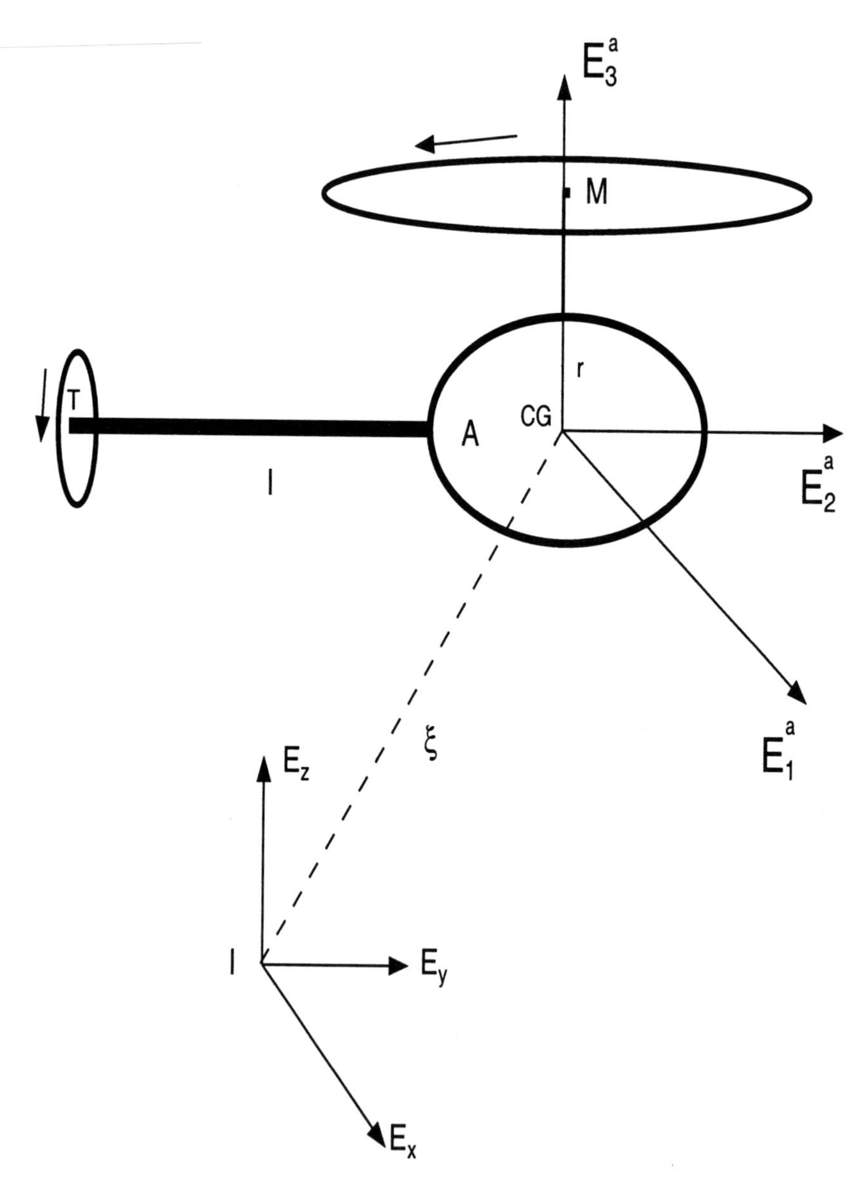


Figure 4.8: Geometric model of the helicopter

$$R(\phi,\theta,\psi) = \begin{bmatrix} c_\phi & -s_\phi & 0 \\ s_\phi & c_\phi & 0 \\ 0 & 0 & 1 \end{bmatrix} \begin{bmatrix} c_\theta & 0 & s_\theta \\ 0 & 1 & 0 \\ -s_\theta & 0 & c_\theta \end{bmatrix} \begin{bmatrix} 1 & 0 & 0 \\ 0 & c_\psi & -s_\psi \\ 0 & s_\psi & c_\psi \end{bmatrix} \quad (4.96)$$

$$R(\phi,\theta,\psi) = \begin{pmatrix} c_\theta c_\phi & s_\psi s_\theta c_\phi - c_\psi s_\phi & c_\psi s_\theta c_\phi + s_\psi s_\phi \\ c_\theta s_\phi & s_\psi s_\theta s_\phi + c_\psi c_\phi & c_\psi s_\theta s_\phi - s_\psi c_\phi \\ -s_\theta & s_\psi c_\theta & c_\psi c_\theta \end{pmatrix} \quad (4.97)$$

where the following shorthand notation is used

$$c_\beta = \cos(\beta), \qquad s_\beta = \sin(\beta) \quad (4.98)$$

The generalized coordinates for the helicopter H are

$$q = (x, y, z, \phi, \theta, \phi) \in I\!R^6 \quad (4.99)$$

We can divide the generalized coordinates into translational and rotational coordinates respectively:

$$\xi = (x, y, z) \in I\!R^3, \qquad \eta = (\phi, \theta, \psi) \in I\!R^3 \quad (4.100)$$

The translational kinetic energy of the helicopter is

$$T_{trans} = \frac{M}{2}\left\langle \dot{\xi}, \dot{\xi} \right\rangle = \frac{M}{2}(\dot{x}^2 + \dot{y}^2 + \dot{z}^2) \quad (4.101)$$

An angular velocity in the body fixed frame $\mathcal{A}$ is related to the generalized velocities $(\dot{\phi}, \dot{\theta}, \dot{\psi})$ [59].

$$\Omega_a = \begin{pmatrix} \dot{\psi} - \dot{\phi} s_\theta \\ \dot{\theta} c_\psi + \dot{\phi} c_\theta s_\psi \\ \dot{\phi} c_\theta c_\psi - \dot{\theta} s_\psi \end{pmatrix} \quad (4.102)$$

which can also be written

$$\Omega_a = W_\eta \dot{\eta} \quad (4.103)$$

where

$$W_\eta = \begin{pmatrix} -s_\theta & 0 & 1 \\ c_\theta s_\psi & c_\psi & 0 \\ c_\theta c_\psi & -s_\psi & 0 \end{pmatrix} \quad (4.104)$$

Therefore

$$\dot{\eta} = \begin{pmatrix} \dot{\phi} \\ \dot{\theta} \\ \dot{\psi} \end{pmatrix} = W_{\eta}^{-1}\Omega_a \tag{4.105}$$

The total kinetic energy of the system is given by

$$T = T_{trans.} + T_{rot.}^{\mathcal{A}} = \frac{1}{2}M\left\langle \dot{\xi}, \dot{\xi} \right\rangle + \frac{1}{2}\left\langle \Omega_a, \mathcal{I}_{\mathcal{A}}\Omega_a \right\rangle \tag{4.106}$$

where $M$ is the total mass of the helicopter, g is the gravitation constant and $\mathcal{I}_{\mathcal{A}}$ denote the inertia of the helicopter airframe $\mathcal{A}$.

$$\mathbf{I}_{\mathcal{A}} = \begin{bmatrix} I_1 & 0 & 0 \\ 0 & I_2 & 0 \\ 0 & 0 & I_3 \end{bmatrix} \tag{4.107}$$

The potential energy is

$$U = Mgz \tag{4.108}$$

and thus the Lagrangian function is defined as

$$L = T - U \tag{4.109}$$

which satisfies the Euler-Lagrange equations

$$\frac{d}{dt}\left(\frac{\partial L}{\partial \dot{q}}\right) - \frac{\partial L}{\partial q} = F_T \tag{4.110}$$

where $F_T = (F, \tau)$, with $F = uE_3^a \in \mathcal{I}$. Since the Lagrangian contains no cross terms in the kinetic energy combining $\dot{\xi}$ with $\dot{\eta}$ , the Euler-Lagrange equation can be partitioned into dynamics for $\xi$ coordinates and the $\eta$ coordinates.

$$L_{trans} = \frac{M}{2}(\dot{x}^2 + \dot{y}^2 + \dot{z}^2) - Mgz \tag{4.111}$$

The Euler-Lagrange equation for the translation motion is

$$\frac{d}{dt}\left[\frac{\partial L_{trans}}{\partial \dot{q}}\right] - \frac{\partial L_{trans}}{\partial q} = F(\xi) \tag{4.112}$$

where

$$F(\xi) = \begin{bmatrix} M\ddot{x} \\ M\ddot{y} \\ M\ddot{z} - Mg \end{bmatrix} \tag{4.113}$$

As for the $\eta$ coordinates we can rewrite

$$\frac{d}{dt}\left[(\Omega_a)^T \mathbf{I}_{\mathcal{A}} \frac{\partial \Omega_a}{\partial \dot{q}}\right] - (\Omega_a)^T \mathcal{I}_{\mathcal{A}} \frac{\partial \Omega_a}{\partial q} = \tau \tag{4.114}$$

$$\frac{\partial \Omega_a}{\partial \dot{q}} = \begin{bmatrix} -s_\theta & 0 & 1 \\ c_\theta s_\psi & c_\psi & 0 \\ c_\theta c_\psi & -s_\psi & 0 \end{bmatrix} \tag{4.115}$$

$$(\Omega_a)^T \mathcal{I}_{\mathcal{A}} \frac{\partial \Omega_a}{\partial \dot{q}} = \begin{bmatrix} b_1 & b_2 & b_3 \end{bmatrix} \tag{4.116}$$

$$\begin{aligned} b_1 &= -I_1(\dot{\psi} s_\theta - \dot{\phi} s_\theta^2) + I_2(\dot{\theta} c_\theta s_\psi c_\psi + \dot{\phi} c_\theta^2 s_\psi^2) \\ &\quad + I_3(\dot{\phi} c_\theta^2 c_\psi^2 - \dot{\theta} c_\theta s_\psi c_\psi) \\ b_2 &= I_2(\dot{\theta} c_\psi^2 + \dot{\phi} c_\theta s_\psi c_\psi) - I_3(\dot{\phi} c_\theta s_\psi c_\psi - \dot{\theta} s_\psi^2) \\ b_3 &= I_1(\dot{\psi} - \dot{\phi} s_\theta) \end{aligned} \tag{4.117}$$

$$\frac{d}{dt}\left[(\Omega_a)^T \mathcal{I}_{\mathcal{A}} \frac{\partial \Omega_a}{\partial \dot{q}}\right] = \begin{bmatrix} d_1 & d_2 & d_3 \end{bmatrix} \tag{4.118}$$

$$\begin{aligned} d_1 &= -I_1(\ddot{\psi} s_\theta + \dot{\psi}\dot{\theta} c_\theta - \ddot{\phi} s_\theta^2 - 2\dot{\phi}\dot{\theta} s_\theta c_\theta) \\ &\quad + I_2(\ddot{\theta} c_\theta s_\psi c_\psi - \dot{\theta}^2 s_\theta s_\psi c_\psi - \dot{\theta}\dot{\psi} c_\theta s_\psi^2 + \dot{\theta}\dot{\psi} c_\theta c_\psi^2 \\ &\quad + \ddot{\phi} c_\theta^2 s_\psi^2 - 2\dot{\phi}\dot{\theta} s_\theta c_\theta s_\psi^2 + 2\dot{\phi}\dot{\psi} c_\theta^2 s_\psi c_\psi) \\ &\quad + I_3(\ddot{\phi} c_\theta^2 c_\psi^2 - 2\dot{\phi}\dot{\theta} s_\theta c_\theta c_\psi^2 - 2\dot{\phi}\dot{\psi} c_\theta^2 s_\psi c_\psi \\ &\quad - \ddot{\theta} c_\theta s_\psi c_\psi + \dot{\theta}^2 s_\theta s_\psi c_\psi + \dot{\theta}\dot{\psi} c_\theta s_\psi^2 - \dot{\theta}\dot{\psi} c_\theta c_\psi^2) \\ d_2 &= I_2(\ddot{\theta} c_\psi^2 - 2\dot{\theta}\dot{\psi} s_\psi c_\psi + \ddot{\phi} c_\theta s_\psi c_\psi - \dot{\phi}\dot{\theta} s_\theta s_\psi c_\psi + \dot{\phi}\dot{\psi} c_\theta c_\psi^2 \\ &\quad - \dot{\phi}\dot{\psi} c_\theta s_\psi^2) - I_3(\ddot{\phi} c_\theta s_\psi c_\psi - \dot{\phi}\dot{\theta} s_\theta s_\psi c_\psi - \dot{\phi}\dot{\psi} c_\theta s_\psi^2 + \dot{\phi}\dot{\psi} c_\theta c_\psi^2 \\ &\quad - \ddot{\theta} s_\psi^2 - 2\dot{\theta}\dot{\psi} s_\psi c_\psi) \\ d_3 &= I_1(\ddot{\psi} - \ddot{\phi} s_\theta - \dot{\phi}\dot{\theta} c_\theta) \end{aligned} \tag{4.119}$$

$$\frac{\partial \Omega_a}{\partial q} = \begin{bmatrix} 0 & -\dot{\phi} c_\theta & 0 \\ 0 & -\dot{\phi} s_\theta s_\psi & -\dot{\theta} s_\psi + \dot{\phi} c_\theta c_\psi \\ 0 & -\dot{\phi} s_\theta c_\psi & -\dot{\phi} c_\theta s_\psi - \dot{\theta} c_\psi \end{bmatrix} \tag{4.120}$$

$$(\Omega_a)^T \mathcal{I}_{\mathcal{A}} \frac{\partial \Omega_a}{\partial q} = \begin{bmatrix} h_1 & h_2 & h_3 \end{bmatrix} \tag{4.121}$$

$$\begin{aligned} h_1 &= 0 \\ h_2 &= -I_1(\dot{\phi}\dot{\psi} c_\theta - \dot{\phi}^2 s_\theta c_\theta) \\ &\quad - I_2(\dot{\phi}\dot{\theta} s_\theta s_\psi c_\psi + \dot{\phi}^2 s_\theta c_\theta s_\psi^2) \\ &\quad - I_3(\dot{\phi}^2 s_\theta c_\theta c_\psi^2 - \dot{\phi}\dot{\theta} s_\theta s_\psi c_\psi) \\ h_3 &= I_2(-\dot{\theta}^2 s_\psi c_\psi - \dot{\phi}\dot{\theta} c_\theta s_\psi^2 + \dot{\phi}\dot{\theta} c_\theta c_\psi^2 + \dot{\phi}^2 c_\theta^2 s_\psi c_\psi) \\ &\quad + I_3(-\dot{\phi}^2 c_\theta^2 s_\psi c_\psi + \dot{\phi}\dot{\theta} c_\theta s_\psi^2 - \dot{\phi}\dot{\theta} c_\theta c_\psi^2 + \dot{\theta}^2 s_\psi c_\psi) \end{aligned} \tag{4.122}$$

The Euler-Lagrange equation for the torques is

$$\tau = \begin{bmatrix} \tau_1 \\ \tau_2 \\ \tau_3 \end{bmatrix} = \begin{bmatrix} d_1 - h_1 \\ d_2 - h_2 \\ d_3 - h_3 \end{bmatrix} \tag{4.123}$$

$$\begin{aligned}
\tau_1 &= -I_1(\ddot{\psi} s_\theta + \dot{\psi}\dot{\theta} c_\theta - \ddot{\phi} s_\theta^2 - 2\dot{\phi}\dot{\theta} s_\theta c_\theta) \\
&\quad + I_2(\ddot{\theta} c_\theta s_\psi c_\psi - \dot{\theta}^2 s_\theta s_\psi c_\psi - \dot{\theta}\dot{\psi} c_\theta s_\psi^2 + \dot{\theta}\dot{\psi} c_\theta c_\psi^2 \\
&\quad + \ddot{\phi} c_\theta^2 s_\psi^2 - 2\dot{\phi}\dot{\theta} s_\theta c_\theta s_\psi^2 + 2\dot{\phi}\dot{\psi} c_\theta^2 s_\psi c_\psi) \\
&\quad + I_3(\ddot{\phi} c_\theta^2 c_\psi^2 - 2\dot{\phi}\dot{\theta} s_\theta c_\theta c_\psi^2 - 2\dot{\phi}\dot{\psi} c_\theta^2 s_\psi c_\psi \\
&\quad - \ddot{\theta} c_\theta s_\psi c_\psi + \dot{\theta}^2 s_\theta s_\psi c_\psi + \dot{\theta}\dot{\psi} c_\theta s_\psi^2 - \dot{\theta}\dot{\psi} c_\theta c_\psi^2) \\
\tau_2 &= I_1(\dot{\phi}\dot{\psi} c_\theta - \dot{\phi}^2 s_\theta c_\theta) \\
&\quad + I_2(\ddot{\theta} c_\psi^2 - 2\dot{\theta}\dot{\psi} s_\psi c_\psi + \ddot{\phi} c_\theta s_\psi c_\psi + \dot{\phi}\dot{\psi} c_\theta c_\psi^2 \\
&\quad - \dot{\phi}\dot{\psi} c_\theta s_\psi^2 + \dot{\phi}^2 s_\theta c_\theta s_\psi^2) - I_3(\ddot{\phi} c_\theta s_\psi c_\psi - \dot{\phi}^2 s_\theta c_\theta c_\psi^2 \\
&\quad - \dot{\phi}\dot{\psi} c_\theta s_\psi^2 + \dot{\phi}\dot{\psi} c_\theta c_\psi^2 - \ddot{\theta} s_\psi^2 - 2\dot{\theta}\dot{\psi} s_\psi c_\psi) \\
\tau_3 &= I_1(\ddot{\psi} - \ddot{\phi} s_\theta - \dot{\phi}\dot{\theta} c_\theta) \\
&\quad - I_2(-\dot{\theta}^2 s_\psi c_\psi - \dot{\phi}\dot{\theta} c_\theta s_\psi^2 + \dot{\phi}\dot{\theta} c_\theta c_\psi^2 + \dot{\phi}^2 c_\theta^2 s_\psi c_\psi) \\
&\quad - I_3(-\dot{\phi}^2 c_\theta^2 s_\psi c_\psi + \dot{\phi}\dot{\theta} c_\theta s_\psi^2 - \dot{\phi}\dot{\theta} c_\theta c_\psi^2 + \dot{\theta}^2 s_\psi c_\psi)
\end{aligned} \tag{4.124}$$

The nonlinear model can be written as

$$M(\eta)\ddot{\eta} + C(\eta, \eta)\,\dot{\eta} = \tau \tag{4.125}$$

where

$$M(\eta) = \begin{bmatrix} I_1 s_\theta^2 + I_2 c_\theta^2 s_\psi^2 + I_3 c_\theta^2 c_\psi^2 & I_2 c_\theta s_\psi c_\psi - I_3 c_\theta s_\psi c_\psi & -I_1 s_\theta \\ I_2 c_\theta s_\psi c_\psi - I_3 c_\theta s_\psi c_\psi & I_2 c_\psi^2 + I_3 s_\psi^2 & 0 \\ -I_1 s_\theta & 0 & I_1 \end{bmatrix} \tag{4.126}$$

$$C(\eta, \dot{\eta}) = \begin{bmatrix} c_{11} & c_{12} & c_{13} \\ c_{21} & c_{22} & c_{23} \\ c_{31} & c_{32} & c_{33} \end{bmatrix} \tag{4.127}$$

$$
\begin{aligned}
c_{11} &= I_1\dot\theta s_\theta c_\theta - I_2\dot\theta s_\theta c_\theta s_\psi^2 + I_2\dot\psi c_\theta^2 s_\psi c_\psi \\
&\quad -I_3\dot\theta s_\theta c_\theta c_\psi^2 - I_3\dot\psi c_\theta^2 s_\psi c_\psi \\
c_{12} &= I_1\dot\phi s_\theta c_\theta - I_2\dot\phi s_\theta c_\theta s_\psi^2 - I_2\dot\theta s_\theta s_\psi c_\psi - I_2\dot\psi c_\theta s_\psi^2 + I_2\dot\psi c_\theta c_\psi^2 \\
&\quad -I_3\dot\phi s_\theta c_\theta c_\psi^2 + I_3\dot\theta s_\theta s_\psi c_\psi + I_3\dot\psi c_\theta s_\psi^2 - I_3\dot\psi c_\theta c_\psi^2 \\
c_{13} &= -I_1\dot\theta c_\theta + I_2\dot\phi c_\theta^2 s_\psi c_\psi - I_3\dot\phi c_\theta^2 s_\psi c_\psi \\
c_{21} &= -I_1\dot\phi s_\theta c_\theta + I_2\dot\phi s_\theta c_\theta s_\psi^2 + I_3\dot\phi s_\theta c_\theta c_\psi^2 \\
c_{22} &= -I_2\dot\psi s_\psi c_\psi + I_3\dot\psi s_\psi c_\psi \\
c_{23} &= I_1\dot\phi c_\theta - I_2\dot\phi c_\theta s_\psi^2 + I_2\dot\phi c_\theta c_\psi^2 - I_2\dot\theta s_\psi c_\psi \\
&\quad +I_3\dot\theta s_\psi c_\psi + I_3\dot\phi c_\theta s_\psi^2 - I_3\dot\phi c_\theta c_\psi^2 \\
c_{31} &= -I_2\dot\phi c_\theta^2 s_\psi c_\psi + I_3\dot\phi c_\theta^2 s_\psi c_\psi \\
c_{32} &= -I_1\dot\phi c_\theta + I_2\dot\phi c_\theta s_\psi^2 - I_2\dot\phi c_\theta c_\psi^2 + I_2\dot\theta s_\psi c_\psi \\
&\quad -I_3\dot\theta s_\psi c_\psi - I_3\dot\phi c_\theta s_\psi^2 + I_3\dot\phi c_\theta c_\psi^2 \\
c_{33} &= 0
\end{aligned}
\tag{4.128}
$$

Note that $M$ is positive definite and

$$
\begin{aligned}
\dot M_{11} &= 2I_1\dot\theta s_\theta c_\theta - 2I_2\dot\theta s_\theta c_\theta s_\psi^2 + 2I_2\dot\psi c_\theta^2 s_\psi c_\psi \\
&\quad -2I_3\dot\theta s_\theta c_\theta c_\psi^2 - 2I_3\dot\psi c_\theta^2 s_\psi c_\psi \\
\dot M_{12} &= -I_2\dot\theta s_\theta s_\psi c_\psi + I_2\dot\psi c_\theta c_\psi^2 - I_2\dot\psi c_\theta s_\psi^2 \\
&\quad +I_3\dot\theta s_\theta s_\psi c_\psi - I_3\dot\psi c_\theta c_\psi^2 + I_3\dot\psi c_\theta s_\psi^2 \\
\dot M_{13} &= -I_1\dot\theta c_\theta \\
\dot M_{21} &= -I_2\dot\theta s_\theta s_\psi c_\psi + I_2\dot\psi c_\theta c_\psi^2 - I_2\dot\psi c_\theta s_\psi^2 \\
&\quad +I_3\dot\theta s_\theta s_\psi c_\psi - I_3\dot\psi c_\theta c_\psi^2 + I_3\dot\psi c_\theta s_\psi^2 \\
\dot M_{22} &= -2I_2\dot\psi s_\psi c_\psi + 2I_3\dot\psi s_\psi c_\psi \\
\dot M_{23} &= 0 \\
\dot M_{31} &= -I_1\dot\theta c_\theta \\
\dot M_{32} &= 0 \\
\dot M_{33} &= 0
\end{aligned}
\tag{4.129}
$$

The matrix

$$P = \dot M - 2C \tag{4.130}$$

is given by

$$
\begin{array}{rcl}
P_{11} & = & 0 \\
P_{12} & = & -2I_1\dot{\phi}s_\theta c_\theta + 2I_2\dot{\phi}s_\theta c_\theta s_\psi^2 + I_2\dot{\theta}s_\theta s_\psi c_\psi + I_2\dot{\psi}c_\theta s_\psi^2 \\
 & & -I_2\dot{\psi}c_\theta c_\psi^2 + 2I_3\dot{\phi}s_\theta c_\theta c_\psi^2 - I_3\dot{\theta}s_\theta s_\psi c_\psi + I_3\dot{\psi}c_\theta c_\psi^2 - I_3\dot{\psi}c_\theta s_\psi^2 \\
P_{13} & = & I_1\dot{\theta}c_\theta - 2I_2\dot{\phi}c_\theta^2 s_\psi c_\psi + 2I_3\dot{\phi}c_\theta^2 s_\psi c_\psi \\
P_{21} & = & 2I_1\dot{\phi}s_\theta c_\theta - 2I_2\dot{\phi}s_\theta c_\theta s_\psi^2 - I_2\dot{\theta}s_\theta s_\psi c_\psi - I_2\dot{\psi}c_\theta s_\psi^2 + I_2\dot{\psi}c_\theta c_\psi^2 \\
 & & -2I_3\dot{\phi}s_\theta c_\theta c_\psi^2 + I_3\dot{\theta}s_\theta s_\psi c_\psi - I_3\dot{\psi}c_\theta c_\psi^2 + I_3\dot{\psi}c_\theta s_\psi^2 \\
P_{22} & = & 0 \\
P_{23} & = & -2I_1\dot{\phi}c_\theta + 2I_2\dot{\phi}c_\theta s_\psi^2 - 2I_2\dot{\phi}c_\theta c_\psi^2 + 2I_2\dot{\theta}s_\psi c_\psi \\
 & & -2I_3\dot{\theta}s_\psi c_\psi - 2I_3\dot{\phi}c_\theta s_\psi^2 + 2I_3\dot{\phi}c_\theta c_\psi^2 \\
P_{31} & = & -I_1\dot{\theta}c_\theta + 2I_2\dot{\phi}c_\theta^2 s_\psi c_\psi - 2I_3\dot{\phi}c_\theta^2 s_\psi c_\psi \\
P_{32} & = & 2I_1\dot{\phi}c_\theta - 2I_2\dot{\phi}c_\theta s_\psi^2 + 2I_2\dot{\phi}c_\theta c_\psi^2 - 2I_2\dot{\theta}s_\psi c_\psi \\
 & & +2I_3\dot{\theta}s_\psi c_\psi + 2I_3\dot{\phi}c_\theta s_\psi^2 - 2I_3\dot{\phi}c_\theta c_\psi^2 \\
P_{33} & = & 0
\end{array}
\tag{4.131}
$$

which is skew-symmetric, and this result will be proved to hold for general Lagrangian systems in the next chapter. By defining $V = \dot{\eta}^T M \dot{\eta}$ , it follows that $\dot{V} = \dot{\eta}^T \tau$. Finally the operator $h : \tau \rightharpoonup \dot{\eta}$ is passive. The translational part of the helicopter (4.113) can not be proved to be passive. This is due to the effect of the potential energy of gravitation that does not satisfy the requirements of lemma 5.1. As an exercise the reader may wish to try to prove (e.g. calculate the available storage and use theorem 4.4) the dissipativity of a simple mass subject to gravity: $\ddot{q} = -g$. This system is **not** passive. The addition of viscous friction does not modify the conclusion.

# Chapter 5

# Dissipative Physical Systems

In this chapter we shall present a class of dissipative systems which correspond to models of physical systems and hence embed in their structure the conservation of energy (first principle of thermodynamics) and the interaction with their environment through pairs of conjugated variables with respect to the power. Firstly, we shall recall three different definitions of systems obtained by an energy based modeling: controlled Lagrangian, input-output Hamiltonian systems and port controlled Hamiltonian systems. We shall illustrate and compare these definitions on some very simple examples. Secondly we shall treat a class of systems which gave rise to numerous stabilizing control using passivity theory and corresponds to models of robotic manipulators. In each worked case we show how the main functions associated to a dissipative system (the available storage, the required supply, storage functions) can be computed analytically and related to the energy of the physical system.

## 5.1 Lagrangian control systems

Lagrangian systems arose from variational calculus and gave a first general analytical definition of physical dynamical systems in Analytical Mechanics [2] [93] [101]. However they also make it possible to describe the dynamics of various engineering systems as electromechanical systems or electrical circuits. They also gave rise to intensive work in control in order to derive different control laws by taking into account the structure of the system's dynamics derived from energy based modeling [173] [176]. We have already studied the passivity properties of a Lagrangian system (a simplified helicopter model) in chapter 4. In this section we shall present several definitions of Lagrangian systems with inputs and outputs and in which way they represent the interaction of physical systems with their environment.

### 5.1.1 General Lagrangian control systems

In this section we shall briefly recall the definition of Lagrangian systems with external forces on $\mathbb{R}^n$ and the definition of Lagrangian control systems derived from it.

**Definition 5.1 (Lagrangian systems with external forces)** Consider a *configuration manifold* $Q = \mathbb{R}^n$ whose points are denoted by $q \in \mathbb{R}^n$ and are called *generalized coordinates*. Denote by $TQ = \mathbb{R}^{2n}$ its tangent bundle and its elements by $(q, \dot{q}) \in \mathbb{R}^{2n}$ where $\dot{q}$ is called *generalized velocity*. A *Lagrangian system with external forces* on the configuration manifold $Q = \mathbb{R}^n$ is defined by a real function $L(q, \dot{q})$, from the tangent bundle $TQ$ to $\mathbb{R}$ called *Lagrangian function* and the *Lagrangian equations*:

$$\frac{d}{dt}\left(\frac{\partial L}{\partial \dot{q}}(q, \dot{q})\right) - \frac{\partial L}{\partial q}(q, \dot{q}) = F \tag{5.1}$$

where $F \in \mathbb{R}^n$ is the vector of *generalized forces* acting on the system. ♠♠

**Remark 5.1** In this definition the configuration space is the real vector space $\mathbb{R}^n$ to which we shall restrict ourselves hereafter, but in general one may consider a differentiable manifold as configuration space [101]. Considering real vector spaces as configuration manifolds corresponds actually to consider a *local* definition of a Lagrangian system.

If the vector of external forces $F$ is the vector of control inputs, then the Lagrangian control system is fully actuated. Such models arise for instance for fully actuated kinematic chains [142].

**Example 5.1 (Harmonic oscillator with external force)** Let us consider the very simple example of the linear mass-spring system consisting in a mass attached to a fixed frame through a spring and suject to a force $F$. The coordinate $q$ of the system is the position of the mass with respect to the fixed frame and the Lagrangian function is given by $L(q, \dot{q}) = K(\dot{q}) - U(q)$ where $K(\dot{q}) = \frac{1}{2}m\dot{q}^2$ is the kinetic co-energy of the mass and $U(q) = \frac{1}{2}kq^2$ is the potential energy of the spring. Then the Lagrangian system with external force is:

$$m\ddot{q} + kq = F \tag{5.2}$$

♠♠

Lagrangian systems with external forces satisfy, by construction, a power balance equation that leads to some passivity property.

**Lemma 5.1 (Losslessness of Lagrangian systems)** A Lagrangian system with external forces (5.1) satisfies the following *power balance equation*

$$F^T \dot{q} = \frac{dH}{dt} \tag{5.3}$$

where the real function $H$ is obtained by the *Legendre transformation* of the Lagrangian function $L(q, \dot{q})$ with respect to the generalized velocity $\dot{q}$ and is defined by:

$$H(q,p) = \dot{q}^T p - L(q, \dot{q}) \tag{5.4}$$

where $p$ is the vector of *generalized momenta*:

$$p(q, \dot{q}) = \left( \frac{\partial L}{\partial \dot{q}}(q, \dot{q}) \right) \tag{5.5}$$

and the Lagrangian function is assumed to be *hyperregular* [101] in such a way that the map from the generalized velocities $\dot{q}$ to the generalized momenta $p$ is bijective.
If moreover the function $H$ is bounded from below, then the Lagrangian system with external forces is lossless with respect to the supply rate: $F^T \dot{q}$ with storage function $H$. ♠♠

**Proof** Let us first compute the power balance equation by computing $F^T \dot{q}$ using the Lagrangian equation (5.1) and the definition of the generalized momentum (5.5).

$$\begin{aligned} \dot{q}^T F &= \dot{q}^T \left[ \tfrac{d}{dt} \left( \tfrac{\partial L}{\partial \dot{q}}(q, \dot{q}) \right) - \tfrac{\partial L}{\partial q}(q, \dot{q}) \right] \\ &= \dot{q}^T \tfrac{d}{dt} p - \dot{q}^T \tfrac{\partial L}{\partial q} \\ &= \tfrac{d}{dt} \left( \dot{q}^T p \right) - \ddot{q}^T p + \ddot{q}^T \tfrac{\partial L}{\partial \dot{q}} - \tfrac{d}{dt} L(q, \dot{q}) \\ &= \tfrac{dH}{dt} \end{aligned} \tag{5.6}$$

Then, using as outputs the generalized velocities and assuming that the function $H$ is bounded from below, according to the definition 4.10, the Lagrangian system with external forces is passive and lossless with storage function $H$. ♠♠

**Remark 5.2** The name *power balance equation* for (5.3) comes from the fact that for physical systems, the supply rate is the power going into the system due to the external force $F$ and that the function $H$ is equal to the total energy of the system.

♠♠

**Example 5.2** Consider again the example 5.1 of the harmonic oscillator. In this case the supply rate is the mechanical power ingoing the system and the storage function is $H = K(p) + U(q)$ and is the total energy of the system, i.e. the sum of the elastic potential and kinetic energy.

Actually the definition of Lagrangian systems with external forces may be too restrictive, as, for instance, the external forces $F$ may not correspond to actual inputs. For example they may be linear functions of the inputs $u$:

$$F = J^T(q)u \tag{5.7}$$

where $J(q)$ is a $p \times n$ matrix depending on the generalized coordinates $q$. This is the case when for instance the dynamics of a robot is described in generalized coordinates for which the generalized velocities are not collocated to the actuators' forces and torques. Then the matrix $J(q)$ is the Jacobian of the geometric relations between the actuators' displacement and the generalized coordinates [142]. This system remains lossless with storage function $H(q,p)$ defined in (5.4) by choosing the outputs: $y = J(q)\dot{q}$.
In order to cope with such situations, a more general definition of Lagrangian systems with external controls is given and consists in considering that the input is directly modifying the Lagrangian function [173] [176].

**Definition 5.2 (Lagrangian control system)** Consider a *configuration manifold* $Q = I\!R^n$ and its tangent bundle $TQ = I\!R^{2n}$, an input vector space $U = I\!R^p$. A *Lagrangian control systems* is defined by a real function $L(q, \dot{q}, u)$ from $TQ \times U$ to $I\!R$, and the equations:

$$\frac{d}{dt}\left(\frac{\partial L}{\partial \dot{q}}(q, \dot{q}, u)\right) - \frac{\partial L}{\partial q}(q, \dot{q}, u) = 0 \tag{5.8}$$

♠♠

This definition includes the Lagrangian systems with external forces (5.1) by choosing the Lagrangian function to be:

$$L_1(q, \dot{q}, F) = L(q, \dot{q}) + q^T F \tag{5.9}$$

It includes as well the case when the the external forces are given by (5.7) as a linear function of the inputs where the matrix $J(q)$ is the Jacobian of some geometric function $C(q)$ from $I\!R^n$ to $I\!R^p$:

$$J(q) = \frac{\partial C}{\partial q}(q) \tag{5.10}$$

Then the Lagrangian function is given by:

$$L_1(q, \dot{q}, F) = L(q, \dot{q}) + C(q)^T u \tag{5.11}$$

But it also encompasses Lagrangian systems where the inputs do not appear as forces as may be seen on the following example.

**Example 5.3** Consider the harmonic oscillator, but assume now that the spring is no more attached to a fixed basis but to a moving basis with its position $u$ considered as an input. Let us choose as coordinate $q$, the position of the mass with respect to the fixed frame. The displacement of the spring becomes then $q - u$ and the potential energy becomes: $U(q, u) = \frac{1}{2}k(q-u)^2$ and the Lagrangian becomes:

$$L(q, \dot{q}, u) = \frac{1}{2}m\dot{q}^2 - \frac{1}{2}k(q-u)^2 \tag{5.12}$$

The Lagrangian control systems becomes then:

$$m\ddot{q} + kq = ku \tag{5.13}$$

It is interesting to see that the Lagrangian equations of the preceding example may also be interpreted as a Lagrangian system with external forces being linear function of the input, as in (5.7), where however the linear map does not correspond to the Jacobian of a kinematic relation. However, modifying slightly the definition of the system, one obtains a different type of system where the external force is a (linear) function of the derivative of the input as may be seen on the following example (taken from [176]).

**Example 5.4** Consider again the harmonic oscillator with the spring attached to a moving basis, but take now as input the *velocity* $u$ of the basis. Let us choose as coordinate $q$, *the displacement of the spring.* The velocity of the mass with respect to the fixed frame becomes $\dot{q} + u$ and the kinetic energy becomes: $K(\dot{q}, u) = \frac{1}{2}m(\dot{q}+u)^2$. Hence the Lagrangian becomes:

$$L(q, \dot{q}, u) = K(\dot{q}, u) - U(q) \tag{5.14}$$

The Lagrangian control system becomes then:

$$m\ddot{q} + kq = m\dot{u} \tag{5.15}$$

Lagrangian control systems also allow us to consider more inputs that the number of generalized velocities as may be seen on the next example.

**Example 5.5** Consider again the harmonic oscillator, assuming that the basis of the spring is moving with velocity $u_1$ and that there is a force $u_2$ exerted on the mass. Then considering the Lagrangian function:

$$L(q, \dot{q}, u) = \frac{1}{2}m(\dot{q}+u_1)^2 - \frac{1}{2}kq + qu_2 \tag{5.16}$$

one obtains the Lagrangian control system:

$$m\ddot{q} + kq = m\dot{u}_1 + u_2 \tag{5.17}$$

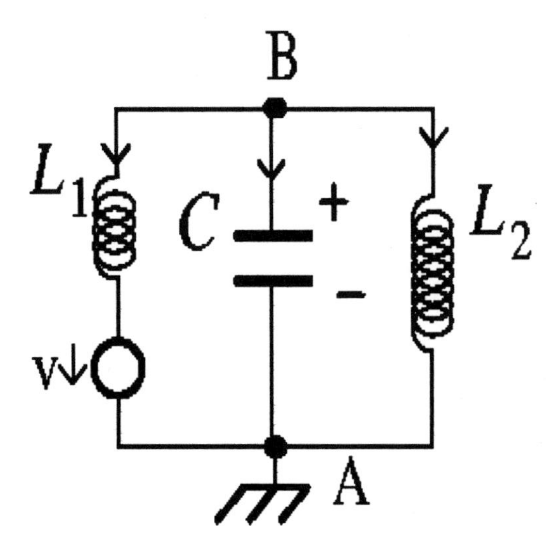


Figure 5.1: LC circuit.

Lagrangian control systems were derived firstly to treat mechanical control systems, as robots for example, but they may also be derived for other types of systems like electrical circuits or electromechanical systems [78]. In this case the Lagrangian system is derived from physical models and, in general, the generalized coordinates are related to some set of variables defining the energy or co-energy of the system or part of it and the Lagrangian function is derived from these energy functions. We shall illustrate such a construction with the following example.

**Example 5.6 (An LC circuit of order 3)** Consider the LC circuit depicted in figure 5.1. Using the procedure proposed by Chua and McPherson [40], the following Lagrangian control system describes its dynamics. Denoting the edges by the element which they connect, the circuit may be partitioned into the spanning tree: $\Gamma = \Gamma_1 \cup \Gamma_2 = \{C\} \cup \{S_u\}$ and its cotree: $\Lambda = \Lambda_1 \cup \Lambda_2 = \{L_1\} \cup \{L_2\}$. Hence one may choose as vector of generalized velocities the voltages of the capacitors in the tree $\Gamma_1$ and the currents of the inductors in the cotree $\Lambda_2$:

$$\dot{q} = \begin{pmatrix} v_C \\ i_{L_2} \end{pmatrix} \tag{5.18}$$

where $v_C$ denotes the voltage at the port of the capacitor and $i_{L_2}$ denotes the current in the inductor labeled $L_2$. And the vector of generalized coordinates is hence obtained by integration of the vector of generalized velocities:

$$q = \begin{pmatrix} \phi_C \\ Q_{L_2} \end{pmatrix} \tag{5.19}$$

Note that this definition of the variables is somewhat unnatural as it amounts to associate flux type variables with capacitors and charge-like variables with inductors (see the discussions in [204] [128]).
The Lagrangian function is constructed as the sum of four terms:

$$L(q, \dot{q}, u) = \hat{\mathcal{E}}(\dot{q}) - \mathcal{E}(q) + \mathcal{C}(q, \dot{q}) + \mathcal{I}(q, u) \tag{5.20}$$

The function $\hat{\mathcal{E}}(\dot{q})$ is the sum of the electric coenergy of the capacitors in the tree $\Gamma_1$ and the magnetic coenergy of the inductors in the cotree $\Lambda_2$ which is, in this example, in the case of linear elements:

$$\hat{\mathcal{E}}(\dot{q}) = \frac{1}{2} C {v_C}^2 + \frac{1}{2} L_2 {i_{L_2}}^2 = \frac{1}{2} C \dot{q}_1^2 + \frac{1}{2} L_2 \dot{q}_2^2 \tag{5.21}$$

The function $\mathcal{E}(q)$ is the sum of the magnetic energy of the inductors in the cotree $\Lambda_1$ and the electric energy of the capacitors in the tree $\Gamma_2$ which is:

$$\mathcal{E}(q) = \frac{1}{2L_1} {\phi_{L_1}}^2 = \frac{1}{2L_1} (q_1 + q_{10})^2 \tag{5.22}$$

where the relation between the flux $\phi_{L_1}$ of the inductor $L_1$ was obtained by integrating the Kirchhoff's mesh law on the mesh consisting of the capacitor $C$ and the inductor $L_1$ yielding $\phi_{L_1} = (q_1 + q_{10})$ and $q_{10}$ denotes some real constant which may be chosen to be null.
The function $\mathcal{C}(q, \dot{q})$ accounts for the coupling between the capacitors in the tree $\Gamma_1$ and inductors in the cotree $\Lambda_2$ depending on the topological interconnection between them and is:

$$\mathcal{C}(q, \dot{q}) = i_{L_2} \phi_C = \dot{q}_2 q_1 \tag{5.23}$$

And the function $\mathcal{I}(q, u)$ is an interaction potential function describing the action of the source element and is:

$$\mathcal{I}(q, u) = q_{L_2} u = q_2 u \tag{5.24}$$

The Lagrangian control system is then :

$$C\ddot{q}_1 - \dot{q}_2 + \frac{1}{L_1}(q_1 + q_{10}) = 0 \tag{5.25}$$

$$L_2 \ddot{q}_2 + \dot{q}_1 - u = 0 \tag{5.26}$$

Note that this system is of order 4 (it has 2 generalized coordinates) which does not correspond to the order of the electrical circuit which, by topological inspection, would be 3; indeed one may choose a maximal tree containing the capacitor and having a cotree containing the 2 inductors. We shall come back to this remark and expand it in the sequel when we shall treat the same example as a port controlled Hamiltonian system. ♠♠

This example illustrates that, although the derivation of Lagrangian system is based on the determination of some energy functions and other physical properties of the system, its structure may not agree with the physical insight. Indeed the Lagrangian control systems are defined on the state space $TQ$, the tangent bundle to the configuration manifold. This state space has a very special structure; it is endowed with a symplectic form which is used to give an intrinsic definition of Lagrangian systems [101]. A very simple property of this state space is that its dimension is even (there are as many generalized coordinates as generalized velocities). Already this property may be in contradiction with the physical structure of the system.
Lagrangian control systems as the Lagrangian systems with external forces, satisfy, by construction, a power balance equation and losslessness passivity property [26].

**Lemma 5.2 (Losslessness of Lagrangian control systems)** A Lagrangian control system, ( definition 5.2), satisfies the following *power balance equation*

$$u^T z = \frac{dE}{dt} \tag{5.27}$$

where:

$$z_i = -\sum_{i=1}^{n} \frac{\partial^2 H}{\partial q_j \partial u_i} \frac{\partial H}{\partial p_j} + \sum_{i=1}^{n} \frac{\partial^2 H}{\partial p_j \partial u_i} \frac{\partial H}{\partial q_j} \tag{5.28}$$

and the real function $E$ is obtained by the *Legendre transformation* of the Lagrangian function $L(q, \dot{q})$ with respect to the generalized velocity $\dot{q}$ and the inputs and is defined by:

$$E(q, p, u) = H(q, p, u) - u^T \frac{\partial H}{\partial u} \tag{5.29}$$

with

$$H(q, p, u) = \dot{q}^T p - L(q, \dot{q}, u) \tag{5.30}$$

where $p$ is the vector of *generalized momenta*:

$$p(q, \dot{q}, u) = \left( \frac{\partial L}{\partial \dot{q}}(q, \dot{q}) \right) \tag{5.31}$$

and the Lagrangian function is assumed to be *hyperregular* [101] in such a way that the map from the generalized velocities $\dot{q}$ to the generalized momenta $p$ is bijective for any $u$.

If moreover the Hamiltonian (5.30) is affine in the inputs (hence the function $E$ is independent of the inputs), the controlled Lagrangian system will be called *affine Lagrangian control system.* Assuming that $E(q,p)$ is bounded from below, then the Lagrangian system with external forces is lossless with respect to the supply rate $u^T z$ with storage function $E(q,p)$. ♠♠

As we have seen above, the affine Lagrangian control systems are lossless with respect to the storage function $E(q,p)$ which in physical systems may be chosen to be equal to the internal energy of the system. However in numerous systems, dissipation has to be included. For instance for robotic manipulator, the dissipation will be due to the friction at the joints and in the actuators. This may be done by modifying the definition of Lagrangian control systems and including dissipating forces as follows.

**Definition 5.3 (Lagrangian control system with dissipation)**
Consider a *configuration manifold* $Q = \mathbb{R}^n$ and its tangent bundle $TQ = \mathbb{R}^{2n}$, an input vector space $U = \mathbb{R}^p$. A *Lagrangian control systems with dissipation* is defined by a Lagrangian function $L(q,\dot{q},u)$ from $TQ \times U$ to $\mathbb{R}$, a function $R(\dot{q})$ from $TQ$ to $\mathbb{R}$, called *Rayleigh dissipation function* and which satisfies:

$$\dot{q}^T \frac{\partial R}{\partial \dot{q}}(\dot{q}) \geq 0 \tag{5.32}$$

and the equations:

$$\frac{d}{dt}\left(\frac{\partial L}{\partial \dot{q}}(q,\dot{q},u)\right) - \frac{\partial L}{\partial q}(q,\dot{q},u) + \frac{\partial R}{\partial \dot{q}} = u \tag{5.33}$$

♠♠

Consider a second example of a magnetically levitated iron ball.

**Example 5.7** Consider the example of the vertical motion of a magnetically levitated ball as depicted in figure 5.2. There are three types of energy involved: the magnetic energy, the kinetic energy of the ball and its potential energy. The vector of generalized coordinates may be chosen as a vector in $\mathbb{R}^2$ where $q_1$ denotes a primitive of the current in the inductor (according to the procedure described in the example 5.6), $q_2 = z$ is the altitude of the sphere.
The Lagrangian function may then be chosen as the sum of three terms:

$$L(q,\dot{q},u) = \hat{\mathcal{E}}_m(q,\dot{q}) + \hat{\mathcal{E}}_k(\dot{q}) - \mathcal{U}(q) + \mathcal{I}(q,u) \tag{5.34}$$

The function $\hat{\mathcal{E}}_m(q,\dot{q})$ is the magnetic coenergy of the inductor and depends on the currents in the coil as well on the altitude of the sphere:

$$\hat{\mathcal{E}}_m(q,\dot{q}) = \frac{1}{2}L(q_2)\,\dot{q}_1^2 \tag{5.35}$$

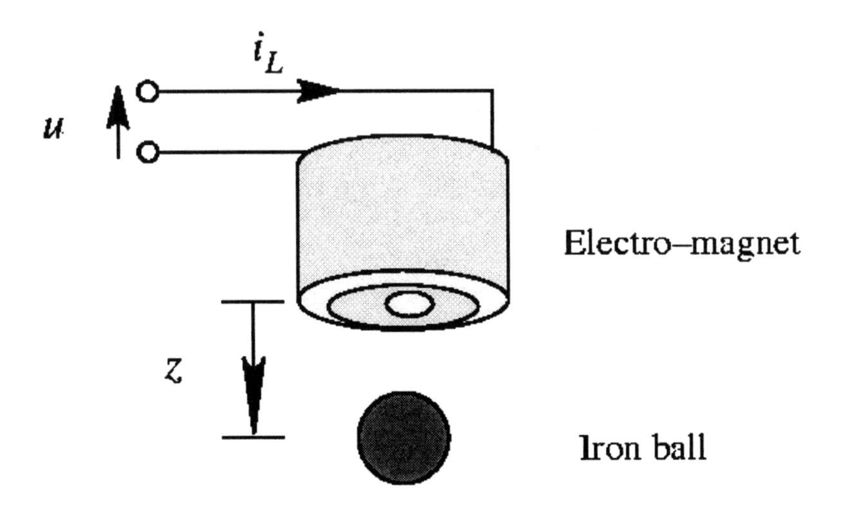


Figure 5.2: Magnetic sphere.

where:

$$L(q_2) = L_0 + \frac{k}{q_2 - z_0} \tag{5.36}$$

The function $\hat{\mathcal{E}}_k(\dot{q})$ is the kinetic coenergy of the ball:

$$\hat{\mathcal{E}}_k(\dot{q}) = \frac{1}{2} m \dot{q}_2^2 \tag{5.37}$$

The function $\mathcal{U}(q)$ denotes the potential energy due to the gravity:

$$\mathcal{U}(q) = g q_2 \tag{5.38}$$

The interaction potential is:

$$\mathcal{I}(q, u) = q_1 u \tag{5.39}$$

In order to take into account the dissipation represented by the resistor $R$, one also define the following Rayleigh potential function:

$$\mathcal{R}(\dot{q}) = \frac{1}{2} R \dot{q}_1^2 \tag{5.40}$$

This leads to the following Lagrangian control system with dissipation:

$$L(q_2)\ddot{q}_1 + \frac{\partial L}{\partial q_2}(q_2)\dot{q}_2\dot{q}_1 + R\dot{q}_1 - u = 0 \tag{5.41}$$

$$m\ddot{q}_2 - \frac{1}{2}\frac{\partial L}{\partial q_2}(q_2)\dot{q}_1^2 + g = 0 \tag{5.42}$$

### 5.1.2 Simple mechanical systems

An important subclass of Lagrangian contol systems is given by the so-called simple mechanical systems where the Lagrangian function takes a particular form.

**Definition 5.4 (Mechanical systems with external forces)** The Lagrangian system for a simple mechanical system is a Lagrangian system with external forces according to definition 5.1 with Lagrangian function:

$$L(q, \dot{q}) = T(q, \dot{q}) - U(q) \tag{5.43}$$

where $U(q)$ is a real function from the configuration space $Q$ on $\mathbb{R}$ and is called *potential energy* and $T(q, \dot{q})$ is a real function from $TQ$ on $\mathbb{R}$, called *kinetic energy* and is defined by:

$$T(q, \dot{q}) = \frac{1}{2}\dot{q}^T M(q)\dot{q} \tag{5.44}$$

where the matrix $M(q) \in \mathbb{R}^{n \times n}$ is positive definite and is called *the inertia matrix.*

Considering the special form of the Lagrangian function, the Lagrangian equations (5.1) may be written in some special form which is particularly useful for deriving stabilizing controllers as will be presented in the subsequent chapters.

**Lemma 5.3 (Simple mechanical systems)** The Lagrangian equations (5.1) for a simple mechanical system may be written:

$$M(q)\ddot{q} + C(q, \dot{q})\dot{q} + g(q) = F \tag{5.45}$$

where $g(q) = \frac{dU}{dq}(q) \in \mathbb{R}^n$,

$$C(q, \dot{q}) = \sum_{k=1}^{n} \Gamma_{ijk}\dot{q}_k \tag{5.46}$$

and $\Gamma_{ijk}$ are called the *Christoffel's symbols* associated with the inertia matrix $M(q)$ and are defined by:

$$\Gamma_{ijk} = \frac{1}{2}\left(\frac{\partial M_{ij}}{\partial q_k} + \frac{\partial M_{ik}}{\partial q_j} - \frac{\partial M_{kj}}{\partial q_i}\right) \tag{5.47}$$

♠♠

A property of Christoffel's symbols which is easily derived but is of great importance for the derivation of stabilizing control laws, is given below [190].

**Lemma 5.4** The Christoffel's symbols (5.47) satisfy the following property: the matrix $\dot{M}(q) - 2C(q,\dot{q})$ is skew-symmetric. ♠♠

**Remark 5.3** A consequence of the lemma is that:

$$\dot{q}^T\left(\dot{M}(q) - 2C(q,\dot{q})\right)\dot{q} = 0 \tag{5.48}$$

hence the reflects that the generalized inertial forces $\left[\dot{M}(q) - 2C(q,\dot{q})\right]\dot{q}$ do not work. This may be seen as follows:

$$\begin{aligned} \tau^T\dot{q} = \tfrac{dH}{dt}(q,p) &= \dot{q}^T M(q)\ddot{q} + \tfrac{1}{2}\dot{q}^T\dot{M}(q)\dot{q} + g(q) \\ &= \dot{q}^T\left[-C(q,\dot{q})\dot{q} - g(q) + \tau\right] + \tfrac{1}{2}\dot{q}^T\dot{M}(q)\dot{q} + g(q) \\ &= \dot{q}^T\tau + \tfrac{1}{2}\dot{q}^T\left[\dot{M}(q) - 2C(q,\dot{q})\right]\dot{q} \end{aligned} \tag{5.49}$$

from which (5.48) follows. Such forces are sometimes called *gyroscopic* [135]. It is noteworthy that (5.48) does not mean that the matrix $\dot{M}(q) - 2C(q,\dot{q})$ is skew-symmetric. Skew-symmetry is true only for the particular definition of the matrix $C(q,\dot{q})$ using Christoffel's symbols.

**Remark 5.4** The definition of a positive definite symmetric inertia matrix for simple mechanical systems, may be expressed in some coordinate independent way by using so-called *Riemannian manifolds* [2]. In ([180], chap. 4) the properties of the Christoffell's symbols, that shall be used in the sequel for the synthesis of stabilizing controllers, may also be related to properties of Riemannian manifolds.

A class of systems which typically may be represented in this formulation is the dynamics of multibody systems, for which systematic derivation procedures were obtained (see [142] and the references herein).

## 5.2 Hamiltonian control systems

In this section we shall treat an alternative formulation of the dynamics of physical controlled systems using the Hamiltonian formalism. This formalism has arised from the Lagrangian one in the end of the nineteenth century and has now become the fundamental structure of the mathematical description of physical systems [2] [101]. In particular it allowed one to deal with symmetry and reduction and also to describe the extension of classical mechanics to quantum mechanics.

### 5.2.1 Input-output Hamiltonian systems

Lagrangian systems may be transformed to standard Hamiltonian systems by using the Legendre transformation [2] [101].

**Lemma 5.5 (Legendre transformation)** Consider a Lagrangian system with external forces and define the vector of *generalized momenta*:

$$p(q,\dot{q}) = \left(\frac{\partial L}{\partial \dot{q}}(q,\dot{q})\right) \in I\!R^n \tag{5.50}$$

assume that the map from generalized velocities to generalized momenta is invertible, and consider the *Legendre transformation with respect to* $\dot{q}$ of the Lagrangian function, called *Hamiltonian function*:

$$H_0(q,p) = \dot{q}^T p - L(q,\dot{q}) \tag{5.51}$$

then the Lagrangian system with external forces is equivalent to the following *standard Hamiltonian system*:

$$\begin{aligned} \dot{q} &= \tfrac{\partial H_0}{\partial p} \\ \dot{p} &= -\tfrac{\partial H_0}{\partial q} + F \end{aligned} \tag{5.52}$$

♠♠

There is an alternative way of writing these equations as follows:

$$\begin{pmatrix} \dot{q} \\ \\ \dot{p} \end{pmatrix} = J_s \begin{pmatrix} \frac{\partial H_0}{\partial q} \\ \\ \frac{\partial H_0}{\partial p} \end{pmatrix} + \begin{pmatrix} 0_n \\ \\ I_n \end{pmatrix} F \tag{5.53}$$

where $J_s$ is the following matrix, called *symplectic matrix*:

$$J_s = \begin{pmatrix} 0_n & I_n \\ -I_n & 0_n \end{pmatrix} \tag{5.54}$$

where $I_n$ denotes the identity matrix of order $n$ and $0_n$, the zero matrix of order $n$. This symplectic matrix is the local representation, in canonical coordinates, of the symplectic Poisson tensor field which defines the geometric structure of the state space of standard Hamiltonian systems. (The interested reader may find an precise exposition to symplectic geometry in [101].)
In the same way as an Lagrangian system with external forces may be expressed as a control Lagrangian system (for which the inputs are an argument of the Lagrangian function), the standard Hamiltonian system with external forces (5.53) may be expressed as Hamiltonian system where the Hamiltonian function depends on the inputs: $H(q,p,u) = H_0(q,p) - q^T F$ which yields:

$$\begin{pmatrix} \dot{q} \\ \dot{p} \end{pmatrix} = J_s \left( \begin{pmatrix} \frac{\partial H_0}{\partial q} \\ \frac{\partial H_0}{\partial p} \end{pmatrix} + \begin{pmatrix} -I_n \\ 0_n \end{pmatrix} F \right) = J_s \begin{pmatrix} \frac{\partial H_0}{\partial q} \\ \frac{\partial H_0}{\partial p} \end{pmatrix} + \begin{pmatrix} 0_n \\ I_n \end{pmatrix} F \tag{5.55}$$

As the simplest example let us consider the harmonic oscillator with an external force.

**Example 5.8 (Harmonic oscillator with external force)** Firstly let us recall that in its Lagrangian representation (see example 5.1), the state space is given by the position of the mass (with respect to the fixed frame) and its velocity. Its Lagrangian is: $L(q,\dot{q},F) = \frac{1}{2}m\dot{q}^2 - \frac{1}{2}kq^2 + q^T F$. Hence the (generalized) momentum is: $p = \frac{\partial L}{\partial \dot{q}} = m\dot{q}$. The Hamiltonian function, obtained through the Legendre transformation is $H(q,p,F) = H_0(q,p) - q^T F$ where the Hamiltonian function $H_0$ represents the total internal energy $H_0(q,p) = K(p) + U(q)$, the sum of the kinetic energy $K(p) = \frac{1}{2}\frac{p^2}{m}$ and the potential energy $U(q)$. The Hamiltonian system becomes:

$$\begin{pmatrix} \dot{q} \\ \dot{p} \end{pmatrix} = \begin{pmatrix} 0 & 1 \\ -1 & 0 \end{pmatrix} \begin{pmatrix} kq \\ \frac{p}{m} \end{pmatrix} + \begin{pmatrix} 0 \\ 1 \end{pmatrix} F \tag{5.56}$$

♠♠

Hamiltonian systems with external forces may be generalized to so-called *input-output Hamiltonian systems* [26] for which the Hamiltonian function depends on the inputs. In the sequel we shall restrict ourselves to systems for which the Hamiltonian function depends linearly on the inputs, which actually constitute the basis of the major part of the work dedicated to the system theoretic analysis and the control of Hamiltonian systems [26] [176] [148].

**Definition 5.5 (Input-output Hamiltonian systems)** An *input-output Hamiltonian system* on $\mathbb{R}^{2n}$ is defined by a Hamiltonian function

$$H(x) = H_0(x) - \sum_{i=1}^{m} H_i(x)\, u_i \tag{5.57}$$

composed of the sum of the internal Hamiltonian $H_0(x)$ and a linear combination of $m$ *interaction Hamiltonian functions* $H_i(x)$ and the dynamic equations:

$$\dot{x} = J_s dH_0(x) + \sum_{i=1}^{m} J_s dH_i(x) u_i \tag{5.58}$$

$$\tilde{y}_i = H_i(x), \; i = 1, .., m \tag{5.59}$$

denoting the state by $x^T = (q^T, p^T) \in \mathbb{R}^{2n}$ and the gradient of a function $H$ by $dH = \frac{dH}{dx} \in \mathbb{R}^{2n}$. ♠♠

One may note that an input-output Hamiltonian system (5.59) is a nonlinear system affine in the inputs in the sense of [72] [148]. It is composed of a Hamiltonian drift vector field $J_s dH_0(q,p)$ and the input vector fields $J_s dH_i(q,p)$ are also Hamiltonian and generated by the interaction Hamiltonian functions. The outputs are the Hamiltonian interaction functions and are called *natural outputs* [26]. We may note already here that these outputs although called "natural" are not the outputs conjuguated to the inputs for which the system is passive (see lemma 4.4 and (4.36)) as will be shown in the sequel.

**Example 5.9** Consider again the example 5.5. The state space is given by the displacement of the spring and its velocity. Its Lagrangian is:

$$L(q, \dot{q}, F) = \frac{1}{2} m (\dot{q} + u_1)^2 - \frac{1}{2} k q^2 + q u_2 \tag{5.60}$$

Hence the generalized momentum is: $p = \frac{\partial L}{\partial \dot{q}} = m(\dot{q} + u_1)$. The Hamiltonian function, obtained through the Legendre transformation with respect to $\dot{q}$ is

$$H(q, p, u_1, u_2) = \dot{q}^T p - L(q, \dot{q}, u_1, u_2) = H_0(q,p) - p u_1 - q u_2 \tag{5.61}$$

where the Hamiltonian function $H_0(q,p) = \frac{1}{2}\frac{p^2}{m} + \frac{1}{2} k q^2$ represents, as in the preceding example, the sum of the kinetic and the elastic potential energy. The interaction potentials are the momentum of the mass $H_1(q,p) = p$, for the input $u_1$ which represents the controlled velocity of the basis and the the displacement of the spring $H_2(q,p) = q$ for the input $u_2$ which is the external force exerted on the mass. The dynamic is now described by the following input-output Hamiltonian system:

$$\begin{pmatrix} \dot{q} \\ \dot{p} \end{pmatrix} = \begin{pmatrix} 0 & 1 \\ -1 & 0 \end{pmatrix} \begin{pmatrix} kq \\ \frac{p}{m} \end{pmatrix} + \begin{pmatrix} -1 \\ 0 \end{pmatrix} u_1 + \begin{pmatrix} 0 \\ 1 \end{pmatrix} u_2 \tag{5.62}$$

♠♠

Note that the definition of the generalized momentum $p$ corresponds to a generalized state space transformation involving the input $u_1$. Consequently in the Hamiltonian formulation (5.62) the derivative of the input no longer appears, contrary to the Lagrangian dynamics in (5.17). Moreover, like affine Lagrangian control systems, input-output Hamiltonian systems satisfy a power balance equation, however considering, instead of the natural outputs $\tilde{y}_i$ (5.59), their derivatives.

**Lemma 5.6 (Losslessness of Hamiltonian systems)** An input-output Hamiltonian system (according to definition 5.5), satisfies the following *power balance equation*

$$u^T \dot{\tilde{y}} = \frac{dH_0}{dt} \tag{5.63}$$

If moreover the Hamiltonian function $H_0(x)$ is bounded from below, then the input-output Hamiltonian system is lossless with respect to the supply rate: $u^T \dot{\tilde{y}}$ with storage function $H_0(q,p)$. ♠♠

Let us comment this power balance equation on the example of the harmonic oscillator with moving frame and continue example 5.9.

**Example 5.10** The natural outputs are then the momentum of the system: $\tilde{y}_1 = H_1(q,p) = p$ which is conjuguated to the input $u_1$ (the velocity of the basis of the spring) and the displacement of the spring $\tilde{y}_2 = H_2(q,p) = q$ which is conjuguated to the input $u_2$ (the external force exerted on the mass). The passive outputs defining the supply rate are then:

$$\dot{\tilde{y}}_1 = \dot{p} = -kq + u_2 \tag{5.64}$$

and:

$$\dot{\tilde{y}}_2 = \dot{q} = \frac{p}{m} - u_1 \tag{5.65}$$

Computing the supply rate, the terms in the inputs cancel each other and one obtains:

$$\dot{\tilde{y}}_1 u_1 + \dot{\tilde{y}}_2 u_2 = kqu_1 + u_2 \frac{p}{m} \tag{5.66}$$

This is precisely the sum of the mechanical power supplied to the mechanical system by the source of displacement at the basis of the spring and the source of force at the mass. This indeed is equal to the variation of the total energy of the mechanical system. However it may be noticed that the natural outputs as well as their derivatives are *not* the variables which one uses in order to define the interconnection of this system with some other mechanical system: the force at the basis of the spring which should be used to write a force balance equation at that point and the velocity of the mass $m$ which should be used in order to write the kinematic interconnection of the mass (their dual variables are the input variables). In general input-output Hamiltonian systems (or their Lagrangian counterpart) are not well suited for expressing their interconnection.

**Example 5.11** Consider the LC circuit of order 3 considered in the example 5.6. In the Lagrangian formulation, the generalized velocities were $\dot{q}_1 = V_C$ the voltage of the capacitor, $\dot{q}_2 = i_{L_2}$ the current of the inductor $L_2$ and the generalized coordinates were some primitives denoted by: $q_1 = \phi_C$ and $q_2 = Q_{L_2}$. The Lagrangian function was given by: $L(q, \dot{q}, u) = \hat{\mathcal{E}}(\dot{q}) - \mathcal{E}(q) + \mathcal{C}(q, \dot{q}) + \mathcal{I}(q, u)$ where $\hat{\mathcal{E}}(\dot{q})$ is the sum of the electric coenergy of the capacitor and of the inductor $L_2$, $\mathcal{E}(q)$ is the magnetic energy of the inductor $L_1$, $\mathcal{C}(q, \dot{q})$ is a coupling function between the capacitor and the inductor $L_2$ and $\mathcal{I}(q, u)$ is the interaction potential function.
Let us now define the generalized momenta. The first momentum variable is :

$$p_1 = \frac{\partial L}{\partial \dot{q}_1} = \frac{\partial \hat{\mathcal{E}}}{\partial \dot{q}_1} + \frac{\partial \hat{\mathcal{C}}}{\partial \dot{q}_1} = \frac{\partial \hat{\mathcal{E}}}{\partial \dot{q}_1} = C\dot{q}_1 = Q_C \tag{5.67}$$

and is the *electrical charge of the capacitor*, i.e. its energy variable. The second momentum variable is:

$$p_2 = \frac{\partial L}{\partial \dot{q}_2} = \frac{\partial \hat{\mathcal{E}}}{\partial \dot{q}_2} + \frac{\partial \hat{\mathcal{C}}}{\partial \dot{q}_2} = L_2\dot{q}_2 + q_1 = \phi_{L_2} + \phi_C \tag{5.68}$$

and is *the sum* of the *the total magnetic flux of the inductor* $L_2$ (its energy variable) and of the fictitious flux at the capacitor $\phi_C$. The Hamiltonian function is obtained as the Legendre transformation of
$L(q, \dot{q}, u)$ with respect to $\dot{q}$:

$$H(q, p, u) = \dot{q}_1 p_1 + \dot{q}_2 p_2 - L(q, \dot{q}, u) = H_0(q, p) - H_i(q)u \tag{5.69}$$

where $H_i = q_2$ and $H_0$ is:

$$H_0(q, p) = \frac{1}{2L_1}q_1^2 + \frac{1}{2C}p_1^2 + \frac{1}{2L_2}(p_2 - q_1)^2 \tag{5.70}$$

Note that the function $H_0$ is the total electromagnetic energy of the circuit as the state variables are equal to the energy variables of the capacitors and inductors. Indeed using Kirchhoff's law on the mesh containing the inductor $L_1$ and the capacitor $C$, up to a constant: $q_1 = \phi_C = \phi_{L_1}$ is the magnetic flux in the inductor, by definition of the momenta: $p_1 = Q_C$ is the charge of the capacitor and $p_2 - q_1 = \phi_{L_2}$ is the magnetic flux of the inductor $L_2$. This input-output Hamiltonian system has again order 4 (and not the order of the circuit). But one may note that the Hamiltonian function $H_0$ does not depend on $q_2$. Hence it has a symmetry and the drift dynamics may be reduced to a third order system (the order of the circuit) and in a second step to a second order system [101]. However the interaction Hamiltonian depends on the symmetry variable $q_2$, so the *controlled* system may not be reduced to a lower order input-output Hamiltonian system. The power balance equation (5.63) becomes: $\frac{dH_0}{dt} = u\dot{q}_2 = i_{L_2}u$ which is exactly the power delivered by the source as the current $i_{L_2}$ is also the current flowing in the voltage source.♠♠

The preceding input-output Hamiltonian systems may be extended by considering more general structure matrices than the symplectic structure matrix $J_s$ which appear in the reduction of Hamiltonian systems with symmetries [101]. Indeed one may consider so-called *Poisson structure matrices* that are matrices $J(x)$ depending on $x \in \mathbb{R}^{2n}$ , skew-symmetric and satisfying the *Jacobi identities*:

$$\sum_{k,l=1}^{n} \left( J_{lj} \frac{\partial J_{ik}}{\partial x_l}(x) + J_{li}(x) \frac{\partial J_{kj}}{\partial x_l}(x) + J_{lk} \frac{\partial J_{ji}}{\partial x_l}(x) \right) = 0 \qquad (5.71)$$

**Remark 5.5** These structure matrices are the local definition of Poisson brackets defining the geometrical structure of the state-space [2] [101] of Hamiltonian systems defined on differentiable manifold endowed with a Poisson bracket. Such systems appear for instance in the Hamiltonian formulation of a rigid body spinning around its center of mass (the Euler-Poinsot problem) [101].

**Remark 5.6** Poisson structure matrices may be related to symplectic structure matrices as follows. Note first that, by its skew-symmetry, the rank of the structure matrix of a Poisson bracket at any point is even, say $2n$. (Then one says also that the Poisson bracket has the rank $2n$). Suppose moreover that the structure matrix has constant rank $2n$ in a neighborhood of a point $x_0 \in M$. Then the Jacobi identities (5.71) ensure the existence of *canonical coordinates* $(q, p, r) = (q_1, .., q_n, p_1, .., p_n, r_1, .., r_l)$ where $(2n + l) = m$, such that the $m \times m$ structure matrix $J(q, p, r)$ is given as follows:

$$J(q,p,r) = \begin{pmatrix} 0_n & I_n & 0_{n \times l} \\ -I_n & 0_n & 0_{n \times l} \\ 0_{l \times n} & 0_{l \times n} & 0_{l \times l} \end{pmatrix} \qquad (5.72)$$

One may hence see appear a symplectic matrix associated with the first $2n$ coordinates. The remaining coordinates correspond to so-called distinguished functions or Casimir functions which define an important class of dynamical invariants of the Hamiltonian system [101]. ♠♠

With such structure matrices, the input-output Hamiltonian systems may be generalized to Poisson control systems as follows [148].

**Definition 5.6 (Poisson control systems)** A *Poisson control system* on $\mathbb{R}^n$ is defined by a Poisson structure matrix $J(x)$, a Hamiltonian function $H(x) = H_0(x) - \sum_{i=1}^{m} H_i(x)\, u_i$ composed of the sum of the internal Hamiltonian $H_0(x)$ and a linear combination of $m$ *interaction Hamiltonian functions* $H_i(x)$ *and the dynamic equations:*

$$\dot{x} = J(x)dH_0(x) - \sum_{i=1}^{m} J(x)dH_i(x)\, u_i \qquad (5.73)$$

♠♠

### 5.2.2 Port controlled Hamiltonian systems

As the examples of the LC circuit and of the levitated ball have shown, although the input-output Hamiltonian systems represent the dynamics of physical systems in a way that the conservation of energy is embedded in the model, they fail to represent accurately some other of their structural properties. Therefore another type of Hamiltonian systems, called *port controlled Hamiltonian systems* was introduced which made it possible to represent both the energy conservation as well as some other structural properties of physical systems, mainly related to their internal interconnection structure [124] [180].

**Definition 5.7 (Port controlled Hamiltonian system)** A *port controlled Hamiltonian system* on $\mathbb{R}^n$ is defined by a skew-symmetric structure matrix $J(x)$, a real-valued Hamiltonian function $H_0(x)$, $m$ input vector fields $g_i(x)$ and the dynamic equations:

$$\begin{aligned} \dot{x} &= J(x)dH_0(x) + \textstyle\sum_{i=1}^m g_i(x)U_i \\ y_i &= g_i^T(x)dH_0(x) \end{aligned} \tag{5.74}$$

♠♠

One may note that port controlled Hamiltonian system, as the input-output Hamiltonian systems, are affine with respect to the inputs [72] [148].

**Remark 5.7** The system-theoretic properties of port controlled Hamiltonian systems were investigated in particular concerning the external equivalence, but as this subject goes beyond the scope of this book, the reader is referred to [178] ([180], chap.4). ♠♠

The systems (5.74) have been called *port controlled Hamiltonian system* in allusion to the network concept of the interaction through ports [124] [178] [180]. In this case the Hamiltonian function corresponds to the internal energy of the system, the structure matrix corresponds to the interconnection structure associated with the energy flows in the system [125] [128] [129] and the interaction with the environment of the network is defined through pairs of port variables [124] [178]. Moreover the underlying modeling formalism is a network formalism which provides a practical frame to construct models of physical systems and roots on a firmly established tradition in engineering [23] which found its achievement in the bond graph formalism [154] [24] [124].
Port controlled Hamiltonian systems differ from input-output Hamiltonian systems in three ways which we shall illustrate below on some examples. Firstly, the structure matrix $J(x)$ does not have to satisfy the Jacobi identities (5.71); such structure matrices indeed arise in the reduction of simple mechanical systems with nonholonomic constraints [177]. Secondly the input vector fields are

no more necessarily Hamiltonian, that is they may not derive from an interaction potential function. Thirdly, the definition of the output is changed. The most simple examples of port controlled Hamiltonian system consist in elementary energy storing systems, corresponding for instance to a linear spring or a capacitor.

**Example 5.12 (Elementary energy storing systems)** Consider the following first order port controlled Hamiltonian system:

$$\begin{aligned} \dot{x} &= u \\ y &= \frac{dH_0}{dx}(x) \end{aligned} \tag{5.75}$$

where $x \in \mathbb{R}^n$ is the state variable, $H_0(x)$ is the Hamiltonian function and the structure matrix is equal to 0. In the scalar case, this system represents the *integrator* which is obtained by choosing the Hamiltonian function to be: $H_0 = \frac{1}{2}x^2$. This system represents also a *linear spring*, where the state variable $x$ is the displacement of the spring and the energy function is the elastic potential energy of the spring (for instance $H(x) = \frac{1}{2}k\,q^2$ where $k$ is the stiffness of the spring). In the same way (5.75) represents a *capacitor* with $x$ being the charge and $H_0$ the electrical energy stored in the capacitor, or an *inductance* where $x$ is the total magnetic flux and $H_0$ is the magnetic energy stored in the inductance.
In $\mathbb{R}^3$ such a system represents the point mass in the 3-dimensional Euclidean space with mass $m$ where the state variable $x \in \mathbb{R}^3$ is the momentum vector, the input $u \in \mathbb{R}^3$ is the vector of forces applied on the mass, the output vector $y \in \mathbb{R}^3$ is the velocity vector and the Hamiltonian function is the kinetic energy $H_0(x) = \frac{1}{2m}x^T x$.
It may be noted that such elementary systems may take more involved forms when the state variable belongs to some manifold different from $\mathbb{R}^n$, as it is the case for instance for spatial springs which deform according to rigid body displacements [129] [104] [53] [54]. ♠♠

Like affine Lagrangian control systems and input-output Hamiltonian systems, port controlled Hamiltonian systems satisfy a power balance equation and under some assumption on the Hamiltonian function are lossless.

**Lemma 5.7 (Losslessness)** A port controlled Hamiltonian system (according to definition 5.7), satisfies the following *power balance equation*

$$u^T y = \frac{dH_0}{dt} \tag{5.76}$$

If moreover the Hamiltonian function $H_0(x)$ is bounded from below, then the port controlled Hamiltonian system is lossless with respect to the supply rate $u^T y$ with storage function $H_0(x)$. ♠♠

Again in the case when the Hamiltonian function is the energy, the balance equation corresponds to a power balance expressing the conservation of energy. Let us now consider a slightly more involved example, the LC circuit of order 3 treated here above, in order to comment on the structure of port controlled Hamiltonians sytems as well as to compare it to the structure of input-output and Poisson control systems.

**Example 5.13 (LC circuit of order 3)** Consider again the circuit of example 5.6. According to the partition of the interconnection graph into the spanning tree: $\Gamma = \{C\} \cup \{S_u\}$ and its cotree: $\Lambda = \{L_1\} \cup \{L_2\}$, one may write Kirchhoff's mesh law for the meshes defined by the edges in $\Lambda$ and the node law corresponding to the edges in $\Gamma$ as follows:

$$\begin{pmatrix} i_C \\ v_{L_1} \\ v_{L_2} \\ -i_S \end{pmatrix} = \begin{pmatrix} 0 & -1 & -1 & 0 \\ 1 & 0 & 0 & 0 \\ 1 & 0 & 0 & -1 \\ 0 & 0 & 1 & 0 \end{pmatrix} \begin{pmatrix} v_C \\ i_{L_1} \\ i_{L_2} \\ v_S \end{pmatrix} \tag{5.77}$$

Now, taking as state variables the energy variables of the capacitor (the charge $Q_C$, the total magnetic fluxes $\phi_{L_1}$ and $\phi_{L_2}$ in the two inductors) one identifies immediately the first 3 components of the left hand side in (5.77) as the time derivative of the state vector $x = (Q_C, \phi_{L_1}, \phi_{L_2})^T$. Denoting by $H_C(Q_C)$, $H_{L_1}(\phi_{L_1})$ and $H_{L_2}(\phi_{L_2})$ the electric and magnetic energies stored in the elements, one may identify the coenergy variables as follows: $v_C = \frac{\partial H_C}{\partial Q_C}$, $i_{L_1} = \frac{\partial H_{L_1}}{\partial \phi_{L_1}}$ and $i_{L_2} = \frac{\partial H_{L_2}}{\partial \phi_{L_2}}$. Hence the 3 first components of the vector on the right hand side of the equation (5.77) may be interpreted as the components of the gradient of the total electromagnetic energy of the LC circuit $H_0(x) = H_C(Q_C) + H_{L_1}(\phi_{L_1}) + H_{L_2}(\phi_{L_2})$. Hence the dynamics of the LC circuit may be written as the following port controlled Hamiltonian system:

$$\begin{cases} \dot{x} = J dH_0(x) + gu \\ \\ y = g^T dH_0(x) \end{cases} \tag{5.78}$$

where the structure matrix $J$ and the input vector $g$ are part of the matrix describing Kirchhoff's laws in (5.77) (i.e. part of the fundamental loop matrix associated with the tree $\Gamma$):

$$J = \begin{pmatrix} 0 & -1 & -1 \\ 1 & 0 & 0 \\ 1 & 0 & 0 \end{pmatrix} \text{ and } g = \begin{pmatrix} 0 \\ 0 \\ 1 \end{pmatrix} \tag{5.79}$$

The input is $u = v_S$ and the output is the current with generator sign convention: $y = -i_S$. In this example the power balance equation (5.76) is simply interpreted as the time derivative of the total electromagnetic energy being the

power supplied by the source. Actually this formulation is completely general to LC circuits and it may be found in [128] as well as the comparison with the formulation in terms of Lagrangian or input-output Hamiltonian systems [20] [128].
The port controlled Hamiltonian formulation of the dynamics of the LC circuit may be compared with the input-output formulation derived in the example 5.11. Firstly, one may notice that in the port controlled Hamiltonian formulation, the information on the topology of the circuit and the information about the elements (i.e. the energy) is represented in two *different* objects: the structure matrix and the input vector on the one side and the Hamiltonian function on the other side. In the input-output Hamiltonian formulation this information is captured solely in the Hamiltonian function (with interaction potential), in the same way as in the Lagrangian formulation in example 5.6. Secondly the port controlled Hamiltonian system is defined with respect to a non-symplectic structure matrix and its order coincides with the order of the circuit, whereas the input-output system is given (by definition) with respect to a symplectic (even order) structure matrix of order larger than the order of the circuit. Thirdly, the definition of the state variables in the port controlled system corresponds simply to the energy variables of the different elements of the circuit whereas in the input-output Hamiltonian system, they are defined for the total circuit and for instance the flux of capacitor $L_2$ does not appear as one of them. Finally, although the two structure matrices of the port controlled and the input-output Hamiltonian systems may be related by projection of the dynamics using the symmetry in $q_2$ of the input-output Hamiltonian system, the *controlled* systems remain distinct. Indeed, consider the input vector $g$; it is clear that it is not in the image of the structure matrix $J$. Hence there exist no interaction potential function which generates this vector and the port controlled Hamiltonian formulation *cannot* be formulated as an input-output Hamiltonian system or Poisson control system. ♠♠

In order to illustrate a case where the energy function defines some inter-domain coupling, let us consider the example of the iron ball in magnetic levitation. This example may be seen as the one-dimensional case of general electromechanical coupling arising in electrical motors or actuated multibody systems.

**Example 5.14** Consider again the example of the vertical motion of a magnetically levitated ball as treated in the example 5.7 (see figure 5.2). Following a bond graph modeling approach, one defines the state space as being the variables defining the energy of the system. Here the state vector is then $x = (\phi, z, p_b)^T$ where $\phi$ is the magnetic flux in the coil, $z$ is the altitude of the sphere and $p_b$ is the kinetic momentum of the ball. The total energy of the system is composed of 3 terms: $H_0(x) = H_{mg}(\phi, z) + \mathcal{U}(z) + H_{kin}(p_b)$ where

$H_{mg}(\phi, z)$ denotes the magnetic energy of the coil and is:

$$H_{mg}(\phi, z) = \frac{1}{2}\frac{1}{L(z)}\phi^2 \tag{5.80}$$

where $L(z)$ is given in (5.36), $\mathcal{U}(z) = gz$ is the gravitational potential energy and $H_{kin}(p_b) = \frac{1}{2m}p^2$ is the kinetic energy of the ball. Hence the gradient of the energy function $H_0$ is the vector of the coenergy variables: $\frac{\partial H_0}{\partial x} = (v_L, f, v_b)$ where $v_L$ is the voltage at the coil:

$$v_L = \frac{\partial H_{mg}}{\partial \phi} = \frac{\phi}{L(z)} \tag{5.81}$$

The sum of the gravity force and the electromagnetic force is given by $f = g - f_{mg}$:

$$f_{mg} = \frac{1}{2}\frac{\phi^2}{L^2(z)}\frac{\partial L}{\partial z}(z) \tag{5.82}$$

and $v_b = \frac{p_b}{m}$ is the velocity of the ball. Then from Kirchhoff's laws and the kinematic and static relations in the system, it follows that the dynamics may be expressed as a port controlled Hamiltonian system (5.74) where the structure matrix is constant:

$$J = \begin{pmatrix} 0 & 0 & 0 \\ 0 & 0 & 1 \\ 0 & -1 & 0 \end{pmatrix} \tag{5.83}$$

and the input vector is constant:

$$g = \begin{pmatrix} 1 \\ 0 \\ 0 \end{pmatrix} \tag{5.84}$$

Note that the structure matrix is already in canonical form. In order to take into account the dissipation represented by the resistor $R$, one also defines the following dissipating force $v_R = -Ri_R = -Ri_L$ which may be expressed in a Hamiltonian like format as a *Hamiltonian-system with dissipation* [44].
Let us compare now the port controlled Hamiltonian formulation with the Lagrangian or input-output Hamiltonian formulation. Therefore recall first the input-output Hamiltonian system obtained by the Legendre transformation of the Lagrangian system of the example 5.7. The vector of the momenta is:

$$p = \frac{\partial L}{\partial \dot{q}}(q, \dot{q}) = \begin{pmatrix} \phi \\ p_b \end{pmatrix} \tag{5.85}$$

and the Hamiltonian function obtained by Legendre transformation of the Lagrangian function, defined in the example 5.7, is:

$$H(q, p) = H_0(x) - q_1 u \tag{5.86}$$

Hence the state space of the input-output representation is the state space of the port controlled system augmented with the variable $q_1$ (the primitive if the current in the inductor). Hence the order of the input output Hamiltonian system is 4 and larger than 3, the *natural* order of the system (a second order mechanical system coupled with a first order electrical circuit), which is precisely the order of the port controlled Hamiltonian system. Moreover the state variable "in excess" is $q_1$ and is precisely the symmetry variable of the internal Hamiltonian function $H_0(x)$ in $H(q,p)$. In an analogous way as in the LC circuit example above, this symmetry variable defines the interaction Hamiltonian, hence the *controlled* input-output Hamiltonian system may not be reduced. One may notice again that the input vector $g$ does not belong to the image of the structure matrix $J$, hence cannot be generated by any interaction potential function. ♠♠

Now we shall compare the definitions of the outputs for input output Hamiltonian or Poisson control systems and port controlled Hamiltonian systems. Consider the port controlled system (5.74) and assume that the input vector fields are Hamiltonian, i.e. there exists interaction Hamiltonian functions such that: $g_i(x) = J(x)dH_i(x)$. The port conjugated outputs are then: $y_i = dH_0^T(x)g_i(x) = dH_0^T(x)J(x)dH_i(x)$. The natural outputs are: $\tilde{y}_i = H_i(x)$. Using the drift dynamics in (5.74), their derivatives are computed as:

$$\dot{\tilde{y}}_i = dH_i^T(x)\dot{x} = y_i + \sum_{j=1, j\neq i}^{m} u_j dH_i^T(x)J(x)dH_j(x) \tag{5.87}$$

Hence the passive outputs of both systems differ, in general, by some skew symmetric terms in the inputs. This is related to the two versions of the Kalman-Yakubovich-Popov lemma where the output includes or not a skew symmetric feedthrough term.

**Example 5.15 (Mass-spring system with moving basis)** Consider again the mass-spring system with moving basis and its input-output model treated in the examples 5.9 and 5.10. The input vector fields are Hamiltonian, hence we may compare the definition of the passive outputs in the input-output Hamiltonian formalism and in the port controlled Hamiltonian formalism. The derivatives of the natural outputs derived in the example 5.10 are: $\dot{\tilde{y}}_2 = \dot{q} = \frac{p}{m} - u_1$ and $\dot{\tilde{y}}_1 u_1 + \dot{\tilde{y}}_2 u_2 = u_1(kq) + u_2\frac{p}{m}$. The port conjugated outputs are: $y_1 = (-1, 0)\begin{pmatrix} kq \\ \frac{p}{m} \end{pmatrix} = -kq$ and $y_2 = (0,1)\begin{pmatrix} kq \\ \frac{p}{m} \end{pmatrix} = \frac{p}{m}$. These outputs, contrary to the natural outputs and their derivatives, are precisely the interconnection variables needed to write the kinematic and static relation for interconnecting this mass-spring system to some other mechanical systems. ♠♠

The mass-spring example shows how the different definitions of the pairs of input-output variables for input-output and port controlled Hamiltonian systems, although both define a supply rate for the energy function as storage function, are fundamentally different with respect to the interconnection of the system with its environment. One may go one step further and investigate the *interconnection* of Hamiltonian and Lagrangian systems which preserve their structure. It was shown that the port controlled Hamiltonian systems may be interconnected in a structure preserving way by so-called *power continuous interconnections* [44] [130]. Therefore a generalization of port controlled Hamiltonian systems to *implicit* port controlled Hamiltonian systems (encompassing constrained systems) was used [178] [44] [130] [180]. However this topics beyond the scope of this section and we shall only discuss the interconnection of Lagrangian and Hamiltonian systems on the example of the ball in magnetic levitation.

**Example 5.16 (Levitated ball)** Often, the model of the levitated ball is given as a set of coupled differential equations. The first one is a Lagrangian-like equation for the mechanical part (equation (5.42)). The second one is a first order system representing the dynamics of the electrical part which is expressed with respect to the state $\chi_1 = \dot{q}_1 = v_L$ (according to (5.42) and not including the dissipation part):

$$\begin{aligned} &L\left(q_2\right)\dot{\chi}_1 + \tfrac{\partial L}{\partial q_2}\left(q_2\right)\dot{q}_2\chi_1 - u = 0 \\ &m\ddot{q}_2 - \tfrac{1}{2}\tfrac{\partial L}{\partial q_2}\left(q_2\right)\chi_1^2 + g = 0 \end{aligned} \tag{5.88}$$

It appears that the complete third order system may not of course be put in Lagrangian form but it may neither be considered as the passive coupling of the two subsystems defined by the equations. Indeed, considering the first subsystem, the electrical part, is first order and cannot be expressed in Lagrangian form. It can neither be considered as passive with respect to input $u$ as the inductance defining the energy, is a function of an external variable $q_2$. And the mechanical part is a Lagrangian system (where the electromagnetic force has to be accounted as an external force). Moreover this electromagnetic force corresponds to a coupling between both subsystems which, however, is not simply expressed as a passivity preserving coupling between these two systems. We have seen that the dynamics of the levitated ball may be formulated as a third order port controlled Hamiltonian system where the coupling between the potential and kinetic energy is expressed in the structure matrix (the symplectic coupling) and the coupling through the electromagnetic energy in the Hamiltonian function. But it also allows us to express this system as the coupling, through a passivity preserving interconnection, of two port controlled Hamiltonian systems. Therefore one may conceptually split the physical properties of the iron ball into purely electric and purely mechanical ones. Then the electromechanical energy transduction is represented by a second order port controlled Hamiltonian system:

$$\begin{pmatrix} \dot{\phi} \\ \dot{z} \end{pmatrix} = \begin{pmatrix} 0 & 0 \\ 0 & 0 \end{pmatrix} \begin{pmatrix} \frac{\partial H_{mg}}{\partial \phi} \\ \frac{\partial H_{mg}}{\partial z} \end{pmatrix} + \begin{pmatrix} 1 \\ 0 \end{pmatrix} u + \begin{pmatrix} 0 \\ 1 \end{pmatrix} u_1 \tag{5.89}$$

with output equations:

$$i_S = (1,0) \begin{pmatrix} \frac{\partial H_{mg}}{\partial \phi} \\ \frac{\partial H_{mg}}{\partial z} \end{pmatrix} \tag{5.90}$$

$$\tag{5.91}$$

$$y_1 = f_{mg} = (0,1) \begin{pmatrix} \frac{\partial H_{mg}}{\partial \phi} \\ \frac{\partial H_{mg}}{\partial z} \end{pmatrix} \tag{5.92}$$

The second subsystem simply represents the dynamics of a ball in vertical translation submitted to the action of an external force $u_2$:

$$\begin{pmatrix} \dot{q} \\ \dot{p} \end{pmatrix} = \begin{pmatrix} 0 & 1 \\ -1 & 0 \end{pmatrix} \begin{pmatrix} \frac{\partial H_2}{\partial q} \\ \frac{\partial H_2}{\partial p} \end{pmatrix} + \begin{pmatrix} 0 \\ 1 \end{pmatrix} u_2 \tag{5.93}$$

where the Hamiltonian $H_2$ is the sum of the kinetic and the potential energy of the ball: $H_2(q,p) = \frac{1}{2m}p^2 + gq$ and the conjugated output is the velocity of the ball:

$$y_2 = (0,1) \begin{pmatrix} \frac{\partial H_2}{\partial q} \\ \frac{\partial H_2}{\partial p} \end{pmatrix} \tag{5.94}$$

Consider the interconnection defined by:

$$u_1 = y_2 \tag{5.95}$$

$$u_2 = -y_1 \tag{5.96}$$

It is clear that this interconnection satisfies a power balance: $u_1 y_1 + u_2 y_2 = 0$. Hence it may be proved [44] [130] [180] that the interconnection of the two port controlled Hamiltonian systems leads to a port controlled Hamiltonian system (actually much more general interconnection relations may be considered, involving also constraints). In this example a simple elimination of the variables involved in the interconnection leads to the port controlled Hamiltonian system with Hamiltonian function $H_{tot} = H_{mg} + H_2$ and structure matrix:

$$J_{tot} = \begin{pmatrix} 0 & 0 & 0 & 0 \\ 0 & 0 & 0 & 1 \\ 0 & 0 & 0 & 1 \\ 0 & -1 & -1 & 0 \end{pmatrix} \tag{5.97}$$

Considering the lines 2 and 3 of the structure matrix, one deduces that the variations of $z$ and $q$ satisfy:

$$\dot{z} - \dot{q} = 0 \tag{5.98}$$

This is precisely a Casimir function, i.e. a dynamical invariant of any Hamiltonian system defined with respect to the structure matrix $J_{tot}$. Hence it is possible to identify (up to an arbitrary constant) the two positions $z$ and $q$, thus to reduce this system to the 3-dimensional port controlled Hamiltonian system presented here above. It is clear that this splitting is not possible using the input-output Hamiltonian system or Poisson control systems as the subsystem 1 in (5.89) has a non symplectic (null) structure matrix and the input vector hence are not Hamiltonian (else they would be null too).

Until now we have considered examples pertaining to electrical, mechanical or electromechanical systems. We shall present an example of port controlled Hamiltonian systems which differs radically from the mechanical and electromechanical systems and pertains to thermodynamics.

**Example 5.17 (Two gases in thermal interaction)** Consider two gases, indexed by 1 and 2, which are contained in two closed and rigid reservoir, thermically isolated from the environment but in thermic interaction through a heat conducting wall obeying Fourier's conduction law. Assume furthermore that they are in thermal interaction with two heat sources (this model may be found as a submodel of the model of a heat exchanger, see [127] for more details). The total energy of the system is the sum of the internal energy of both gases:

$$H(S) = U_1(S_1) + U_2(S_2) \tag{5.99}$$

where $S_1$ and $S_2$ are the entropy of each gas which constitute the state vector of the system: $S = (S_1, S_2)^T$. The coenergy variables are then the gradient of the internal energies and define the temperatures:

$$dH(S) = \begin{pmatrix} T_1 \\ T_2 \end{pmatrix} \tag{5.100}$$

For each gas, for instance gas 1, one may write a heat flow balance:

$$T_1 \dot{S}_1 = R(T_2 - T_1) + T_1 u_1 \tag{5.101}$$

where $R$ is the heat conduction coefficient and $u_1$ is the entropy flow delivered by the heat source in contact with gas 1. For gas 2 the same heat balance equation holds permuting the indices 1 and 2. Using the definition of the variables and the heat flow balance equations, one may derive the dynamics of this system as the following port controlled Hamiltonian system:

$$\frac{dS}{dt} = J(T)dH(S) + g_1 u_1 + g_2 u_2 \tag{5.102}$$

where the matrix $J(T)$ is the skew-symmetric structure matrix:

$$J(T) = R\left(\frac{1}{T_1} - \frac{1}{T_2}\right)\begin{pmatrix} 0 & 1 \\ -1 & 0 \end{pmatrix} \tag{5.103}$$

and the input vectors are:

$$g_1 = \begin{pmatrix} 1 \\ 0 \end{pmatrix}, \; g_2 = \begin{pmatrix} 0 \\ 1 \end{pmatrix} \tag{5.104}$$

The matrix $J(S)$ is skew-symmetric and satisfies the Jacobi identities (5.71) (it is a simple consequence of its order 2), hence it defines a Poisson bracket. Its rank is 2 everywhere except at: $T_1 = T_2$, hence it is not symplectic. Consequently the *constant* input vectorfields are not Hamiltonian (else they would become zero at the singular points). Therefore there exists no interaction Hamiltonian function that generate them and this system may not be expressed as an input-output Hamiltonian or Poisson control system.
Actually this degeneracy of the Poisson structure matrix plays a fundamental role in this system for the analysis of the equilibrium. Indeed as the differential $dH(S)$ is the vector of temperatures (which takes strictly positive values), it follows that the equilibria of the system are solely due to degeneracy of the Poisson bracket and actually correspond to the thermal equilibrium of the system: $T_1 = T_2$. This property makes this thermal system quite different from the input-output Hamiltonian systems (for instance the simple mechanical systems) where the equilibria are defined solely as the extrema of the Hamiltonian function. ♠♠

As a conclusion to this section we shall present an extension of lossless port control Hamiltonian systems to dissipative system, called *port controlled Hamiltonian systems with dissipation* introduced in [44]. The main difference is that the skew-symmetry of the structure matrix $J$ is no more required, hence the structure matrix is in general an addition of a skew-symmetric matrix and a symetric positive matrix.

**Definition 5.8 (Port controlled Hamiltonian system)** A *port controlled Hamiltonian system* on $\mathbb{R}^n$ is defined by a skew-symmetric structure matrix $J(x)$, a symmetric positive matrix $R(x)$, a real-valued Hamiltonian function $H_0(x)$, $m$ input vector fields $g_i(x)$ and the dynamic equations:

$$\begin{cases} \dot{x} = (J(x), R(x))\, dH_0(x) + \sum_{i=1}^{m} g_i(x) U_i \\ y_i = g_i^T(x) dH_0(x) \end{cases} \tag{5.105}$$

♠♠

Of course such a system is no more lossless, but it still satisfies a power balance equation and under some assumption on the Hamiltonian system, a passivity property.

**Lemma 5.8 (Dissipativity)** A port controlled Hamiltonian system with dissipation (according to definition 5.8) satisfies the following *power balance equation*

$$u^T y = \frac{dH_0}{dt} + dH_0^T(x)R(x)dH_0(x) \tag{5.106}$$

If moreover the Hamiltonian function $H_0(x)$ is bounded from below, then the port controlled Hamiltonian system with dissipation is dissipative with respect to the supply rate: $u^T y$ with storage function $H_0(x)$. ♠♠

As an example recall the levitated ball as the interconnection of 2 subsystems.

**Example 5.18** Consider firstly the magnetic part. Considering the losses in the coil amounts to add to the skew symmetric structure matrix defined in (5.89) the symmetric positive matrix:

$$R = \begin{pmatrix} -R & 0 \\ 0 & 0 \end{pmatrix} \tag{5.107}$$

Then the total system becomes also a port controlled Hamiltonian system with a symmetric matrix: $R_{tot} = \text{diag}(-R, 0_3)$, where $O_3 \in I\!R^3$ is the zero matrix. ♠♠

## 5.3 Rigid joint-rigid link manipulators

In this section and in the next ones we shall recall the simple models corresponding to electromechanical systems, which motivated numerous results on passivity-based control. We shall recall and derive their passivity properties, and we illustrate some concepts introduced in the previous sections and chapters. Actually the results in the next sections of the present chapter will serve as a basis for introducing the control problem in chapter 6. Our aim now is to show how one can use the passivity properties of the analyzed processes, to construct globally stable control laws. We shall insist on the calculation of storage functions, and it will be shown at some places (see for instance section 6.2) that this can be quite useful to derive Lyapunov functions for closed-loop systems.

The dynamics of the mechanism constituting the mechanical part of a robotic manipulator is given by a simple mechanical system according to definition 5.4 and lemma 5.3:

$$M(q)\ddot{q} + C(q,\dot{q})\dot{q} + g(q) = \tau \tag{5.108}$$

From lemma 5.2, it follows that they are lossless systems with respect to the supply rate $\tau^T \dot{q}$ with storage function $E(q,\dot{q}) = \frac{1}{2}\dot{q}^T M(q)\dot{q} + V(q)$ and $g(q) = \frac{\partial V}{\partial q}$ is the gradient of the gravitation potential energy $V(q)$.

### 5.3.1 The available storage

We have seen that storage functions play an important role in the dissipativity theory. In particular the dissipativity of a system can be characterized by the available storage $V_a(q,\dot{q})$ and the required supply $V_r(q,\dot{q})$ functions. Let us focus now on the calculation of the available storage function (see definition 4.8), which represents the maximum internal energy contained in the system that can be extracted from it. More formally recall that we have

$$\begin{aligned} V_a(q_0,\dot{q}_0) &= -\inf_{\tau:(0,q_0,\dot{q}_0)\to} \int_0^t \tau^T \dot{q} ds \\ &= \sup_{\tau:(0,q_0,\dot{q}_0)\to} -\int_0^t \tau^T \dot{q} ds \end{aligned} \tag{5.109}$$

The notation $\inf\limits_{\tau:(0,q_0,\dot{q}_0)\to}$ means that one performs the infinimization over all trajectories of the system on intervals $[0,t]$, $t \geq 0$, starting from the extended state $(0,q_0,\dot{q}_0)$, with $(q_0,\dot{q}_0) = (q(0),\dot{q}(0))$, with admissible inputs (at least the closed-loop system must be shown to be well-posed). In other words the infinimization is done over all trajectories $\phi(t;0,q_0,\dot{q}_0,\tau)$, $t \geq 0$. From (5.109) one obtains:

$$\begin{aligned} V_a(q_0,\dot{q}_0) &= \sup_{\tau:(0,q_0,\dot{q}_0)\to} -\left\{ \left[\frac{1}{2}\dot{q}^T M(q)\dot{q}\right]_0^t + U(q(t)) - U(q(0)) \right\} \\ &= \tfrac{1}{2}\dot{q}(0)^T M(q(0))\dot{q}(0) + U(q(0)) \\ &= E(q_0,\dot{q}_0) \end{aligned} \tag{5.110}$$

It is not surprizing that the available storage is just the total initial mechanical energy of the system (but we shall see in a moment that for certain systems this is not so evident).

**Remark 5.8** We might have deduced that the system is dissipative since $V_a(q,\dot{q}) < +\infty$ for any bounded state, see theorem 4.4. On the other hand $V_a(q,\dot{q})$ must be bounded since we already know that the system is dissipative with respect to the chosen supply rate.

**Remark 5.9** In section 5.1 we saw that the addition of Rayleigh dissipation enforces the dissipativity property of the system. Let us recalculate the available storage of a rigid joint-rigid link manipulator when the dynamics are given by:

$$M(q)\ddot{q} + C(q,\dot{q})\dot{q} + g(q) + \frac{\partial R}{\partial \dot{q}} = \tau \tag{5.111}$$

One has:

$$
\begin{aligned}
V_a(q_0,\dot{q}_0) &= \sup_{\tau:(0,q_0,\dot{q}_0)\to} -\int_0^t \tau^T \dot{q} ds \\
&= \sup_{\tau:(0,q_0,\dot{q}_0)\to} \left\{ -\left[\frac{1}{2}\dot{q}^T M(q)\dot{q}\right]_0^t - [U_g(q)]_0^t - \int_0^t \dot{q}^T \frac{\partial R}{\partial \dot{q}} ds \right\} \\
&= \tfrac{1}{2}\dot{q}(0)^T M(q(0))\dot{q}(0) + U_g(q(0)) \\
&= E(q_0,\dot{q}_0)
\end{aligned}
\tag{5.112}
$$

since $\dot{q}^T \frac{\partial R}{\partial \dot{q}} \geq \delta \dot{q}^T \dot{q}$ for some $\delta > 0$. One therefore concludes that the dissipation does not modify the available storage function, which is a logical feature from the intuitive physical point of view (the dissipation and the storage are defined independently).

### 5.3.2 The required supply

Let us now compute the required supply $V_r(q,\dot{q})$ as in definition 4.9. Recall that it is given in a variational form by:

$$
V_r(q_0,\dot{q}_0) = \inf_{\tau:(-t,q_t,\dot{q}_t)\to(0,q_0,\dot{q}_0)} \int_{-t}^0 \tau^T \dot{q} ds \tag{5.113}
$$

where $(q_t,\dot{q}_t) = (q(-t),\dot{q}(-t))$, $(q_0,\dot{q}_0) = (q(0),\dot{q}(0))$, $t \geq 0$. Thus this time the minimization process is taken over all trajectories of the system, joining the extended states $(-t,q_t,\dot{q}_t)$ and $(0,q_0,\dot{q}_0)$ (i.e. $(q_0,\dot{q}_0) = \phi(0;-t,q_t,\dot{q}_t,\tau)$). For the rigid manipulator case one finds

$$
\begin{aligned}
V_r(q_0,\dot{q}_0) &= \inf_{\tau:(-t,q_t,\dot{q}_t)\to(0,q_0,\dot{q}_0)} [E(q_0,\dot{q}_0) - E(q(-t),\dot{q}(-t))] \\
&= E(0) - E(-t)
\end{aligned}
\tag{5.114}
$$

Notice that $V_r$ hence defined is not necessarily positive. However if we compute it from $(-t,q_t,\dot{q}_t) = (-t,0,0)$ then indeed $V_r \geq 0$ is a storage function. Here one trivially finds that $V_r(q_0,\dot{q}_0) = E(q_0,\dot{q}_0)$.

**Remark 5.10** The system is reachable from any state $(q_0,\dot{q}_0)$ (actually, this system is globally controllable). Similarly to the available storage function property, the system is dissipative with respect to a supply rate if and only if the required supply $V_r \geq -K$ for some $K > -\infty$, see theorem 4.4. Here we can take $K = E(-t)$.

## 5.4 Flexible joint-rigid link manipulators

In this section we consider another class of systems which corresponds to models of manipulators whose joints are no longer assumed to be perfectly rigid, but can be fairly modelled by a linear elasticity. Their simplified dynamics can be written as:

$$\left\{ \begin{array}{l} M(q_1)\ddot{q}_1 + C(q_1,\dot{q}_1)\dot{q}_1 + g(q_1) = K(q_2 - q_1) \\ \\ J\ddot{q}_2 = K(q_1 - q_2) + u \end{array} \right. \tag{5.115}$$

where $q_1 \in I\!R^n$ is the vector of rigid links angles, $q_2 \in I\!R^n$ is the vector of motorschaft angles, $K \in I\!R^{n\times n}$ is the joint stiffness matrix and $J \in I\!R^{n\times n}$ is the motorschaft inertia matrix (both assumed here to be constant and diagonal). It is a simple mechanical system in Lagrangian form (5.45), we can say that $M(q) = \begin{pmatrix} M(q_1) & 0 \\ 0 & J \end{pmatrix}$, $C(q,\dot{q}) = \begin{pmatrix} C(q_1,\dot{q}_1) & 0 \\ 0 & 0 \end{pmatrix}$, $\tau = \begin{pmatrix} 0 \\ u \end{pmatrix}$, $g(q) = \begin{pmatrix} g(q_1) \\ 0 \end{pmatrix} + \begin{pmatrix} K(q_2 - q_1) \\ K(q_1 - q_2) \end{pmatrix}$.
Actually the potential energy is given by the sum of the gravity and the elasticity terms, $U_g(q_1)$ and $U_e(q_1, q_2) = \frac{1}{2}(q_2 - q_1)^T K(q_2 - q_1)$ respectively.
The dynamics of flexible joint-rigid link manipulators can be seen as the interconnection of the simple mechanical system representing the dynamics of the rigid joint-rigid link manipulators with a set of linear Lagrangian systems with external forces representing the inertial dynamics of the rotor, interconnected by the rotational spring representing the compliance of the joints. It may be seen as the power continuous interconnection of the corresponding three port controlled Hamiltonian systems in a way completely similar to the example of the levitated ball (example 5.16). We shall not detail the procedure here but summarize it on the figure 5.4. As a result it follows that the system is passive, lossless with respect to the supply rate $u^T\dot{q}_2$ with storage function being the sum of the kinetic energies and potential energies of the different elements. We shall see in section 5.6 that including actuator dynamics produces similar interconnected systems, but with quite different interconnection terms. These terms will be shown to play a crucial role in the stabilizability properties of the overall system.

**Remark 5.11** The model in (5.115) was proposed by Spong [188] and is based on the assumption that the rotation of the motorschafts due to the link angular motion does not play any role in the kinetic energy of the system, compared to the kinetic energy of the rigid links. In other words the angular part of the kinetic energy of each motorschaft rotor is considered to be due to its own rotation only. This is why the inertia matrix is diagonal. This assumption seems satisfied in practice for most of the manipulators. It is also satisfied (mathematically speaking) for those manipulators whose actuators are all mounted

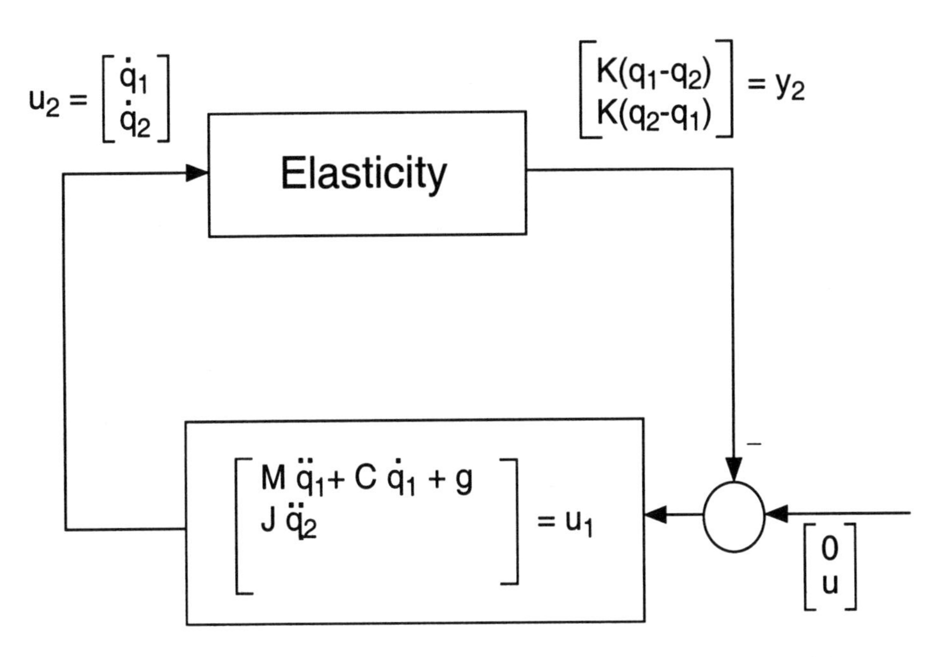


Figure 5.3: Flexible joint-rigid link: interconnection as two passive blocks.

at the base, known as parallel-drive manipulators (the Capri robot presented in chapter 8 is a parallel-drive manipulator). If this assumption is not satisfied [203], the inertia matrix takes the form $M(q) = \begin{pmatrix} M(q_1) & M_{12}(q_1) \\ M_{12}^T(q_1) & J \end{pmatrix}$. ♠♠

The particular feature of the model in (5.115) is that it is static feedback linearizable and possesses a triangular structure [119] that will be very useful when we deal with control.

Let us now prove on some other way that the system is passive (i.e. dissipative with respect to the supply rate $\tau^T \dot{q} = u^T \dot{q}_2$). We get for all $t \geq 0$:

$$\begin{aligned} \int_0^t u(s)^T \dot{q}_2(s) ds &= \int_0^t \left\{ \left[ J\ddot{q}_2 + K(q_2 - q_1) \right]^T \dot{q}_2 \pm (q_2 - q_1)^T K \dot{q}_1 \right\} ds \\ &= \left[ \tfrac{1}{2} \dot{q}_2^T J \dot{q}_2 \right]_0^t + \left[ \tfrac{1}{2} (q_2 - q_1)^T K (q_2 - q_1) \right]_0^t \\ &+ \int_0^t (q_2 - q_1)^T K \dot{q}_1 ds \end{aligned} \tag{5.116}$$

The last integral term can be rewritten as

$$\int_0^t (q_2 - q_1)^T K \dot{q}_1 ds = \int_0^t \dot{q}_1^T \left[ M(q_1)\ddot{q}_1 + C(q_1, \dot{q}_1)\dot{q}_1 + g(q_1) \right] ds \tag{5.117}$$

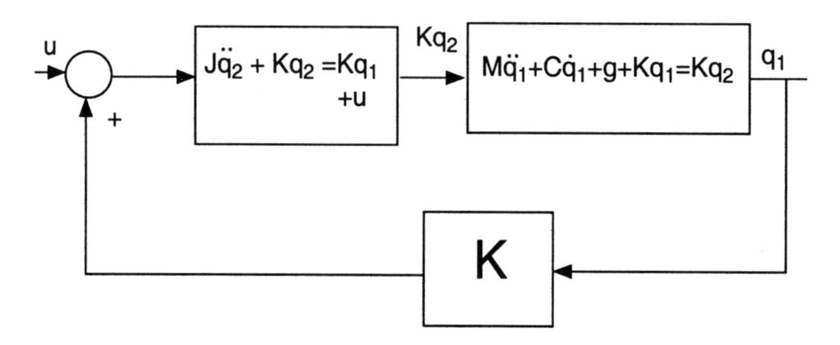


Figure 5.4: Flexible joint-rigid link manipulator.

Looking at the rigid joint-rigid link case, one sees that:

$$\int_0^t (q_2 - q_1)^T K \dot{q}_1 ds = \left[\frac{1}{2}\dot{q}_1^T M(q_1)\dot{q}_1 + U_g(q_1)\right]_0^t \tag{5.118}$$

Therefore grouping (5.116) and (5.118) one obtains:

$$\begin{aligned} \int_0^t u^T \dot{q}_2 ds \quad & \geq -\tfrac{1}{2}\dot{q}_2(0)^T J \dot{q}_2(0) \\ & -\tfrac{1}{2}\dot{q}_1(0)^T M(q_1(0)\dot{q}_1(0) \\ & -\tfrac{1}{2}[q_2(0) - q_1(0)]^T K [q_2(0) - q_1(0)] - U_g(q_1(0)) \end{aligned} \tag{5.119}$$

The results is therefore true whenever $U_g(q_1)$ is bounded from below.

**Remark 5.12** One could have thought of another decomposition of the system as depicted in figure 5.4. In this case the total system is decomposed into two Lagrangian control systems with input being the free end of the springs with respect to each submodel. The subsystem with generalized coordinate $q_1$ (i.e. representing the dynamics of the multibody system of the robot) is analogous to the harmonic oscillator of the example 5.12 and with input $q_2$. The dynamics of the rotors (with generalized coordinates $q_2$) is again analogous to an additional external force $u$. But the interconnection of these two subsystems is defined by : $u_1 = q_2$ and $u_2 = q_1$ involving the *generalized coordinates* which are not passive outputs of the subsystems.

**Remark 5.13** In relationship with the helicopter dynamics analysis at the end of chapter 4, let us point out that manipulators with prismatic joints cannot be passive, except if those joints are horizontal. Hence all those results on open-loop dissipativity hold for revolute joint manipulators only. This will not at all preclude the application of passivity tools for any sort of joints when we deal with feedback control -for instance it suffices to compensate for gravity to avoid this problem-.

### 5.4.1 The available storage

Mimicking the rigid joint-rigid link case, one finds that

$$V_a(q,\dot{q}) = E(q,\dot{q}) \quad = \tfrac{1}{2}\dot{q}_1^T M(q_1)\dot{q}_1 + \tfrac{1}{2}\dot{q}_2^T J\dot{q}_2 + \tfrac{1}{2}[q_1 - q_2]^T K[q_1 - q_2] + U_g(q_1) \tag{5.120}$$

### 5.4.2 The required supply

From subsection 5.3.2 one finds that the energy required from an external source to transfer the system from the extended state $(-t, q_1(-t), q_2(-t), \dot{q}_1(-t), \dot{q}_2(-t)) = (-t, q_{1t}, q_{2t}, \dot{q}_{1t}, \dot{q}_{2t})$ to $(0, q_1(0), q_2(0), \dot{q}_1(0), \dot{q}_2(0)) = (0, q_{10}, q_{20}, \dot{q}_{10}, \dot{q}_{20})$, is given by

$$V_r(q_1(0), q_2(0), \dot{q}_1(0), \dot{q}_2(0)) \quad = E(q_1(0), q_2(0), \dot{q}_1(0), \dot{q}_2(0)) - E(q_1(-t), q_2(-t), \dot{q}_1(-t), \dot{q}_2(-t)) \tag{5.121}$$

**Remark 5.14** One should not think from these examples that the available storage and the required supply are always equal one to each other. See remark 6.6 for a counter-example, where it is is checked that $V_a < V_r$ for nonzero argument.

### 5.4.3 Non-dissipativity for the supply rate $u^T\dot{q}_1$

In all the foregoing developments, we have chosen a physically motivated supply rate that corresponds to the work performed by the generalized torques along a generalized displacement. We have seen that there might be some alternative output variables for which the system is dissipative with respect to the same storage function (the total energy of the system) in the comparison of the outputs of input output and port controlled Hamiltonian systems. In this section, as an illustration on the necessity of such supply rates, we shall consider another candidate supply rate and give a proof that $u^T\dot{q}_1$ cannot be a supply rate for the system in (5.115).

#### The relative degree condition

Recall that there are necessary and sufficient conditions for a *linear* system to be passive (i.e. the transfer function must satisfy some properties in order to be PR or SPR, as explained in chapters 2 and 3). For the nonlinear case of smooth systems affine in the input, there are also some necessary properties: the considered output must have a relative degree equal to one with respect to the input, see section 4.9. We can therefore conclude at once that the model in (5.115) cannot be passive with respect to the supply rate $u^T\dot{q}_1$, because the relative degree $n^*$ between $u$ and $\dot{q}_1$ is equal to three (the number of times on needs to differentiate $\dot{q}_1$ to obtain an expression that contains explicitely the input). Indeed:

$$\begin{aligned}
y &= \dot{q}_1 \\
\dot{y} &= \ddot{q}_1 = M^{-1}(q_1)\left[-C(q_1,\dot{q}_1)\dot{q}_1 - g(q_1) + K(q_2 - q_1)\right] \\
\ddot{y} &= \tfrac{d}{dt}\left\{M^{-1}(q_1)\left[-C(q_1,\dot{q}_1)\dot{q}_1 - g(q_1)\right]\right\} + \tfrac{d}{dt}\left[M^{-1}(q_1)\right] \\
&\quad +M^{-1}(q_1)K(\dot{q}_2 - \dot{q}_1) \\
&\triangleq f(q_1,\dot{q}_1,q_2) + M^{-1}(q_1)K\dot{q}_2 \\
y^{(3)} &= \tfrac{d}{dt}f(q_1,\dot{q}_1,q_2) + M^{-1}(q_1)K\ddot{q}_2
\end{aligned} \tag{5.122}$$

so that $y^{(3)} = g(q_1,q_2,\dot{q}_1,\dot{q}_2) + M^{-1}(q_1)KJ^{-1}u$ which proves that $n^* = 3$.

### The KYP lemma is not satisfied

Recall from the positive-real (or Kalman-Yacubovich-Popov) lemma 4.4 that a system of the form

$$\begin{aligned}
\dot{x} &= f(x) + g(x)u \\
y &= h(x)
\end{aligned} \tag{5.123}$$

is passive (dissipative with respect to the supply rate $u^Ty$) if and only if there exists at least one function $V(t,x) \geq 0$ such the following conditions are satisfied:

$$\begin{aligned}
h^T(x) &= \tfrac{\partial V}{\partial x}(x)g(x) \\
\tfrac{\partial V}{\partial x}(x)f(x) &\geq 0
\end{aligned} \tag{5.124}$$

The **if** part of this lemma tells us that an unforced system that is Lyapunov stable with Lyapunov function $V$ is passive when the output has the particular form in (5.124). The **only if** part tells us that given an output function, then passivity holds only if the searched $V$ does exist.

Now let us assume that the potential function $U_g(q_1)$ is finite for all $q \in \mathcal{C}$. Then it follows that the available storage calculated in (5.120) is a storage function, hence it satisfies the conditions in (5.124) when $y = JJ^{-1}\dot{q}_2 = \dot{q}_2$ and

$u$ is defined in (5.115). More explicitly the function $E(q,\dot{q})$ in (5.120) satisfies the partial differential equations (in (5.115) one has $g^T(x) = (0,0,0,J^{-1})$)

$$\begin{cases} \frac{\partial E}{\partial \dot{q}_2}^T J^{-1} = \dot{q}_2^T \\ \\ \frac{\partial E}{\partial q_1}^T \dot{q}_1 + \frac{\partial E}{\partial \dot{q}_1}^T M(q_1)^{-1}\left[-C(q_1,\dot{q}_1)\dot{q}_1 - g(q_1) + K(q_2-q_1)\right] \\ \\ +\frac{\partial E}{\partial q_2}^T \dot{q}_2 + \frac{\partial E}{\partial \dot{q}_2} J^{-1}\left[K(q_1-q_2)\right] = 0 \end{cases} \tag{5.125}$$

Now if the system was passive with respect to the supply rate $u^T\dot{q}_1$ it would mean that there exists a positive function $V(q,\dot{q})$ such that $V$ satisfies the second equation in (5.125) and

$$\frac{\partial V}{\partial \dot{q}_2}^T J^{-1} = \dot{q}_1^T \tag{5.126}$$

Such a function does not exist, for if it did the system would be dissipative with respect to the supply rate $u^T\dot{q}_1$ and it is not as we know from the preceding paragraph and the next one. The direct conclusion from (5.126) can be drawn if one considers for instance a $V$ that is quadratic in each state component. More generally (5.126) can hold only if $\frac{\partial V}{\partial \dot{q}_2} = J\dot{q}_1$, i.e. $V(q,\dot{q}) = \dot{q}_2 J\dot{q}_1 + \bar{V}(q_1,q_2,\dot{q}_1)$ for a certain function $\bar{V}$. We could go ahead by examing the form of $\bar{V}$ but we prefer to stop here. Actually the first and third paragraphs of this subsection prove to be more efficient to prove that the system is not dissipative with respect to the chosen supply rate. The positive real lemma is essentially useful to construct an output such that a Lyapunov stable system is passive. But as soon as the linear invariant systems case is left, its "only if" part becomes too cumbersome to use because it amounts to solving partial differential equations (or inequations).

### Unboundedness of the available storage

Let us now propose a rather less elegant proof, but closer to the intuitive idea one has of dissipative systems. As we saw above another necessary and sufficient condition for a system to be dissipative is that its available storage be bounded for any state from which the system is reachable. Let us compute the available storage of the system with supply rate $u^T\dot{q}_1$, i.e. $V_a(q(0),\dot{q}(0)) = \sup_{\tau:(0,q_0,\dot{q}_0)\rightarrow} -\int_0^t u^T\dot{q}_1 ds$. Some calculations yield

$$V_a(q(0),\dot{q}(0)) = \sup_{\tau:(0,q_0,\dot{q}_0)\rightarrow} -\int_0^t \dot{q}_1^T\left[J\ddot{q}_2 + K(q_1-q_2)\right] ds \tag{5.127}$$

Our goal is now to show that there exists an admissible $u$ that drives $(q(0),\dot{q}(0))$ to $(q(t),\dot{q}(t))$ and such that $V_a(q(0),\dot{q}(0)) = +\infty$. In order to enable us to make calculations, let us confine ourselves to the linear case for which the dynamics are given by:

$$\begin{cases} m\ddot{q}_1 + kq_1 = kq_2 \\ \\ J\ddot{q}_2 + kq_2 = kq_1 + \tau \end{cases} \tag{5.128}$$

Choosing $\tau = k(q_2 - q_1) + v$ we get

$$\begin{cases} m\ddot{q}_1 + kq_1 = kq_2 \\ \\ J\ddot{q}_2 = v \end{cases} \tag{5.129}$$

We thus aim at calculating

$$-\int_0^t [\dot{q}_1 v + \dot{q}_1(q_1 - q_2)]\, ds \tag{5.130}$$

and we would like to find a $v$ such that the supremum of this integral is not bounded for all $t \geq 0$. Let us consider the input $v = Ja$ where $a \in I\!R$ is a constant. After some manipulations one finds that

$$\begin{aligned} q_1(t) &= \beta \cos\left(\sqrt{\tfrac{k}{m}} + \varphi\right) + kq_2(0) + k\dot{q}_2(0)t + \tfrac{Ja}{2}t^2 - \tfrac{mJa}{k} \\ \\ \dot{q}_1(t) &= -\beta\sqrt{\tfrac{k}{m}} \sin\left(\sqrt{\tfrac{k}{m}} + \varphi\right) + k\dot{q}_2(0) + Jat \\ \\ q_2(t) &= q_2(0) + \dot{q}_2(0)t + \tfrac{Ja}{2}t^2 \\ \\ \dot{q}_2(t) &= \dot{q}_2(0) + Jat \end{aligned} \tag{5.131}$$

with $\beta = \sqrt{\frac{m}{k}\dot{q}_1(0) + q_1(0)}$ and $\varphi = \text{Arctan}\left(-\sqrt{\frac{m}{k}}\frac{\dot{q}_1(0)}{q_1(0)}\right)$. To simplify the calculations further let us choose $J = m = k = 1$. Let us develop the computations for the three terms in (5.130). The first one is given by:

$$\int_0^t \dot{q}_1(s)v(s)ds = aq_1(t) - aq_1(0) \tag{5.132}$$

that is bounded for bounded states, the second one is:

$$\int_0^t \dot{q}_1(s)q_1(s)ds = \frac{1}{2}q_1^2 - \frac{1}{2}q_1(0)^2 \tag{5.133}$$

and hence is also bounded for bounded states. The last term is given by:

$$\begin{aligned} -\int_0^t \dot{q}_1(s)q_2(s)ds &= -\int_0^t -\beta\left(q_2(0) + \dot{q}_2(0)s + \tfrac{a}{2}s^2\right)\sin(s+\varphi)ds \\ &\quad -\int_0^t (\dot{q}_2(0) + as)\left(q_2(0) + \dot{q}_2(0)s + \tfrac{a}{2}s^2\right)ds \end{aligned} \tag{5.134}$$

The first integral term in (5.134) (i.e. $\int_0^t \beta q_2(0)\sin(s+\varphi)ds$) is bounded. But the second and the third one yield:

$$\beta\dot{q}_2(0)\int_0^t s\sin(s+\varphi)ds = -\beta\dot{q}_2(0)t\cos(t+\varphi) + [\sin(t+\varphi)]_0^t \tag{5.135}$$

and

$$\begin{aligned}\beta\tfrac{a}{2}\int_0^t s^2\sin(s+\varphi)ds &= \beta\tfrac{a}{2}\Big\{\left[-s^2\cos(s+\varphi)\right]_0^t \\ &+2[-s\cos(s+\varphi)]_0^t + 2\int_0^t \cos(s+\varphi)ds\Big\}\end{aligned} \tag{5.136}$$

Both those terms may grow unbounded as $t \to +\infty$. Now the second integral in (5.134) contains a term in $t^3$ with coefficient $\frac{a^2}{6}+\frac{a\dot{q}_2(0)}{2}$. As a consequence it follows that for some $q_1(0), q_2(0), \dot{q}_1(0), \dot{q}_2(0)$ there exists $a$ such that the integral in (5.130) tends to infinity. Hence its supremum $V_a(q_1(0), q_2(0), \dot{q}_1(0), \dot{q}_2(0))$ is not finite for all $t \geq 0$. Since the system is reachable from any bounded initial state, one concludes that it cannot be dissipative with respect to the chosen supply rate. Clearly such an input that yields an unbounded $V_a$ cannot be found for the supply rate $\tau\dot{q}_2$.

**Remark 5.15** Let us end these calculations by recalling that concluding that a system is not dissipative with respect to a certain supply rate, does not hamper its dissipativity with respect to another supply rate. For instance a one degree-of-freedom mechanical system (hence a second order system) will be passive between the applied force and the velocity. It is not passive between the force and the position. The system has not physically changed because a wrong supply rate has been chosen: the designer has merely not chosen the right "output" function. The KYP lemma shows how such a function has to be taken.

## 5.5 A bouncing system

We may conclude from the preceding examples that in general, for mechanical systems, the total mechanical energy is a storage function. However the calculation of the available storage may not always be so straightforward as the following example shows. Let us consider a one degree-of-freedom system composed of a mass striking a compliant obstacle modelled as a spring-dashpot system. The dynamical equations for contact and non-contact phase are given by

$$m\ddot{q} = \tau + \begin{cases} -f\dot{q} - kq & \text{if } q > 0 \\ 0 & \text{if } q \leq 0 \end{cases} \tag{5.137}$$

It is noteworthy that the system in (5.137) is nonlinear since the switching condition depends on the state. Moreover existence of a solution with $q$ continuously differentiable is proved in [152] when $\tau$ is a Lipschitz continuous function of time, $q$ and $\dot{q}$. The control objective is to stabilize the system at rest in contact with the obstacle. To this aim let us choose the input

$$\tau = -\lambda_2\dot{q} - \lambda_1(q - q_d) + v \tag{5.138}$$

with $q_d > 0$ constant, $\lambda_1 > 0$, $\lambda_2 > 0$ and $v$ is an auxiliary signal. The input in (5.138) is a PD controller but can be also interpreted as an input transformation. Let us now consider the equivalent closed-loop system with input $v$ and output $\dot{q}$, and supply rate $w = v\dot{q}$. The available storage function is given by

$$V_a(x_0, \dot{x}_0) = \sup_{\tau:(0,q_0,\dot{q}_0)\rightarrow} -\int_{t_0}^{t} v(s)\dot{q}(s)ds \tag{5.139}$$

Due to the system's dynamics in (5.137) we have to consider two cases:

- $q_0 \leq 0$:

  Let us denote $\Omega_{2i} = [t_{2i}, t_{2i+1}]$ the time intervals such that $q(t) \leq 0$, and $\Omega_{2i+1} = [t_{2i+1}, t_{2i+2}]$ the intervals such that $q(t) > 0$, $i \in I\!N$. From (5.138) and (5.137) one has

$$\begin{aligned}
&V_a(q_0, \dot{q}_0) = \\
&= \sup_{\tau:(0,q_0,\dot{q}_0)\rightarrow} -\sum_{i\geq 0}\left\{\int_{\Omega_{2i}} (m\ddot{q}(s) + \lambda_2\dot{q}(s) + \lambda_1 q(s) - \lambda_1 q_d)\dot{q}(s)ds\right\} \\
&-\sum_{i\geq 0}\left\{\int_{\Omega_{2i+1}} (m\ddot{q}(s) + \lambda_2\dot{q}(s) + (\lambda_1 + k)q(s) - \lambda_1 q_d)\dot{q}(s)ds\right\} \\
&= \sup_{\tau:(0,q_0,\dot{q}_0)\rightarrow} \sum_{i\geq 0}\left\{-\left[m\frac{\dot{q}^2}{2}\right]_{t_{2i}}^{t_{2i+1}} - \left[\frac{\lambda_1}{2}(x - x_d)^2\right]_{t_{2i}}^{t_{2i+1}} - \lambda_2\int_{\Omega_{2i}} \dot{q}^2(t)dt\right\} \\
&+\sum_{i\geq 0}\left\{-\left[m\frac{\dot{q}^2}{2} - \frac{\lambda_1 + k}{2}\left(q - \frac{\lambda_1 q_d}{\lambda_1 + k}\right)^2\right]_{t_{2i+1}}^{t_{2i+2}} - (\lambda_2 + f)\int_{\Omega_{2i+1}} \dot{q}^2(t)dt\right\}
\end{aligned} \tag{5.140}$$

  In order to maximize the terms between brackets it is necessary that the integrals $-\int_{\Omega_i} \dot{q}^2(t)dt$ be zero and that $\dot{q}(t_{2i+1}) = 0$. In view of the system's controllability, there exists an impulsive input $v$ that fullfills these requirements [79] (let us recall that this impulsive input is applied while the system evolves in a free-motion phase, hence has linear dynamics). In order to maximize the second term $-\left[\frac{\lambda_1}{2}(q - q_d)^2\right]_{t_0}^{t_1}$ it is also necessary that $q(t_1) = 0$. Using similar arguments, it follows that $\dot{q}(t_{2i+2}) = 0$ and that $q(t_2) = \frac{\lambda_1 q_d}{\lambda_1 + k}$. This reasoning can be iterated to obtain the optimal path which is $(q_0, \dot{q}_0) \to (0,0) \to (\frac{\lambda_1 q_d}{\lambda_1+k}, 0)$ where all the transitions are instantaneous. This leads us to the following available storage function

$$V_a(q_0, \dot{q}_0) = m\frac{\dot{q}_0^2}{2} + \frac{\lambda_1 q_0^2}{2} - \lambda_1 q_d q_0 + \frac{\lambda_1^2 q_d^2}{2(\lambda_1 + k)} \tag{5.141}$$

- $q_0 > 0$:

  Using a similar reasoning one obtains

$$V_a(q_0, \dot{q}_0) = m\frac{\dot{q}_0^2}{2} + \frac{(\lambda_1 + k)}{2}\left(q_0 - \frac{\lambda_1 q_d}{\lambda_1 + k}\right)^2 \tag{5.142}$$

Notice that the two functions in (5.141) and (5.142) are not equal. Their concatenation yields a positive definite function of $(\tilde{q}, \dot{q}) = (0,0)$ with $\tilde{q} = q - \frac{\lambda_1 q_d}{\lambda_1 + k}$, that is continuous at $q = 0$.

**Remark 5.16** Let us now consider the following systems

$$m\ddot{q} + \lambda_2 \dot{q} + \lambda_1 (q - q_d) = v \tag{5.143}$$

and

$$m\ddot{q} + (\lambda_2 + f)\dot{q} + \lambda_1 (q - q_d) + kq = v \tag{5.144}$$

that represent the persistent free motion and the persistent contact motion dynamics respectively. The available storage function for the system in (5.143) is given by (see remark 5.9)

$$V_a(q, \dot{q}) = \frac{1}{2}m\dot{q}^2 + \frac{1}{2}\lambda_1 (q - q_d)^2 \tag{5.145}$$

whereas it is given for the system in (5.144) by

$$V_a(q, \dot{q}) = \frac{1}{2}m\dot{q}^2 + \frac{1}{2}\lambda_1 (q - q_d)^2 + \frac{1}{2}kq^2 \tag{5.146}$$

It is clear that the functions in (5.141) and (5.145), (5.142) and (5.146), are respectively not equal. Notice that this does not preclude that the concatenation of the functions in (5.145) and (5.146) yield a storage function for the system (in which case it must be larger than the concatenation of the functions in (5.141) and (5.142) for all $(q, \dot{q})$). In fact an easy inspection shows that the functions in (5.145) and (5.146) are obtained by adding $\frac{1}{2}\frac{\lambda_1 k q_d^2}{\lambda_1 + k}$ to those in (5.141) and (5.142) respectively. Thus their concatenation indeed yields a storage function for the system in (5.137) with input (5.138). ♠♠

An open issue is to study the conditions under which the available storage function of the piecewise continuous system

$$\begin{array}{lll} \dot{x} = f_i(x,u) & \text{if} & C_i x \geq 0 \\ \dot{x} = g_i(x,u) & \text{if} & C_i x < 0 \\ i \in \{1,\cdots,m\} & & \end{array} \tag{5.147}$$

can be deduced as a concatenation of the available storages of the independent systems $\dot{x} = f_i(x,u)$ and $\dot{x} = g_i(x,u)$. See also remark 6.26.

## 5.6 Including actuator dynamics

### 5.6.1 Armature-controlled DC motors

In all the foregoing examples it has been assumed that the control is directly provided by the generalized torque $\tau$. In reality the actuators possess their own dynamics, and the torque is just the output of a dynamical system. In practice the effect of neglecting those dynamics may deteriorate the closed-loop performance [35]. In other words, the dynamics in (5.45) are replaced by a more accurate armature-controlled DC motor model as:

$$\left\{ \begin{array}{l} M(q)\ddot{q} + C(q,\dot{q})\dot{q} + g(q) = \tau = K_t I \\ RI + L\frac{dI}{dt} + K_t \dot{q} = u \end{array} \right. \tag{5.148}$$

where $R, L, K_t$ are diagonal constant matrices with strictly positive entries, $R \in I\!R^{n\times n}$ is a matrix whose $j$th entry is the resistance of the $j$th motor armature circuit, $L \in I\!R^{n\times n}$ has entries which represent the inductances of the armature, $K_t \in I\!R^{n\times n}$ represents the torque constants of each motor, $u \in I\!R^n$ is the vector of armature voltage, $I \in I\!R^n$ is the vector of armature currents. For the sake of simplicity we have assumed that all the gear ratios that might relate the various velocities are equal to one. Moreover the inertia matrix $M(q)$ is the sum of the manipulator and the motorschaft inertias. The new control input is therefore $u$, see figure 5.5. For the moment we are interested in deriving the passivity properties of this augmented model. We shall see further that the (disconnected) dynamics of the motor are strictly output passive with respect to the supply rate $u^T I$.

**Remark 5.17** One may consider this system as the interconnection of two subsystems as in figure 5.6. One notes at once a strong similarity between the model in (5.148) and the example of the magnetic ball in example 5.14. The difference is that there is no coupling through the energy function (no state variable in common) but that the simple mechanical system, representing the dynamics of the mechanical part, is non-linear. The interconnection structure is best seen on the formulation using port controlled Hamiltonian systems as follows and illustrated in the figure 5.6. The Legendre transformation of the simple mechanical system leads to the definition of the momentum vector $p = \frac{\partial L}{\partial \dot{q}} = M(q)\dot{q}$, the Hamiltonian function $H(q,p) = \frac{1}{2}p^T M^{-1}(q)p + V(q)$ and the following port controlled Hamiltonian system:

$$\begin{pmatrix} \dot{q} \\ \dot{p} \end{pmatrix} = \begin{pmatrix} 0_n & I_n \\ -I_n & 0_n \end{pmatrix} \begin{pmatrix} \frac{\partial H}{\partial q} \\ \frac{\partial H}{\partial p} \end{pmatrix} + \begin{pmatrix} 0_n \\ I_n \end{pmatrix} \tau \tag{5.149}$$

$$\tag{5.150}$$

$$y_{mech} = (0_n, I_n) \begin{pmatrix} \frac{\partial H}{\partial q} \\ \frac{\partial H}{\partial p} \end{pmatrix} = \dot{q} \tag{5.151}$$

where the input $\tau$ represents the electromechanical forces. The dynamics of the motors is described by the following port controlled Hamiltonian system with dissipation with state variable being the total magnetic flux $\phi = LI$ and the magnetic energy being $H_{mg} = \frac{1}{2L}\phi^2$:

$$\dot{\phi} = -R\frac{\partial H_{mg}}{\partial \phi} + u + u_{mg} \tag{5.152}$$

$$y_{mg} = \frac{\partial H_{mg}}{\partial \phi} = I \tag{5.153}$$

where $u_{mg}$ represents the electromotive forces. Note that the structure matrix consists only in a negative definite part, thus is purely energy dissipating system. The interconnection between the two subsystems is defined by the following power continuous interconnection:

$$\tau = K_t y_{mg} \tag{5.154}$$

$$u_{mg} = -K_t y_{mech} \tag{5.155}$$

A simple elimination leads to the following port controlled Hamiltonian system with dissipation:

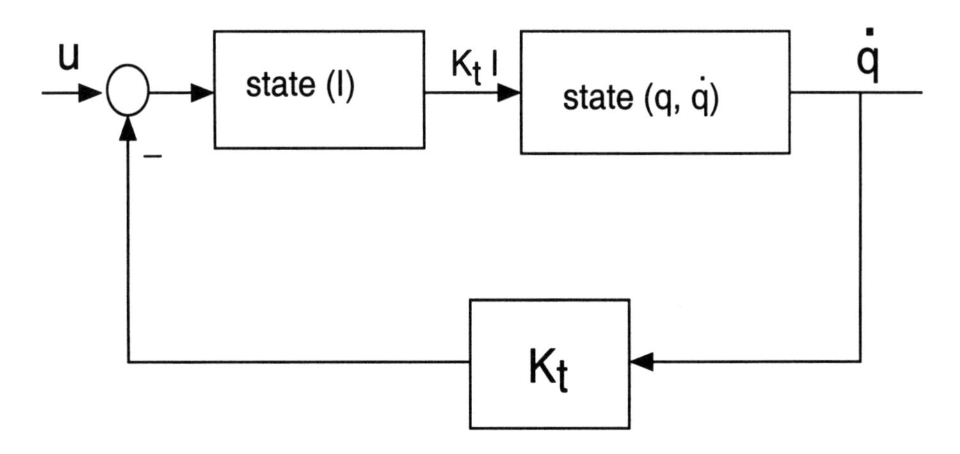


Figure 5.5: Manipulator + armature-controlled DC motor.

$$\begin{pmatrix} \dot{q} \\ \dot{p} \\ \dot{\phi} \end{pmatrix} = \left[ \begin{pmatrix} 0_n & I_n & 0_n \\ -I_n & 0_n & K_t \\ 0_n & -K_t & 0_n \end{pmatrix} + \begin{pmatrix} 0_{2n} & 0_{2n\times n} \\ 0_{n\times 2n} & -R \end{pmatrix} \right] \begin{pmatrix} \frac{\partial H}{\partial q} \\ \frac{\partial H}{\partial p} \\ \frac{\partial H_{mg}}{\partial \phi} \end{pmatrix} + \begin{pmatrix} 0_n \\ 0_n \\ I_n \end{pmatrix} u \tag{5.156}$$

$$y = (0_n, 0_n, I_n) \begin{pmatrix} \frac{\partial H}{\partial q} \\ \frac{\partial H}{\partial p} \\ \frac{\partial H_{mg}}{\partial \phi} \end{pmatrix} = I \tag{5.157}$$

From this formulation of the system as interconnected port controlled Hamiltonian with dissipation, the interconnected system is seen to be passive with supply rate $u^T I$ and storage function $H(q,p) + H_{mg}(\phi)$.

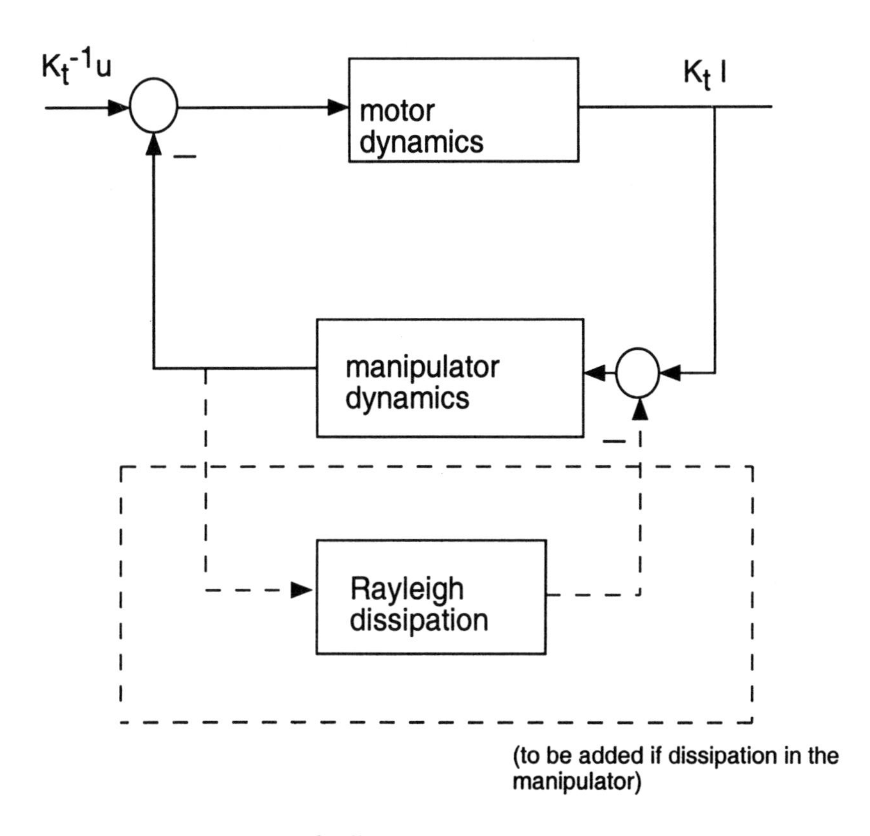


Figure 5.6: Negative feedback interconnection in two dissipative blocks.

### Passivity with respect to the supply rate $u^T I$

Let us calculate directly the value of $u^T I$, where the choice of this supply rate is motivated by an (electrical) energy expression:

$$\begin{aligned} u^T I &= \int_0^t I^T \left[RI + L\frac{dI}{dt} + K_v \dot{q}\right] \\ &= \int_0^t I(s)^T RI(s)ds + \tfrac{1}{2}\left[I(s)^T LI(s)\right]_0^t \\ &+\tfrac{1}{2}\left[\dot{q}(s)^T M(q(s))\dot{q}(s)\right]_0^t + [U_g(q(s))]_0^t \\ &\geq \int_0^t I(s)^T RI(s)ds - \tfrac{1}{2}I(0)^T LI(0) \\ &-\tfrac{1}{2}\dot{q}(0)^T M(q(0))\dot{q}(0) - U_g(q(0)) \end{aligned} \tag{5.158}$$

where we used the fact that $R > 0$, $L > 0$. One sees that the system in (5.148) is even strictly output passive when the output is $y = K_t I$. Indeed $I^T RI \geq \lambda_{\min}(R) y^T y$ where $\lambda_{\min}(R)$ denotes the minimum eigenvalue of $R$.

### Available storage and required supply

Using the same supply rate as in subsection 5.6.1, one gets:

$$\begin{aligned} V_a(q, \dot{q}, I) &= \tfrac{1}{2}I^T LI + \tfrac{1}{2}\dot{q}^T M(q)\dot{q} + U_g(q) \\ &= V_r(q, \dot{q}, I) \end{aligned} \tag{5.159}$$

### Necessity and sufficiency for the supply rate to be $u^T I$

The supply rate $u^T I$ has been chosen according to the definition of conjugated port variables of port controlled Hamiltonian systems. In the sequel, we shall prove that no other form on the port variables may be chosen to define a supply rate for another storage function. Therefore let us introduce a more general supply rate of the form $u^T A^T BI$ for some constant matrices $A$ and $B$ of suitable dimensions. Our goal is to show that if the system is dissipative with respect to this new supply rate, then necessarily (and sufficiently) $A = \frac{1}{\alpha}U^{-1}K_t^{-1}$ and $B = \alpha K_t U$, where $\alpha \neq 0$ and $U$ is a full-rank symmetric matrix. Let us compute the available storage associated to this supply rate, i.e.

$$\begin{aligned} V_a(q_0,\dot{q}_0,I_0) &= \sup_{u_2:(0,q_0,\dot{q}_0,I_0)} -\int_0^t u^T A^T BI ds \\ &= \sup_{u_2:(0,q_0,\dot{q}_0,I_0)} -\left\{\frac{1}{2}[I^T LA^T BI]_0^t + \int_0^t I^T RA^T BI ds \right. \\ &\left. + \int_0^t \dot{q}^T K_t A^T B K_t^{-1} \left[M(q)\ddot{q} + C(q,\dot{q})\dot{q} + g(q)\right] ds\right\} \end{aligned} \quad (5.160)$$

It follows that necessary conditions for $V_a(q,\dot{q},I)$ to be bounded are that $LA^TB \geq 0$ and $RA^TB \geq 0$. Moreover the last integral concerns the dissipativity of the rigid joint-rigid link manipulator dynamics. We know storage functions for this dynamics, from which it follows that an output of the form $K_t^{-1}B^TAK_t\dot{q}$ does not satisfy the (necessary) Kalman-Yakubovic-Popov property, except if $K_t^{-1}B^TAK_t = I_n$. One concludes that the only supply rate with respect to which the system is dissipative must satisfy

$$\begin{aligned} &K_t^{-1}B^TAK_t = I_n \\ &LA^TB \geq 0 \\ &RA^TB \geq 0 \end{aligned} \quad (5.161)$$

Hence $A = \frac{1}{\alpha}U^{-1}K_t^{-1}$ and $B = \alpha K_t U$ for some $\alpha \neq 0$ and some full-rank matrix $U = U^T$.

### 5.6.2 Field-controlled DC motors

Now consider the model of rigid joint-rigid link manipulators actuated by field-controlled DC motors:

$$\begin{cases} L_1\frac{dI_1}{dt} + R_1 I_1 = u_1 \\ L_2\frac{dI_2}{dt} + R_2 I_2 + K_t(I_1)\dot{q} = u_2 \\ M(q)\ddot{q} + C(q,\dot{q})\dot{q} + g(q) + K_v\dot{q} = \tau = K_t(I_1)I_2 \end{cases} \quad (5.162)$$

where $I_1$, $I_2$ are the vectors of currents in the coils of the $n$ motors actuating the manipulator, $L_1$ and $L_2$ denote their inductances, $R_1$ and $R_2$ are the resistors representing the losses in the coils. The matrix $K_t(I_1)$ represent the electromechanical coupling and is defined by a constant diagonal matrix $K_t$ as follows:

$$K_t(I_1) = \text{diag}(k_{t1}I_{11}, \cdots, k_{tn}I_{1n}) = K_t I_1 \quad (5.163)$$

with $k_{ti} > 0$ The last equation is the Lagrangian control system representing the dynamics of the manipulator with $n$ degrees of freedom defined in (5.108) where the diagonal matrix $K_v$ is definite positive and represents the mechanical losses in the manipulator.
In order to reveal the passive structure of the system, we shall again, like in the preceding case, assemble it as the interconnection of 2 passive port controlled Hamiltonian systems. Therefore let us split this system in two parts: the magnetic part and the mechanical part and interconnect them through a power continuous interconnection. The first port controlled Hamiltonian system with dissipation represents the magnetic energy storage and the electromechanical energy transduction. The state variables are the total magnetic fluxes in the coils $\phi = (\phi_1, \phi_2)^T$ defining the magnetic energy: $H_{mg} = \frac{1}{2}\left(\frac{1}{L_1}\phi^2 + \frac{1}{L_1}\phi^2\right)$ and becomes:

$$\dot{\phi} = \begin{pmatrix} -R_1 & 0_n \\ 0_n & -R_2 \end{pmatrix} \begin{pmatrix} \frac{\partial Hmg}{\partial \phi_1} \\ \frac{\partial Hmg}{\partial \phi_2} \end{pmatrix} + \begin{pmatrix} 1 \\ 0 \end{pmatrix} u_1 + \begin{pmatrix} 0 \\ 1 \end{pmatrix} u_2 + \begin{pmatrix} 0 \\ K_t \frac{\phi_1}{L_1} \end{pmatrix} u_{mg} \tag{5.164}$$

with the conjugated outputs associated to the voltages $u_1$ and $u_2$:

$$y_1 = (1,0) \begin{pmatrix} \frac{\partial Hmg}{\partial \phi_1} \\ \frac{\partial Hmg}{\partial \phi_2} \end{pmatrix} = I_1 \text{and } y_2 = (0,1) \begin{pmatrix} \frac{\partial Hmg}{\partial \phi_1} \\ \frac{\partial Hmg}{\partial \phi_2} \end{pmatrix} = I_2 \tag{5.165}$$

and the output conjuguated to the electromotive force $u_{mg}$ is:

$$y_{mg} = \left(0, K_t \frac{\phi_1}{L_1}\right) \begin{pmatrix} \frac{\partial Hmg}{\partial \phi_1} \\ \frac{\partial Hmg}{\partial \phi_2} \end{pmatrix} \tag{5.166}$$

where the two conjugated port variables $u_{mg}$ and $y_{mg}$ define the interconnection with the mechanical system. The second port controlled Hamiltonian system with dissipation represents the dynamics of the manipulator and was presented above:

$$\begin{pmatrix} \dot{q} \\ \dot{p} \end{pmatrix} = \begin{pmatrix} 0_n & I_n \\ -I_n & -K_v \end{pmatrix} \begin{pmatrix} \frac{\partial H}{\partial q} \\ \frac{\partial H}{\partial p} \end{pmatrix} + \begin{pmatrix} 0_n \\ I_n \end{pmatrix} u_{mech} \tag{5.167}$$

$$y_{mech} = (0_n, I_n) \begin{pmatrix} \frac{\partial H}{\partial q} \\ \frac{\partial H}{\partial p} \end{pmatrix} = \dot{q} \tag{5.168}$$

where one notes that the dissipation defined by the matrix $K_t$ was included in the structure matrix. The interconnection of the two subsystems is defined as a elementary negative feedback interconnection:

$$u_{mech} = y_{mg} \tag{5.169}$$

$$u_{mg} = -y_{mech} \tag{5.170}$$

Again a simple elimination of the interconnection variables leads to the port controlled Hamiltonian system with dissipation, with Hamiltonian being the sum of the Hamiltonian of the subsystems: $H_{tot}(\phi, q, p) = H_{mg}(\phi) + H(q, p)$ and structure matrice with skew-symmetric part:

$$J_{tot} = \begin{pmatrix} 0_n & 0_n & 0_n \\ 0_n & 0_n & -K_t\frac{\phi_1}{L_1} \\ 0_n & K_t\frac{\phi_1}{L_1} & 0_n \end{pmatrix} \tag{5.171}$$

and symmetric positive structure matrix:

$$R_{tot} = \mathrm{diag}(-R_1, -R_2, -K_v) \tag{5.172}$$

hence the complete system is passive with respect to the supply rate of the remaining port variables: $u_1 y_1 + y_2 u_2$ and with storage function being the total energy $H_{tot}$.

### Passivity of the manipulator plus field-controlled DC motor

Motivated by the preceding physical analysis of the field-controlled DC motor, using the integral formulation of the passivity, let us prove the dissipativity with respect to the supply rate $u_1^T I_1 + u_2^T I_2$:

$$\begin{aligned} u_1^T I_1 + u_2^T I_2 \quad & \geq -\tfrac{1}{2} I_1(0)^T L_1 I_1(0) + \int_0^t I_1^T R_1 I_1 ds \\ & -\tfrac{1}{2} I_2(0)^T L_2 I_2(0) + \int_0^t I_2^T R_2 I_2 ds + \int_0^t \dot{q}^T K_t(I_1) I_2 \\ & \geq -\tfrac{1}{2} I_1(0)^T L_1 I_1(0) - \tfrac{1}{2} I_2(0)^T L_2 I_2(0) \\ & -\tfrac{1}{2}\dot{q}(0)^T M(q(0))\dot{q}(0) - U_g(q(0)) \end{aligned} \tag{5.173}$$

which proves the statement.

**Remark 5.18 (Passivity of the motors alone)** The dynamics of a field-controlled DC motor are given by

$$\begin{cases} L_1 \frac{dI_1}{dt} + R_1 I_1 = u_1 \\ L_2 \frac{dI_2}{dt} + R_2 I_2 + K_v(I_1)\dot{q} = u_2 \\ J\ddot{q} = K_t(I_1) I_2 - K_{vt}\dot{q} \end{cases} \tag{5.174}$$

where $J \in \mathbb{R}^{n \times n}$ is the rotor inertia matrix. It follows that the (disconnected) actuator is passive with respect to the supply rate $u_1^T I_1 + u_2^T I_2$. Actually we could have started by showing the passivity of the system in (5.174) and then proceeded to showing the dissipativity properties of the overall system in (5.162) using a procedure analog to the interconnection of subsystems. Similar conclusions hold for the armature-controlled DC motor whose dynamics are given by

$$\begin{cases} J\ddot{q} = K_t I \\ RI + L\frac{dI}{dt} + K_t\dot{q} = u \end{cases} \tag{5.175}$$

and which is dissipative with respect to $u^T I$. This dynamics is even output strictly passive (the output is $y = I$ or $y = \begin{pmatrix} I_1 \\ I_2 \end{pmatrix}$) due to the resistance.

### The available storage

The available storage function of the system in (5.162) with respect to the supply rate $u_1^T I_1 + u_2^T I_2$ is found to be after some calculations

$$V_a(I_1, I_2, q, \dot{q}) = \frac{1}{2} I_1^T L_1 I_1 + \frac{1}{2} I_2^T L_2 I_2 + \frac{1}{2}\dot{q}^T M(q)\dot{q} + U_g(q) \tag{5.176}$$

This is a storage function and a Lyapunov function of the unforced system in (5.162).

**Remark 5.19** Storage functions for the disconnected DC motors are given by $V_{adc}(I, q, \dot{q}) = \frac{1}{2}\dot{q}^T J\dot{q} + \frac{1}{2} I^T L I$ and $V_{fdc}(I_1, I_2, q, \dot{q}) = \frac{1}{2}\dot{q}^T J\dot{q} + \frac{1}{2} I_1^T L_1 I_1 + \frac{1}{2} I_2^T L_2 I_2$. Notice that they are not positive definite functions of the state $(q, \dot{q}, I)$ but they are positive definite functions of the partial state $(\dot{q}, I)$. Hence the fixed point $(\dot{q}, I) = (0, 0)$ (or $(\dot{q}, I_1, I_2) = (0, 0, 0)$) is asymptotically stable. ♠♠

Notice that the actuator dynamics in (5.174) with input $(u_1, u_2)$ and output $(I_1, I_2)$ (which are the signals from which the supply rate is calculated, hence the storage functions) is zero-state detectable: $((u_1, u_2) \equiv (0, 0)$ and $I_1 = I_2 = 0) \longrightarrow \dot{q} = 0$ (but nothing can be concluded on $q$), and is strictly output passive. From lemma 4.6 one may conclude at once that any function satisfying the Kalman-Yacubovic-Popov conditions is indeed positive definite.

**Remark 5.20** The model of field-controlled DC motors in (5.174) is similar to that of induction motors, that may be given in some reference frame by (here we show the model for one motor whereas in (5.174) the dynamics represent a system composed of $n$ motors):

$$\begin{aligned} L\dot{z} + C(z,u_3)z + Rq &= E\begin{pmatrix} u_1 \\ u_2 \end{pmatrix} + d \\ y &= L_{sr}(I_2I_3 - I_1I_4) \end{aligned} \tag{5.177}$$

where $z^T = [I_1, I_2, I_3, I_4, \dot{q}] \in \mathbb{R}^5$, $u^T = [u_1, u_2, u_3] \in \mathbb{R}^3$, $d^T = [0,0,0,0,d_5]$, $L = \text{diag}(L_e, vJ) \in \mathbb{R}^{5\times 5}$, $R = \text{diag}(R_e, vb) \in \mathbb{R}^{5\times 5}$, $C(z,u_3) = \begin{bmatrix} C_e(u_3,\dot{q}) & -c(\dot{q}) \\ c^T(\dot{q}) & 0 \end{bmatrix} \in \mathbb{R}^{5\times 5}$, $E = \begin{bmatrix} I_2 \\ 0_{3\times 2} \end{bmatrix} \in \mathbb{R}^{5\times 2}$. $L_e \in \mathbb{R}^{4\times 4}$ is a matrix of inductance, $v \in \mathbb{R}$ is the number of pole pairs, $J \in \mathbb{R}$ is the rotor inertia, $R_e \in \mathbb{R}^{4\times 4}$ is the matrix of resistance, $b \in \mathbb{R}$ is the coefficient of motor damping, $u_1$ and $u_2$ are stator voltages, $u_3$ is the primary frequency, $I_1$ and $I_2$ are stator currents, $I_3$ and $I_4$ are rotor currents, $\dot{q}$ is the rotor angular velocity, $d_5 = -vy_l$ where $y_L$ is the load torque. Finally $y \in \mathbb{R}$ is the generated torque, where $L_{sr} \in \mathbb{R}$ is the mutual inductance.
It can be shown that this model shares properties with the Euler-Lagrange dynamics. In particular [150] the matrix $C(z,u_3)$ satisfies the skew-symmetry requirement for a certain choice of its definition (which is not unique), and $z^TC(z,u_3)z = 0$ (similarly to workless forces). Also this system is strictly passive with respect to the supply rate $I_1u_1 + I_2u_2$, with storage function $H(z) = \frac{1}{2}z^TLz$ and function $S(z) = z^TRz$ (see definition 4.10).

## 5.7 Interconnection terms

In the preceding section we have seen that the supply rate of a system composed of dissipative physical systems was a part of the supply rate of the disconnected subsystems. The storage function of the total system was the sum of the storage functions of the subsystems. Using the port controlled Hamiltonian formulation, this was derived systematically by considering the subsystems as port controlled Hamiltonian systems interconnected through power continuous interconnection relations on the port variables of the subsystems. In this section we shall in some sense invert the problem and investigate the passivity preserving interconnections on a particular class of systems which includes the models of manipulators presented above. From a general point of view, it is clear that the output one chooses in the supply rate must satisfy the relative degree and the zero-dynamics requirements for passivity: this is a first answer to the question why the supply rate chosen for the complete system, looks like the supply rate of the subsystem within which the control enters.

In this section we analyze the role of the interconnection between both subsystems in the overall stabilization problem using dissipativity tools. To this end we primarily focus on manipulator dynamics including actuator dynamics, i.e. we shall consider a system of the form

$$\begin{cases} M(q)\ddot{q} + C(q,\dot{q})\dot{q} + g(q) = K_t(x) \\ \dot{x} = f(x,\dot{q}) + g(x)u_2 \\ y = h(x) \end{cases} \tag{5.178}$$

with $q \in \mathbb{R}^n$, $x = \begin{pmatrix} I_1 \\ I_2 \end{pmatrix} \in \mathbb{R}^{2n}$, $g(x) = \mathrm{diag}(g_1(x), g_2(x)) \in \mathbb{R}^{2n\times 2n}$, $u_2 = \begin{pmatrix} u_{21} \\ u_{22} \end{pmatrix} \in \mathbb{R}^{2n}$, $f(x,\dot{q}) = \begin{pmatrix} f_1(x) \\ f_2(x) \end{pmatrix} + A(x)\dot{q}$, and $A(x) = \begin{pmatrix} 0_{n\times n} \\ A_2(x) \end{pmatrix} \in \mathbb{R}^{2n\times n}$ full-rank matrices. All functions are smooth enough to assure the well-posedness of the open-loop system. Let us further assume that the overall system in (5.178) is controllable through $u_2$, and that the matrix $(A^\dagger)^T K_t(x)$ is full-rank for all $x$, where $A^\dagger = (A^T A)^{-1} A^T \in \mathbb{R}^{n\times 2n}$. It is supposed to be the torque vector delivered by the actuators is given by $h(x,\dot{q})$. Let us do the following

**Assumption 5.1** .

- The second disconnected subsystem is passive with respect to the supply rate $h(x)^T A^\dagger(x) g(x) u_2 + u_{21}^T I_1 = h(x)^T A_2^{-1}(x) g_2(x) u_{22} + u_{21}^T g_1^T(x) I_1$.
- $h(x)^T A^\dagger(x)$ is bounded for all $x$ ($\Rightarrow h(x)^T A_2^{-1}(x)$ is bounded as well).
- $h(x)^T A_2^{-1}(x)$ does not depend on $I_1$. ♠♠

The novelty with respect to (5.162) is that we do not assume that $h(x) = \tau = K_t(x)$ where $\tau$ is the torque input for the manipulator dynamics. Let us compute the available storage function:

$$\begin{aligned} V_a(q,\dot{q},x) &= \sup_{u:(0,q_0,\dot{q}_0,x_0)\to} -\int_0^t \left\{h(x)^T A^\dagger(x) g(x) u_2 + u_{21}^T I_1\right\} ds \\ &= \sup_{u:(0,q_0,\dot{q}_0,x_0)\to} -\int_0^t \left\{I_2^T A_2^{-1}(x) g_2(x) u_{22} + u_{21}^T I_1\right\} ds \\ &= \sup_{u:(0,q_0,\dot{q}_0,x_0)\to} -\int_0^t \{h(x)^T A_2^{-1}(x) g_2(x) u_{22} + u_{21}^T I_1\} ds \\ &= \sup_{u:(0,q_0,\dot{q}_0,x_0)\to} -\int_0^t \Big\{h(x)^T A_2^{-1}(x) \dot{I}_2 \\ &\quad -h(x)^T A_2^{-1}(x) f_2(x) - h(x)^T \dot{q} + I_1^T I_1 - I_1^T f_1(x)\Big\} ds \end{aligned} \tag{5.179}$$

Since the dynamics of the field-controlled DC motor is supposed to be passive with respect to this supply rate, the first, second and last terms under the integral sign do not pose any problem. In other words from assumption 5.1, $h(x)^T A_2^{-1}(x) = \frac{dV_{i2}}{dI_2}(I_2)$ for some positive definite function $V_{i2}(\cdot)$, and $h(x)^T A_2^{-1}(x) f_2(x) \leq -x^T R_1 x$, $I_1^T f_1(x) \leq -x^T R_2 x$ for some matrices $R_1 \geq 0$, $R_2 \geq 0$. The only bothering term is the third one in which the coupling between both subsystems appear. A sufficient condition for the overall system to be dissipative (i.e. for $V_a$ to be bounded for any bounded state) is therefore that $h(x) = \alpha K_t(x)$ for some real $\alpha > 0$. If one desires to get an available storage that is the sum of the available storages of each subsystem, this interconnection structure even becomes necessary (this can also be seen as a consequence of the KYP property in (4.38) for the first subsystem with input $\tau = K_t(x)$ and output $\dot{q}$). Then the total available storage is given by

$$V_a(q,\dot{q},I_1,I_2) = \frac{1}{2}\dot{q}^T M(q)\dot{q} + U_g(q) + \frac{1}{2}I_1^T I_1 + V_{i2}(I_2) \tag{5.180}$$

similarly to (5.176). A major conclusion of this brief study is that the dissipativity properties are in a sense stringent since they do not permit a great freedom in the choice of the interconnection terms between two dissipative systems.

**Remark 5.21** Actually, considering port controlled Hamiltonian systems, the class of structure preserving interconnections, and hence passivity preserving, may be characterized as so-called *power continuous interconnections* [44] [130].

## 5.8 Passive environment

In this section we shall briefly treat systems which may be considered as models of manipulators in contact with their environment through their end-effector or some other body (for instance in assembly tasks or in cooperation with other robots). These systems are part of a more general class of constrained dynamical systems or implicit dynamical systems which constitute still an open problem for their simulation and control. More precisely we shall consider simple mechanical systems which are subject to two types of constraints. Firstly, we shall consider ideal, i.e. workless, constraints on the generalized coordinates or velocities which again may be split into integrable constraints which may be expressed on the generalized coordinates and non-holonomic constraints which may solely be expressed in terms of the generalized velocities. Secondly we shall consider the case when the environment itself is a simple mechanical system and hence consider two simple mechanical systems related by some constraints on their generalized coordinates.

### 5.8.1 Systems with holonomic constraints

Let us consider first a robotic manipulator whose motion is constrained by some $m$ bilateral kinematic constraints, for instance following a smooth surface while keeping in contact. Its model may be expressed as a simple mechanical system (5.45) of order $2n$ with $m < n$ *kinematic constraints* of order zero, and defined by some real function $\phi$ from the space of generalized coordinates $\mathbb{R}^n$ in $\mathbb{R}^m$:

$$\phi(q) = 0 \tag{5.181}$$

Let us assume moreover that the Jacobian $J(q) = \frac{\partial \phi}{\partial q}$ is of rank $m$ everywhere and the kinematic constraints (5.181) define a smooth submanifold $Q_c$ of $\mathbb{R}^n$. Then by differentiating the constraints (5.181) one obtains kinematic constraints of order 1, defined on the velocities:

$$J(q)\dot{q} = 0 \tag{5.182}$$

The two sets of constraints (5.181) and (5.182) define now a submanifold $S$ on the state space $T\mathbb{R}^n = \mathbb{R}^{2n}$ of the simple mechanical system (5.45):

$$S = \left\{(q, \dot{q}) \in \mathbb{R}^{2n} : \phi(q) = 0, J(q)\dot{q} = 0\right\} \tag{5.183}$$

The dynamics of the constrained simple mechanical system is then described by the following system:

$$\begin{aligned} & M(q)\ddot{q} + C(q, \dot{q})\dot{q} + g(q) = \tau + J^T(q)\lambda \\ & J(q)\dot{q} = 0 \end{aligned} \tag{5.184}$$

where $\lambda \in \mathbb{R}^m$ is the $m$ dimensional vector of the Lagrangian multiplier associated with the constraint (5.181). They define the reaction forces $F_r = J^T(q)\lambda$ associated with the constraint which enforce the simple mechanical system to remain on the constraint submanifold $S$ defined in(5.183).

**Remark 5.22** Note that the constrained system (5.184) may be viewed as a port controlled Hamiltonian system with conjugated port variables $\lambda$ and $y = J(q)\dot{q}$ interconnected to a power continous constraint relation defined by $y = 0$ and $\lambda \in \mathbb{R}^m$. It may then be shown that this defines an *implicit* port controlled Hamiltonian system [44] [130]. More general definition of kinematic constraints were considered in [129] [132].

**Remark 5.23** Constrained dynamical systems are the subject of numerous works which are impossible to present here in any detail, and we refer the interested reader to [123] for a brief historical presentation and presentation of related Hamiltonian and Lagrangian formulation as well as to [175] for a Hamiltonian formulation in some more system theoretic setting.

**Remark 5.24** Note that the kinematic constraint of order zero (5.181) is not included in the definition of the dynamics (5.184). Indeed it is not relevant to it, in the sense that this dynamics is valid for any constraint $\phi(q) = c$ where $c$ is a constant vector and may be fixed to zero by the appropriate initial conditions. ♠♠

One may reduce the constrained system to a simple mechanical system of order $2(n-m)$ by using an adapted set of coordinates [118]. Using the theorem of implicit functions, one may find, locally, a function $\rho$ from $\mathbb{R}^{n-m}$ to $\mathbb{R}^m$ such that:

$$\phi(\rho(q_2), q_2) = 0 \tag{5.185}$$

Then define the change of coordinates:

$$z = \tilde{\mathcal{Q}}(q) = \begin{pmatrix} q_1 - \rho(q_2) \\ \\ q_2 \end{pmatrix} \tag{5.186}$$

Its inverse is then simply:

$$q = \mathcal{Q}(z) = \begin{pmatrix} z_1 - \rho(z_2) \\ \\ z_2 \end{pmatrix} \tag{5.187}$$

In the new coordinates (5.186), the constrained simple mechanical system becomes:

$$\begin{aligned} &\tilde{M}(z)\ddot{z} + \tilde{C}(z,\dot{z})\dot{z} + \tilde{g}(z) = \tfrac{\partial \mathcal{Q}}{\partial \tilde{q}}^T \tau + \begin{pmatrix} I_m \\ 0_{n-m} \end{pmatrix} \lambda \\ &\dot{z}_1 = (I_m, 0_{n-m})\,\dot{z} = 0 \end{aligned} \tag{5.188}$$

where the inertia matrix is defined by:

$$\tilde{M}(z) = \frac{\partial \mathcal{Q}}{\partial \tilde{q}}^T (\mathcal{Q}(z)) M(\mathcal{Q}(\tilde{q})) \frac{\partial \mathcal{Q}}{\partial \tilde{q}}(\mathcal{Q}(z)) \tag{5.189}$$

and $\tilde{g}(z)$ is the gradient of the potential function $\tilde{V}(\mathcal{Q}(z))$. The kinematic constraint is now expressed in a canonical form in (5.188) or in its integral form $z_1 = 0$. The equations in (5.188) may be interpreted as follows: the second equation corresponds to the motion along the tangential direction to the constraints. It is not affected by the interaction force since the constraints are assumed to be frictionless. It is exactly the reduced-order dynamics that one obtains after having eliminated $m$ coordinates, so that the $n-m$ remaining coordinates $z_2$ are independent. Therefore the first equation must be considered as an algebraic relationship that provides the value of the Lagrange multiplier as a function of the system's state and external forces.
Taking into account the canonical expression of the kinematic constraints, the constrained system may then be reduced to the simple mechanical system of order $2(n-m)$ with generalized coordinates $z_2$, and inertia matrix (defining the kinetic energy) being the submatrix $\tilde{M}_r(z_2)$ obtained by extracting the last $n-m$ colums and rows from $\tilde{M}(z)$ and setting $z_1 = 0$. The input term is obtained by taking into account the expression of $\mathcal{Q}$ and computing its Jacobian:

$$\frac{\partial \mathcal{Q}}{\partial z}^T = \begin{pmatrix} I_m & 0_{m\times(n-m)} \\ -\frac{\partial \rho}{\partial q_2}(\mathcal{Q}(z) & I_{n-m} \end{pmatrix} \tag{5.190}$$

The reduced dynamics is then a simple mechanical system with inertia matrix $M_r$ and is expressed by:

$$\tilde{M}_r(z)\ddot{z} + \tilde{C}_r(z,\dot{z})\dot{z} + \tilde{g}_r(z) = \left(-\frac{\partial \rho}{\partial q_2}(z_2), I_{n-m}\right)\tau \tag{5.191}$$

The port conjuguated output to $\tau$ is then:

$$y_r = \begin{pmatrix} -\frac{\partial \rho}{\partial q_2}(q_2) \\ I_{n-m} \end{pmatrix} \dot{z}_2 \tag{5.192}$$

Hence the restricted system is as passive and lossless with respect to the supply rate $\tau^T y_r$ and storage function being the sum of the kinetic and potential energy of the constrained system.

**Remark 5.25** We have considered the case of simple mechanical systems subject to holonomic kinematic constraints, that means kinematic constraints of order 1 in (5.182), that fulfill some integrability conditions which guarantee the existence of kinematic constraints of order 0 (5.182). If this is not the case, the constraints are said to be *non-holonomic*. This means that the system may no more be reduced to a lower order simple mechanical system. As we shall not treat them in the sequel, we do not give a detailed presentation and give a sketch of the results indicating only some references. These systems may still be reduced by choosing an adapted set of velocities (in the case of a Lagrangian formulation) or momenta in the case of a Hamiltonian formulation) and then projecting the dynamics along a subspace of velocities or momenta [39] [177] [123] [177]. This dynamics cannot be expressed as a controlled Lagrangian systems, however it has been proved that it may still be expressed as a port controlled Hamiltonian system for which the structure matrix does not satisfy the Jacobi identities (5.71) [177] [90].

**Remark 5.26** Concerning the case of *unilateral* or *inequality* constraints, which yield nonsmooth complementary-slackness systems, see remark 6.26.

## 5.8.2 Compliant environment

### The general dynamics

The general dynamical equations of a rigid joint-rigid link manipulator in permanent contact with an obstacle (that is also a Euler-Lagrange system and can be for instance another –uncontrolled– kinematic chain) are given by:

$$\begin{cases} M(q)\ddot{q} + C(q,\dot{q})\dot{q} + g(q) = \tau + F_q \\ \\ M_e(x)\ddot{x} + C_e(x,\dot{x})\dot{x} + \frac{dR_e}{d\dot{x}} + g_e(x) + K_e x = F_x \end{cases} \tag{5.193}$$

where $q \in I\!R^n$, $x \in I\!R^m$, $m < n$, $F_q \in I\!R^n$ and $F_x \in I\!R^m$ represent the generalized interaction force in coordinates $q$ and in coordinates $x$ respectively. In other words if $x = \phi(q)$ for some function $\phi(\cdot)$, then $\dot{x} = \frac{d\phi}{dx}^T(q)\dot{q} = J(q)\dot{q}$, and $F_q = J^T(q)F_x$. If we view the system in (5.193) as a whole, then the interaction force becomes an internal force. The virtual work principle (for the moment let us assume that all contacts are frictionless) tells us that for any virtual displacements $\delta q$ and $\delta x$, one has $\delta x^T F_x = -\delta q^T F_q$. This can also be seen as a form of the principle of mutual actions. Let us further assume that rank$(\phi) = m$ and that $K_e > 0$. Let us note that the relation $x = \phi(q)$ relates the generalized displacements of the controlled subsystem to those of the uncontrolled one, i.e. to the deflexion of the environment. With this in mind, one can define following McClamroch a nonlinear transformation $q = \mathcal{Q}(z)$, $z = \mathcal{Q}^{-1}(q) = \begin{bmatrix} z_1 \\ z_2 \end{bmatrix} = \begin{bmatrix} K_e\phi(q_1,q_2) \\ q_2 \end{bmatrix}$, $\begin{bmatrix} q_1 \\ q_2 \end{bmatrix} = \begin{bmatrix} \Omega(z_1,z_2) \\ z_2 \end{bmatrix}$,
$\dot{q} = T(z)\dot{z}$, with $T(z) = \begin{bmatrix} \frac{\partial\Omega}{\partial z_1}^T & \frac{\partial\Omega}{\partial z_2}^T \\ \\ 0 & I_{n-m} \end{bmatrix}$, where $z_1 \in I\!R^m$, $z_2 \in I\!R^{n-m}$, and $\phi(\Omega(z_1,z_2),z_2) = z_1$ for all $z$ in the configuration space. Notice that from the rank assumption on $\phi(q)$ and due to the procedure to split $z$ into $z_1$ and $z_2$

(using the implicit function theorem), the Jacobian $T(z)$ is full-rank. Moreover $z_2 = q_2$ where $q_2$ are the $n-m$ last components of $q$. In new coordinates $z$ one has $z_1 = x$ and

$$\begin{cases} \bar{M}(z)\ddot{z} + \bar{C}(z,\dot{z})\dot{z} + \bar{g}(z) = \bar{\tau} + \begin{pmatrix} \lambda_{z_1} \\ 0 \end{pmatrix} \\ M_e(z_1)\ddot{z}_1 + C_e(z_1,\dot{z}_1)\dot{z}_1 + \frac{dR_e}{dz_1} + g_e(z_1) + K_e z_1 = -\lambda_{z_1} \end{cases} \tag{5.194}$$

where $\lambda_{z_1} \in I\!R^m$ and $\bar{M}(z) = T(z)^T M(q) T(z)$, and $\bar{\tau} = T^T(z)\tau$. In a sense this coordinate change splits the generalized coordinates into "normal" direction $z_1$ and "tangential" direction $z_2$, similarly as in subsection 5.8.1. The virtual work principle tells us that $\delta z^T F_z = -\delta z_1 \lambda_{z_1}$ for all virtual displacement $\delta z$, hence the form of $F_z$ in (5.194) where the principle of mutual actions clearly appears.

**Remark 5.27** .

- The original system may appear as having $n+m$ degrees of freedom. However since the two subsystems are assumed to be bilaterally coupled, the number of degrees of freedom is $n$. This is clear once the coordinate change in (5.194) has been applied.
- The system in (5.193) once again has a cascade form where the interconnection between both subsystems is the contact interaction force.
- An equivalent representation as two passive blocks is shown in figure 5.7. As an exercise one may consider the calculation of the storage functions associated to each block.

### Dissipativity properties

Let us assume that the potential energy terms $U_g(z)$ and $U_{g_e}(z_1)$ are bounded from below. This assumption is clearly justified by the foregoing developments on passivity properties of Euler-Lagrange systems. Now it is an evident choice that the suitable supply rate is given by $(\bar{\tau}^T + F_z^T)\dot{z} - \lambda_{z_1}^T \dot{z}_1$. Notice that although one might be tempted to reduce this expression to $\bar{\tau}^T \dot{z}$ since $F_z^T \dot{z} = \lambda_{z_1}^T \dot{z}_1$, it is important to keep it since they do represent the outputs and inputs of different subsystems: one refers to the controlled system while the other refers to the uncontrolled obstacle. Let us calculate the available storage of

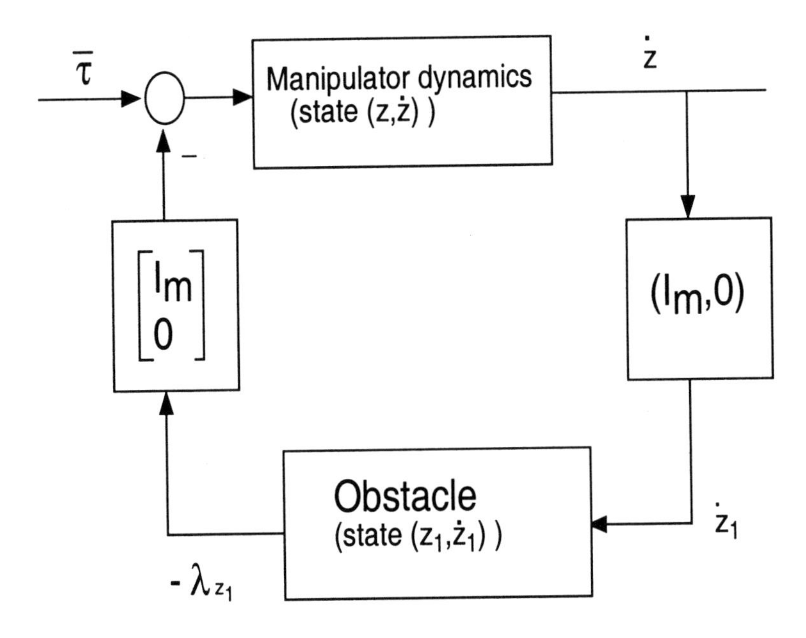


Figure 5.7: Manipulator in bilateral contact with a dynamical passive obstacle.

the total system in (5.194):

$$
\begin{aligned}
V_a(z,\dot z) &= \sup_{\bar\tau:(0,z(0),\dot z(0))\to} -\int_0^t \{(\bar\tau^T + F_z^T)\dot z - \lambda_{z_1}^T \dot z_1\}\, ds \\
&= \tfrac{1}{2}\dot z^T(0)M(z(0))\dot z(0) + \tfrac{1}{2}\dot z_1^T(0)M_e(z_1(0))z_1(0) \\
&\quad + \tfrac{1}{2}z_1^T(0)K_e z_1(0) + U_g(z(0)) + U_{g_e}(z_1(0))
\end{aligned}
\tag{5.195}
$$

Hence the system is dissipative since $V_a$ is bounded for bounded state. Since we introduced some Rayleigh dissipation in the environment dynamics, the system has some strict passivity property.

# Chapter 6

# Passivity-Based Control

This chapter is devoted to investigate how the dissipativity properties of the various systems examined in the foregoing chapter, are related to the stability of the fixed point(s), both in the open-loop (no input) and closed-loop cases. We first start with a classical result of mechanics, which actually is at the base of Lyapunov stability and Lyapunov functions theory. The interest of this result is that its proof hinges on important stability analysis tools, and allows one to make a clear connection between Lyapunov stability and dissipativity theory.

## 6.1 The Lagrange-Dirichlet theorem

In this section we present a stability result that was first stated by Lagrange in 1788 and subsequently proved rigorously by Dirichlet. It provides sufficient conditions for a conservative mechanical system to possess a Lyapunov stable fixed point. The case of Rayleigh dissipation is also presented. The developments base on the dissipativity results of chapter 4.

### 6.1.1 Lyapunov stability

Let us consider the Euler-Lagrange dynamics in (5.1) or that in (5.45). Let us further make the following

**Assumption 6.1** The potential energy $U(q)$ is such that **i)** $\left\{\frac{dU}{dq}(q) = 0 \Leftrightarrow q = q_0\right\}$ and **ii)** $\frac{d^2U}{dq^2}(q_0) > 0$. ♠♠

In other words $U(q)$ is locally convex around $q = q_0$ and $q_0$ is a critical point of the potential energy. Hence the point $(q_0, 0)$ is a fixed point of the dynamics in (5.1). Then it follows that the considered system with input $\tau$, output $\dot{q}$

and state $(q - q_0, \dot{q})$, is zero-state observable (see definition 4.12). Indeed if $\tau \equiv 0$ and $\dot{q} \equiv 0$, it follows from (5.45) that $g(q) = \frac{dU}{dq} = 0$, hence $q = q_0$. The following is then true:

**Theorem 6.1 (Lagrange-Dirichlet)** Let assumption 6.1 hold. Then the fixed point $(q, \dot{q}) = (q_0, 0)$ of the unforced system in (5.1) is Lyapunov stable. ♠♠

**Proof**
First of all notice that the local (strict) convexity of $U(q)$ around $q_0$ precludes the existence of other $q$ arbitrarily close to $q_0$ and such that $\frac{dU}{dq}(q_1) = 0$. This means that the point $(q_0, 0)$ is a strict local minimum for the total energy $E(q, \dot{q})$. We have seen that $E(q, \dot{q})$ is a storage function provided that $U(q)$ remains positive. Now it suffices to define a new potential energy as $U(q) - U(q_0)$ to fullfill this requirement, and at the same time to guarantee that the new $E(q, \dot{q})$ satisfies $E(0, 0) = 0$, and is a positive definite function (locally at least) of $(0, 0)$. Since this is a storage function, we deduce from the dissipation inequality (which is actually here an equality) that for all $\tau \equiv 0$ one gets:

$$E(0) = E(t) - \int_0^t \tau(s)^T \dot{q}(s) ds = E(t) \tag{6.1}$$

Therefore the fixed point of the unforced system is locally Lyapunov stable. Actually we have just proved that the system evolves on a constant energy level (what we already knew) and that the special form of the potential energy implies that the state remains close enough to the fixed point when initialized close enough to it. Notice that (6.1) is of the type (4.37) with $\mathcal{S}(x) = 0$: the system is lossless. All in all, we did not make an extraordinary progress. Before going ahead with asymptotic stability, let us give an illustration of theorem 6.1.

**Example 6.1** Let us consider the dynamics of planar two-link revolute joint manipulator with generalized coordinates the link angles $(q_1, q_2)$ (this notation is not to be confused with that employed for the flexible joint-rigid link manipulators). We do not need here to develop the whole stuff. Only the potential energy is of interest to us. It is given by:

$$U(q) = a_1 \sin(q_1) + a_2 \sin(q_1 + q_2) \tag{6.2}$$

where $a_1 > 0$ and $a_2 > 0$ are constant depending on masses, dimensions and gravity. It is easy to see that $\frac{dU}{dq} = \begin{pmatrix} a_1 \cos(q_1) + a_2 \cos(q_1 + q_2) \\ a_2 \cos(q_1 + q_2) \end{pmatrix} = \begin{pmatrix} 0 \\ 0 \end{pmatrix}$ implies that $q_1 + q_2 = (2n + 1)\frac{\pi}{2}$ and $q_1 = (2m + 1)\frac{\pi}{2}$ for $n, m \in$

$I\!N$. In particular $q_1 = -\frac{\pi}{2}$ and $q_2 = 0$ (i.e. $n = m = -1$) is a point that satisfies the requirements of assumption 6.1. One computes that at this point $\frac{d^2U}{dq^2} = \begin{pmatrix} a_1 + a_2 & a_2 \\ a_2 & a_2 \end{pmatrix}$ that is positive definite since it is symmetric and its determinant is $a_1 a_2 > 0$. Intuitively one notices that global stability is not possible for this example since the unforced system possesses a second fixed point when $q_1 = \frac{\pi}{2}$, $q_2 = 0$. But this one is not stable.

### 6.1.2 Asymptotic Lyapunov stability

Let us now consider the dynamics in (5.33). The following is true:

**Lemma 6.1** Suppose that assumption 6.1 holds. The unforced Euler-Lagrange dynamics with Rayleigh dissipation satisfying $\dot{q}^t \frac{\partial R}{\partial \dot{q}} \geq \delta \dot{q}^T \dot{q}$ for some $\delta > 0$, possesses a fixed point $(q, \dot{q}) = (q_0, 0)$ that is locally asymptotically stable. ♠♠

**Proof**
It is not difficult to prove that the dynamics in definition 5.3 define an OSP system (with the velocity $\dot{q}$ as the output signal). Therefore the system now defines as well an output strictly passive operator $\tau \mapsto \dot{q}$. We could directly compute the derivative of $E(q, \dot{q})$ along the trajectories of the unforced system to attain our target. Let us however use passivity. We know (see remark 4.9) that the dissipation inequality is equivalent to its infinitesimal form, i.e.

$$\frac{dV}{dx}^T f(x, \tau) = \tau^T x_2 - x_2^T \frac{\partial R}{\partial x_2} \tag{6.3}$$

where $x = \begin{pmatrix} x_1 \\ x_2 \end{pmatrix} = \begin{pmatrix} q \\ \dot{q} \end{pmatrix}$, $f(x, u)$ denotes the system vector field in state space notations, and $V(x)$ is any storage function. Let us take $V(x) = E(q, \dot{q})$. We deduce that

$$\dot{E} = \frac{dE}{dx}^T f(x, 0) = -\delta \dot{q}^T \dot{q} \tag{6.4}$$

The only invariant set inside the set $\{(q, \dot{q}) : \dot{q} \equiv 0\}$ is the fixed point $(q_0, 0)$. Resorting to Krasovskii-La Salle invariance theorem one deduces that the trajectories converge asymptotically to this point, provided that the initial conditions are chosen in a sufficiently small neighborhood of it. Notice that we could have used corollary 4.4 to prove the asymptotic stability.

**Remark 6.1** .

- **Convexity:** Convex properties at the core of stability in mechanics: in statics the equilibrium positions of a solid body lying on a horizontal plane, submitted to gravity, are characterized by the condition that the vertical line that passes by its center of mass crosses the convex hull of the contact points of support. In dynamics assumption 6.1 shows that without a convexity property (maybe local), the stability of the fixed point is generically impossible to obtain.

- The case of Amontons-Coulomb friction is more intricate. Indeed this is a nonsmooth hybrid model of friction (here hybrid is to be taken in the dynamical sense, i.e. Amontons-Coulomb friction model has a discrete-event dynamics layer). It is a dissipative model (even more: as shown by Moreau [138], this model satisfies a *maximal dissipation* principle. But this relies on some convex analysis tools which are beyond the scope of this book). The dynamics of systems with Amontons-Coulomb friction is complex and even well-posedness of such systems is still a topic of active research in Applied Mathematics and Mechanics [58]. Asymptotic stability of a unique fixed point (and even its existence) is not assured by Coulomb friction, see [27] pp.207-208.
- The Lagrange-Dirichlet theorem also applies to constrained Euler-Lagrange systems as in (5.188). If Rayleigh dissipation is added and if the potential energy satisfies the required asumptions, then the $(z_2, \dot{z}_2)$ dynamics are asymptotically stable. Thus $\ddot{z}_2$ tends towards zero as well so that $\lambda_{z_1}(t) = \bar{g}_1(z_2(t))$ as $t \to +\infty$.
- It is clear that if assumption 6.1 is strengthened to having a potential energy $U(q)$ that is globally convex, then its minimum point is globally Lyapunov stable.
- Other results generalizing the Lagrange-Dirichlet theorem for systems with cyclic coordinates (i.e. coordinates such that $\frac{\partial T}{\partial q_i}(q) = 0$) were given by Routh and Lyapunov, see [135].

**Remark 6.2** It is a general result that output strict passivity together with zero-state detectability yields under certain conditions asymptotic stability, see corollary 4.4. One basic idea for feedback control may then be to find a control law that renders the closed-loop system strictly output passive with respect to some supply rate, and such that the closed-loop operator is zero-state detectable with respect to the considered output.

**Remark 6.3** Let us come back on the example in section 5.5. As we noted the concatenation of the two functions in (5.141) and (5.142) yields a positive definite function of $(\tilde{q}, \dot{q}) = (0, 0)$ with $\tilde{q} = q - \frac{\lambda_1 q_d}{\lambda_1 + k}$, that is continuous at $q = 0$. The only invariant set for the system in (5.137) with the input in (5.138) is $(q, \dot{q}) = (\frac{\lambda_1 q_d}{\lambda_1 + k}, 0)$. Using the Krasovskii-La Salle invariance theorem one concludes that the point $\tilde{q} = 0$, $\dot{q} = 0$ is globally asymptotically uniformly Lyapunov stable.

### 6.1.3 Invertibility of the Lagrange-Dirichlet theorem

One question that comes to one's mind is that since the strong assumption on which the Lagrange-Dirichlet theorem relies is the existence of a minimum point for the potential energy, what happens if $U(q)$ does not possess a minimum point? Is the equilibrium point of the dynamics unstable in this case? Lyapunov and Chetaev stated the following:

**Theorem 6.2** .

- If at a position of isolated equlibrium $(q, \dot{q}) = (q_0, 0)$ the potential energy does not have a minimum, and, neglecting higher order terms, it can be expressed as a second order polynomial, then the equilibirum is unstable.
- If at a position of isolated equilibrium $(q, \dot{q}) = (q_0, 0)$ the potential energy has a maximum with respect to the variables of smallest order that occur in the expansion of this function, then the equilibrium is unstable.
- If at a position of isolated equilibrium $(q, \dot{q}) = (q_0, 0)$ the potential energy, which is an analytical function, has no minimum, then this fixed point is unstable. ♠♠

Since $U(q) = U(q_0) + \frac{dU}{dq}(q_0)(q-q_0) + \frac{1}{2}(q-q_0)^T \frac{d^2U}{dq^2}(q-q_0) + o[(q-q_0)^T(q-q_0)]$, and since $q_0$ is a critical point of $U(q)$, the first item tells us that the Hessian matrix $\frac{d^2U}{dq^2}$ is not positive definite, otherwise the potential energy would be convex and hence the fixed point would be a minimum. Without going into the details of the proof since we are interested by dissipative systems, not by unstable systems, let us note that the trick consisting of redefining the potential energy as $U(q) - U(q_0)$ in order to get a positive storage function no longer works. Moreover assume there is only one fixed point for the dynamical equations. It is clear at least in the one degree-of-freedom case that if $\frac{d^2U}{dq^2}(q_0) < 0$ then $U(q) \to -\infty$ for some $q$. Hence the available storage function that contains a term equal to $\sup_{\tau:(0,q(0),\dot{q}(0))\to} [U(q(t))]_0^t$ cannot be bounded, assuming that the state space is reachable. Thus the system cannot be dissipative, see theorem 4.4.

## 6.2 Rigid joint-rigid link: state feedback

In this subsection we shall present various feedback controllers that assure some stability properties for the rigid joint-rigid link model in (5.108). We start with the regulation problem and then generalize to the tracking case. In each case we emphasize how the dissipativity properties of the closed-loop systems constitute the basis of the stability properties.

### 6.2.1 PD control

Let us consider the following input:

$$\tau = -\lambda_1 \dot{q} - \lambda_2 \tilde{q} \tag{6.5}$$

where $\lambda_1 > 0$ and $\lambda_2 > 0$ are the constant feedback gains (for simplicity we consider them as being scalars instead of positive definite $n \times n$ matrices, this is not very important for what follows), $\tilde{q} = q - q_d$, $q_d \in I\!R^n$ is a constant desired position. The closed-loop system is given by:

$$M(q)\ddot{q} + C(q, \dot{q})\dot{q} + g(q) + \lambda_1 \dot{q} + \lambda_2 \tilde{q} = 0 \tag{6.6}$$

Two paths are possible: we can search for the available storage function of the closed-loop system in (6.6) which is likely to provide us with a Lyapunov function, or we can try to interpret this dynamics as the negative interconnection of two passive blocks and then use the passivity theorem (more exactly one of its numerous versions) to conclude on stability. To fix the ideas we develop both paths in detail.

#### The closed-loop available storage

First of all notice that in order to calculate an available storage we need a supply rate, consequently we need an input (that will be just an auxiliary signal with no significance). Let us therefore just add a term $u$ in the right-hand-side of (6.6) instead of zero. In other words we proceed as we did for the example in section 5.5: we make an input transformation and the new system is controllable. Let us compute the available storage along the trajectories of this new input-output system, assuming that $U(q)$ is bounded from below, i.e. $U(q) \geq U_{\min} > -\infty$ for all $q \in Q$:

$$\begin{aligned}
V_a(q_0, \dot{q}_0) = & \sup_{u:(0,q_0,\dot{q}_0)\rightarrow} -\int_0^t u^T \dot{q} ds \\
= & \sup_{u:(0,q_0,\dot{q}_0)\rightarrow} -\int_0^t \dot{q}^T \{M(q)\ddot{q} + C(q, \dot{q})\dot{q} + g(q) + \lambda_1 \dot{q} \\
& +\lambda_2 \tilde{q}\} \, ds \\
= & \sup_{u:(0,q_0,\dot{q}_0)\rightarrow} \left\{ -\left[\frac{1}{2}\dot{q}^T M(q)\dot{q}\right]_0^t - [U(q(t))]_0^t - \left[\frac{1}{2}\lambda_2 \tilde{q}^T \tilde{q}\right]_0^t \right. \\
& \left. -\lambda_1 \int_0^t \dot{q}^T \dot{q} ds \right\} \\
= & \tfrac{1}{2}\dot{q}(0)^T M(q(0))\dot{q}(0) + U(q(0)) + \tfrac{1}{2}\lambda_2 \tilde{q}(0)^T \tilde{q}(0)
\end{aligned} \tag{6.7}$$

where we used the fact that $\dot{q}^T[\dot{M}(q,\dot{q}) - 2C(q,\dot{q})]\dot{q} = 0$ for all $q \in Q$ and all $\dot{q} \in T_qQ$, see lemma 5.4. Let us now make a little stop: we want to show some stability property for the unforced system in (6.6), so what is the fixed point of this system? Letting $\dot{q} \equiv 0$ in (6.6) one finds:

$$g(q) + \lambda_2 \tilde{q} = 0 \tag{6.8}$$

Let us state the following:

**Assumption 6.2** The equations in (6.8) possess a finite number of isolated roots $q = q_i$. Moreover the $q_i$s are strict local minima of $U(q)$. ♠♠

Then we have the following:

**Lemma 6.2** Assume that 6.2 is true. The rigid joint-rigid link manipulator dynamics in (5.108) with PD controller in (6.5) has locally asymptotically stable fixed points $(q, \dot{q}) = (q_i, 0)$. ♠♠

**Proof**
From the second part of assumption 6.2 it follows that the available storage $V_a$ in (6.7) is a storage function for the closed-loop system with input $u$ (fictitious) and output $\dot{q}$. Next this also allows us to state that $V_{pd}(q - q_i, \dot{q}) \triangleq V_a(q, \dot{q}) - U(q_i)$, is a Lyapunov function for the unforced system in (6.6): indeed this is a storage function and the conditions of lemma 4.6 are satisfied. Now let us calculate the derivative of this function along the trajectories of (6.6):

$$\begin{aligned} \dot{V}_{pd}(q - q_i, \dot{q}) &= -\lambda_1 \dot{q}^T \dot{q} + \dot{q}^T \left[ -g(q) + \tfrac{dU}{dq} \right] \\ &= -\lambda_1 \dot{q}^T \dot{q} \end{aligned} \tag{6.9}$$

One therefore just has to apply the Krasovskii-La Salle lemma to deduce that the fixed points $(q_i, 0)$ are locally asymptotically Lyapunov stable. Lyapunov second method guarantees that the basin of attraction $B_{r_i}$ of each fixed point has a strictly positive measure.

**Remark 6.1 (Potential energy shaping)** One remarks that asymptotic stability has been obtained in part because the PD control injects some strict output passivity inside the closed-loop system. This may be seen as a forced damping. On the other hand the position feedback may be interpreted as a modification of the potential energy so as to shape it adequately for control purposes. It seems that this technique was first advocated by Takegaki and Arimoto in [200].

**Remark 6.2** The PD control alone cannot compensate for gravity. Hence the system will converge to a configuration that is not the desired one. Clearly increasing $\lambda_2$ reduces the steady-state error. But increasing gains is not always desirable in practice, due to measurement noise in the sensors.

### Equivalent closed-loop interconnections

Since the closed-loop system possesses several equilibrium points, the underlying passivity properties of the complete closed-loop system must be local in nature, i.e. they hold whenever the state remains inside the balls $B_{r_i}$ [165]. It is however possible that each block of the interconnection, when considered separately, possesses global dissipativity properties. But the interconnection does not.

**A first interconnection:** Looking at (6.6) one is tempted to interpret those dynamics as the interconnection of two subsystems with respective inputs $u_1$, $u_2$ and outputs $y_1$ and $y_2$, with $y_1 = u_2$ and $y_2 = -u_1$, and:

$$\begin{cases} u_1 = -\lambda_1 \dot{q} - \lambda_2 \tilde{q} \\ \\ y_1 = \dot{q} \end{cases} \tag{6.10}$$

Evidently this is motivated by the fact that the rigid joint-rigid link manipulator dynamics in (6.6) defines a passive operator between $u_1$ and $y_1$, with state vector $\begin{pmatrix} \tilde{q} \\ \dot{q} \end{pmatrix}$ and dynamics

$$M(q)\ddot{q} + C(q,\dot{q})\dot{q} + g(q) = u_1 \tag{6.11}$$

Let us write this second subsystem in state space form as:

$$\begin{cases} \dot{z}_1 = u_2 \\ \\ y_2 = \lambda_2 z_1 + \lambda_1 u_2 \end{cases} \tag{6.12}$$

with $z_1(0) = q(0) - q_d$. Its transfer matrix is given by $H_{pd}(s) = \frac{\lambda_2 + \lambda_1 s}{s} I_n$ where $I_n$ is the $n \times n$ identity matrix. Thus $H_{pd}(s)$ is PR, see definition 2.3. From theorem 4.7 and corollary 4.1 it follows that $\dot{q} \in L^2(I\!R^+)$. Notice that this is a consequence of the fact that $H_{pd}(s)$ defines an input strictly passive operator, see theorem 2.3 2). We cannot say much more if we do not pick up the storage functions of each subsystem. Now the second subsystem has dynamics such that the associated operator $u_2 \mapsto y_2$ is input strictly passive (hence necessarily of relative degree zero) and with storage function $\frac{\lambda_2}{2} z_1^T z_1$. From the fact that $\dot{z}_1 = \dot{q}$ and due to the choice of the initial data, one has for all $t \geq 0$: $z_1(t) = \tilde{q}(t)$. It is easy to see then that the first subsystem (the rigid joint-rigid links dynamics) has a storage function equal to $\frac{1}{2}\dot{q}^T M(q)\dot{q} + U(q) - U(q_i)$. The sum of both storage functions yields the desired Lyapunov function for the whole system. The interconnection is depicted in figure 6.1.

**Remark 6.4** Looking at the dynamics of both subsystems it seems that the total system order has been augmented. But the interconnection equation

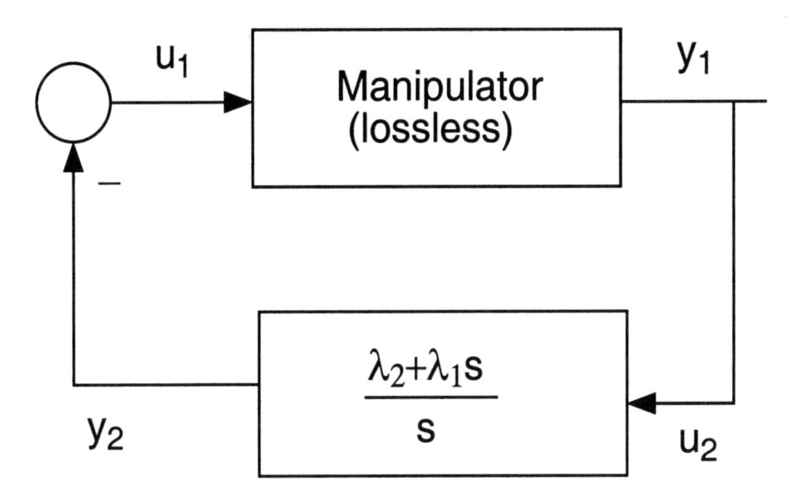


Figure 6.1: The first equivalent representation.

$y_1 = z_1$ may be rewriten as $\dot{z}_1 = \dot{q}$. This defines actually a dynamical invariant $z_1 - q = q_0$, where $q_0 \in I\!R^n$ is fixed by the initial condition $z_1(0) = q(0) - q_d$. Hence the system (6.11) (6.12) may be reduced to the subspace $z_1 - q = -q_d$ and one recovers a system of dimension $2n$ (in other words the space $(q, \dot{q}, z_1)$ is foliated by invariant manifolds $z_1 - q = -q_d$). We refer the reader to section 6.2.3 for further solutions of passive feedback that avoids such drawbacks.

**Remark 6.5** In connection with the remarks at the beginning of this subsubsection, let us note that the fixed points of the first unforced (i.e. $u_1 \equiv 0$) subsystem are given by $\{(q, \dot{q}) : g(q) = 0, \dot{q} = 0\}$, while those of the unforced second subsystem are given by $\{z_1 : \dot{z}_1 = 0 \Rightarrow \tilde{q} = \tilde{q}(0)\}$. Thus the first subsystem has Lyapunov stable fixed points which correspond to its static equilibrium, while the fixed point of the second subsystem corresponds to the desired static position $q_d$. The fixed points of the interconnected blocks are given by the roots of (6.8). If one looks at the system from a pure input-output point of view, such a fixed points problem does not appear. However if one looks at it from a dissipativity point of view, which necessarily implies that the input-output properties are related to the state space properties, then it becomes a necessary step.

**Remark 6.6** $H_{pd}(s)$ provides us with an example of a passive system that is ISP but obviously not asymptotically stable, only stable (see corollary 4.4).

**A second interconnection:** A second possible interpretation of the closed-loop system in (6.6) is made of the interconnection of the two blocks:

$$\begin{cases} u_1 = -\lambda_2 \tilde{q}, \quad y_1 = \dot{q} \\ u_2 = y_1, \qquad y_2 = \lambda_2 \tilde{q} \end{cases} \tag{6.13}$$

The first subsystem then has the dynamics

$$M(q)\ddot{q} + C(q,\dot{q})\dot{q} + g(q) + \lambda_1 \dot{q} = u_1 \tag{6.14}$$

from which one recognizes an output strictly passive system, while the second one has dynamics

$$\begin{aligned} \dot{z}_1 &= u_2 \\ y_2 &= \lambda_2 z_1 \end{aligned} \tag{6.15}$$

with $z_1(0) = q(0) - q_d$. One can check that it is a passive lossless system since $\langle u_2, y_2 \rangle = \int_0^t \lambda_2 \tilde{q}^T(s)\dot{q}(s)ds = \frac{\lambda_2}{2}[\tilde{q}^T\tilde{q}(t) - \tilde{q}^T\tilde{q}(0)]$, with storage function $\frac{\lambda_2}{2}\tilde{q}^T\tilde{q}$. Therefore applying the passivity theorem (see theorem 4.7 and corollary 4.1), one still concludes that $\dot{q} \in \mathcal{L}_2(I\!R^+)$. We however may go a little further with this decomposition. Indeed consider the system with input $u = u_1 + y_2$ and output $y = y_1$. This defines an output strictly passive operator $u \mapsto y$. Setting $u \equiv y \equiv 0$ one obtains that $(q - q_i, \dot{q}) = (0,0)$. Hence this closed-loop system is zero-state observable. Since the storage function (the sum of both storage functions) we have exhibited is positive definite with respect to this error equation fixed point, and since it is proper, it follows that the equilibrium point of the unforced system (i.e. $u \equiv 0$) is globally asymptotically stable. This second interconnection is depicted in figure 6.2.
In conclusion it is not very important whether we associate the strict passivity property to one block or the other. What is important is that we can *systematically* associate to these dissipative subsystems some Lyapunov functions that are *systematically* deduced from their passivity property. This is a fundamental property of dissipative systems that one can calculate Lyapunov functions for them. It has even been originally the main motivation for studying passivity, at least in the field of control and stabilization of dynamic systems.

### 6.2.2 PID control

The PID control is also a feedback controller that is widely used in practice. Let us investigate whether we can redo the above analysis for the PD controller. If we proceed in the same manner, we decompose the closed-loop dynamics

$$M(q)\ddot{q} + C(q,\dot{q})\dot{q} + g(q) + \lambda_1 \dot{q} + \lambda_2 \tilde{q} + \lambda_3 \int_0^t \tilde{q}(s)ds = 0 \tag{6.16}$$

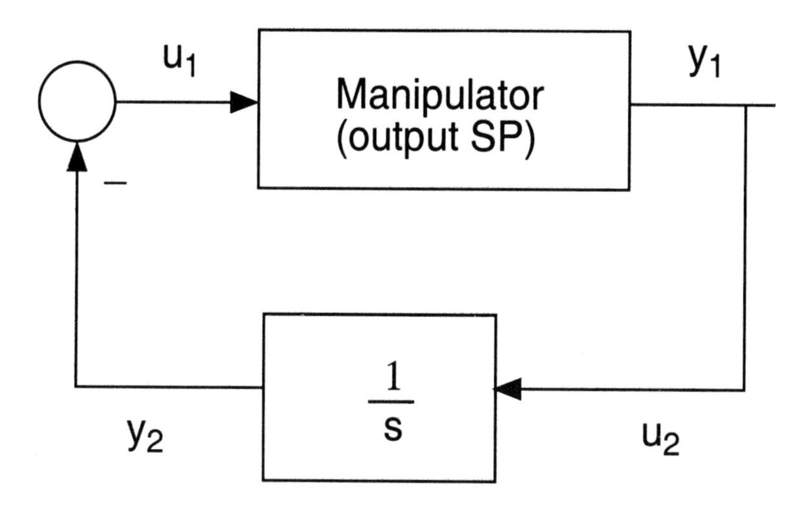


Figure 6.2: The second equivalent representation.

into two subsystems, one of which corresponds to the rigid joint-rigid link dynamics, and the other one to the PID controller itself. The input and output signals of this interconnection are this time chosen to be

$$\begin{cases} u_1 = -\lambda_1 \dot{q} - \lambda_2 \tilde{q} - \lambda_3 \int_0^t \tilde{q}(s)ds = -y_2 \\ \\ y_1 = \dot{q} = u_2 \end{cases} \tag{6.17}$$

The dynamics of the PID block are given by (compare with (6.12)):

$$\begin{cases} \dot{z}_1 = z_2 \\ \\ \dot{z}_2 = u_2 \\ \\ y_2 = \lambda_1 u_2 + \lambda_2 z_2 + \lambda_3 z_1 \end{cases} \tag{6.18}$$

The transfer matrix of this linear operator is given by (compare with (2.64) and (2.65))

$$H_{pid}(s) = \frac{\lambda_1 s^2 + \lambda_2 s + \lambda_3}{s^2} I_n \tag{6.19}$$

Thus it has a a double pole with zero real part and it cannot be a PR transfer matrix, see theorem 2.10. This can also be checked by calculating $\langle u_2, y_2 \rangle$ that contains a term $\int_0^t u_2(s) z_1(s) ds$ which cannot be lower bounded.
If one chooses $u_2' = \tilde{q}$ then the PID block transfer matrix becomes

$$H_{pid}(s) = \frac{\lambda_1 s^2 + \lambda_2 s + \lambda_3}{s} I_n \tag{6.20}$$

which this time is a PR transfer function for a suitable choice of the gains, and one can check that

$$\langle u_2^{'}, y_2 \rangle = \frac{\lambda_1}{2}[\tilde{q}^T\tilde{q}]_0^t + \lambda_2 \int_0^t \tilde{q}^T\tilde{q}(s)ds + \frac{\lambda_3}{2}\left[\int_0^t \tilde{q}^T\tilde{q}(s)ds\right]_0^t \tag{6.21}$$

which shows that the system is even input strict passive (but the transfer function is not SPR otherwise this system would be strictly passive (in the state space sense), see example 4.7, which is is not from inspection of (6.21)). However this change of input is suitable for the PID block, but not for the rigid joint-rigid link block, that we know is not passive with respect to the supply rate $u_1^T\tilde{q}$ because of the relative degree of this output.
As a consequence the dynamics in (6.16) cannot be analyzed through the passivity theorem.

**Remark 6.3** Notice however that the system in (6.16) can be shown to be locally Lyapunov stable [11] with a Lyapunov function $V(z)$, where $z(t) = \begin{pmatrix} \int_0^t \tilde{q}(s)ds \\ \tilde{q}(t) \\ \dot{q}(t)) \end{pmatrix}$. Let us add a fictitious input $\tau$ in the right-hand-side of (6.16) instead of zero. From the KYP lemma we know that there exists an output $y$ (another fictitious signal) such that this closed-loop is passive with respect to the supply rate $\tau^T y$. One has $y = (0,0,1)\frac{\partial V}{\partial z}$.

### 6.2.3 SPR control for stabilization of Euler-Lagrange systems

The foregoing subsections as well as the passivity theorem suggest that any feedback controller that consists of a SPR function with input $u_2 = \dot{q}$ and output $y_2 = -\tau$ is going to stabilize the system in (5.45), provided the potential energy satisfies the requirements that assure the passivity of this dynamics with supply rate $\tau^T\dot{q}$, see lemma 5.1 (in particular the function $E(q,\dot{q}) = K(q,\dot{q}) + U(q)$ should be bounded from below). As we know, asymptotic stability of a fixed point can be obtained if a system is zero-state detectable and output strictly passive, provided its storage function has this fixed point as a minimum, see lemma 4.6 and corollary 4.4. Let us investigate the application of this result for the stabilization of a general Euler-Lagrange system. Let us denote the transfer function of the controller as $H(s)$, and a minimal representation for it as $\dot{z}_1 = Az_1 + Bu_2$, $y_2 = Cz_1 + Du_2$. In the case of a SPR control block, the closed-loop dynamics is:

$$\begin{cases} M(q)\ddot{q} + C(q,\dot{q})\dot{q} + g(q) = Cz_1 + D\dot{q} \\ \dot{z}_1 = Az_1 + B\dot{q} \end{cases} \tag{6.22}$$

where the fixed point is given by $z_1 = 0$, $\dot{q} = 0$ and $g(q^\star) = 0$. In the case of the PR control block one gets:

$$\begin{cases} M(q)\ddot{q} + C(q,\dot{q})\dot{q} + g(q) = Cz_1 + D\dot{q} \\ \\ \dot{z}_1 = B\dot{q} \end{cases} \tag{6.23}$$

and the fixed point is given by $\dot{q} = 0$, $z_1$ constant and $g(q^\star)$ constant.
In order to get an output strictly passive system, one can choose an input strictly passive controller: for instance a transfer $H(s)$ with relative degree $n^* = 0$, and with $A$ strictly Hurwitz (in which case $H(s)$ is SSPR and is input strictly passive) can do the job; for example take $H(s) = \frac{s+a^2}{s+b^2}I_n$. Then the overall system is indeed output strictly passive if the first block is passive, since $\langle u, y\rangle = \langle u_1, y_1\rangle + \langle u_2, y_2\rangle \geq -\gamma_1^2 - \gamma_2^2 + \delta \int_0^t u_2^T u_2 ds$ and $u_2 = y$. It remains to study the zero-state detectability property of the feedback system, i.e. to look at the state $(x, q, \dot{q})$ evolution when $u \equiv y \equiv 0$. First note that $(u_2 \equiv 0) \Rightarrow (z_1 \to 0) \Rightarrow (y_2 \to 0)$. Notice that this result does not hold for the system in (6.12), since in that case $A$ is not Hurwitz (it is the zero matrix). Thus we get $\frac{dU}{dq}(q) = u_1 \to 0$, from which it follows that if $U(q)$ has a unique critical point $q = q_0$, then $q \to q_0$. Hence the closed-loop system with state $(z_1, q - q_0, \dot{q})$ is zero-state detectable. Now a storage function is the sum $E(q, \dot{q}) + z_1^T P z_1$, with $PA + A^T P = -Q$ for some $Q > 0$, and Lyapunov stability of $(z_1, q - q_0, \dot{q}) = (0, 0, 0)$ follows.

**Remark 6.4**

- As pointed out earlier, the potential energy can be shaped by introducing a first feedback $\tau = -\lambda_2(q - q_d) + \tau'$ in order to modify the critical point $q_0$ value.

- The PD transfer function is input strictly passive but it is not SSPR. What did we gain by replacing $H_{pd}(s)$ by a SSPR system?. Actually it can be shown (see 4.1) that since $A$ is strictly Hurwitz, $u_2 \in \mathcal{L}_2(I\!R^+) \Rightarrow (z_1 \to 0 \text{ as } t \to +\infty)$ and $z_1 \in \mathcal{L}_2(I\!R^+) \cap \mathcal{L}_\infty(I\!R^+)$. Hence $y_2$ is the sum of two $\mathcal{L}_2$-bounded signals, thus is $\mathcal{L}_2$-bounded itself. We therefore end up with a closed-loop with all input and output signals $\mathcal{L}_2$-bounded, nothing more. There is little gain with respect to the PD except perhaps for some robustness purposes since one may choose to vary the bandwidth of $H(s)$ depending on noise measurement on $\dot{q}$. The most important improvement is rather related to the zero-state-detectability that the SPR condition introduces, that allows one to get rid of the initial condition choice as discussed in remark 6.4. This may introduce us to the following problem: the design of stabilizing controllers that are based on dissipativity properties is simpler (for mechanical systems) if one has access to the velocity. In practice one would often like to measure the position only, because only displacement sensors are available (actually in practice one often just differentiates the position signal, but for the sake of theoretical completeness, we may want to see how disspativity may be used in case of position measurements only). How can one use dissipativity theory in such framework? We shall present some results in this direction in section 6.3.

- It is clear that this result applies to any passive system, hence to the case of Euler-Lagrange mechanical systems including actuator dynamics. Only the input and the output definitions change.

### 6.2.4 More about Lyapunov functions and the passivity theorem

Before going on with controllers that assure tracking of arbitrary (smooth enough) desired trajectories, let us investigate in more detail the relationships between Lyapunov stable systems and the passivity theorem (which has numerous versions, but is always based on the study of the interconnection of two dissipative blocks). From the study we made about the closed-loop dynamics of the PD and PID controllers, it follows that if one has been able to transform a system (should it be open or closed-loop) as in figure 3.3 and such that both blocks are dissipative, then the sum of the respective storage functions of each block is a suitable Lyapunov function candidate. Now one might like to know whether a Lyapunov stable system possesses some dissipativity properties. More precisely, we would like to know whether a system that possesses a Lyapunov function, can be interpreted as the interconnection of two dissipative subsystems. Let us state the following [112] [31]:

**Lemma 6.3** Let $\mathcal{L}$ denote a set of Lyapunov stable systems with equilibrium point $(x_1, x_2) = (0, 0)$, where $(x_1, x_2)$ generically denotes the state of systems in $\mathcal{L}$. Suppose the Lyapunov function $V(x_1, x_2, t)$ satisfies

1. $$V(x_1, x_2, t) = V_1(x_1, t) + V_2(x_2, t) \tag{6.24}$$

   where $V_1$,$V_2$ are positive definite functions

2. $$\dot{V}(x_1, x_2, t) \leq -\gamma_1\beta_1(\| x_1 \|) - \gamma_2\beta_2(\| x_2 \|) \tag{6.25}$$

   along trajectories of systems in $\mathcal{L}$, where $\beta_1$ and $\beta_2$ are class $K$ functions, and $\gamma_1 \geq 0$, $\gamma_2 \geq 0$.

Suppose there exist functions $F_1$ and $F_2$ such that for all $x_1$, $x_2$ and $t \geq t_0$

$$\frac{\partial V_1}{\partial t} + \frac{\partial V_1}{\partial x_1}^T F_1(x_1, t) \leq -\gamma_1\beta_1(\| x_1 \|) \tag{6.26}$$

$$\frac{\partial V_2}{\partial t} + \frac{\partial V_2}{\partial x_2}^T F_2(x_1, t) \leq -\gamma_2\beta_1(\| x_2 \|) \tag{6.27}$$

and $f_i(0,t) = 0$, dim $x_i'$=dim $x_i$ for $i = 1,2$, for all $t \geq t_0$. Then there exists a set $\mathcal{P}$ of Lyapunov stable systems, with the same Lyapunov function $V(x_1', x_2', t)$, that can be represented as the feedback interconnection of two (strictly) passive subsystems with states $x_1'$ and $x_2'$ respectively. These systems are defined as follows

$$\left\{ \begin{array}{l} \dot{x'}_1 = F_1(x_1', t) + G_1(x_1', x_2', t) u_1 \\ \\ y_1 = G_1^T(x_1', x_2', t) \frac{\partial V_1}{\partial x_1'} \end{array} \right. \tag{6.28}$$

$$\left\{ \begin{array}{l} \dot{x'}_2 = F_2(x_2', t) + G_2(x_1', x_2', t) y_1 \\ \\ y_2 = G_2^T(x_1', x_2', t) \frac{\partial V_2}{\partial x_2'} = -u_1 \end{array} \right. \tag{6.29}$$

where $G_1$ and $G_2$ are arbitrary smooth nonzero functions [1], which can be shown to define the inputs and the outputs of the interconnected systems.♠♠

The proof of lemma 6.3 is straightforward from the KYP property of the outputs of passive systems. Note that lemma 6.3 does not imply any relationship between the system in $\mathcal{L}$ and the system in $\mathcal{P}$ other than the fact that they both have the same Lyapunov function structure. That is why we used different notations for their states $(x_1', x_2')$ and $(x_1, x_2)$. We are now interested in establishing sufficient conditions allowing us to transform a system in $\mathcal{L}$ into a system in $\mathcal{P}$ having the particular form given in (6.28) (6.29). These conditions are discussed next. Suppose $\Sigma_L$ has the following form (notice that this is a *closed-loop* form):

$$\left\{ \begin{array}{l} \dot{x}_1 = F_1(x_1, t) + G_1(x_1, x_2, t) u_1 \\ \\ y_1 = h_1(x_1, t) = u_2 \end{array} \right. \tag{6.30}$$

$$\left\{ \begin{array}{l} \dot{x}_2 = F_2(x_2, t) + G_2(x_1, x_2, t) u_2 \\ \\ y_2 = h_2(x_2, t) = u_1 \end{array} \right. \tag{6.31}$$

From (6.25) we thus have:

$$\begin{array}{rl} \dot{V}(x_1, x_2, t) & = \frac{\partial V_1}{\partial t} + \frac{\partial V_1}{\partial x_1}^T F_1(x_1, t) + \frac{\partial V_2}{\partial t} + \frac{\partial V_2}{\partial x_2}^T F_2(x_2, t) \\ & \\ & + \frac{\partial V_1}{\partial x_1}^T G_1(x_1, x_2, t) h_2(x_2, t) + \frac{\partial V_2}{\partial x_2}^T G_2(x_1, x_2, t) \\ & \\ & \times h_1(x_1, t) \\ & \\ & \leq -\gamma_1 \beta_1(\| x_1 \|) - \gamma_2 \beta_2(\| x_2 \|) \end{array} \tag{6.32}$$

[1] We assume that the considered systems have 0 as a unique equilibrium point.

with inequalities (6.26)(6.27) satisfied for both systems in (6.30) (6.31).
Now let us rewrite $\Sigma_L$ in (6.30) (6.31) as follows (we drop the arguments for convenience; $u_1 = h_2(x_2)$, $u_2 = h_1(x_1)$):

$$\begin{cases} \dot{x_1} = (F_1 + G_1 u_1 - \bar{g}_1 \bar{u}_1) + \bar{g}_1 \bar{u}_1 \\ \bar{y}_1 = \bar{g}_1^T \frac{\partial V_1}{\partial x_1} = \bar{u}_2 \end{cases} \tag{6.33}$$

$$\begin{cases} \dot{x_2} = (F_2 + G_2 u_2 - \bar{g}_2 \bar{u}_2) + \bar{g}_2 \bar{u}_2 \\ \bar{y}_2 = \bar{g}_2^T \frac{\partial V_2}{\partial x_2} = -\bar{u}_1 \end{cases} \tag{6.34}$$

Notice that $\tilde{\Sigma}_L$ in (6.33) (6.34) and $\Sigma_L$ in (6.30) (6.31) strictly represent the same system. We have simply changed the definition of the inputs and of the outputs of both subsystems in (6.30) and (6.31). Then the following lemma is true:

**Lemma 6.4** Consider the closed-loop Lyapunov stable system $\Sigma_L$ in (6.30) (6.31), satisfying (6.32), with $F_1$ and $F_2$ satisfying (6.26) (6.27). A sufficient condition for $\Sigma_L$ to be able to be transformed into a system in $\mathcal{P}$ is that the following two inequalities are satisfied:

1.
$$\frac{\partial V_1}{\partial x_1}^T \left( G_1 h_2 + \bar{g}_1 \bar{g}_2^T \frac{\partial V_2}{\partial x_2} \right) \leq 0 \tag{6.35}$$

2.
$$\frac{\partial V_2}{\partial x_2}^T \left( G_2 h_1 - \bar{g}_2 \bar{g}_1^T \frac{\partial V_1}{\partial x_1} \right) \leq 0 \tag{6.36}$$

for some non-zero, smooth matrices $\bar{g}_1$, $\bar{g}_2$ of appropriate dimensions, and with

$$\begin{aligned} F_1(0,t) + G_1(0,x_2,t)u_1(0,x_2,t) - \bar{g}_1(0,x_2,t)\bar{u}_1(0,x_2,t) = 0 \\ \forall x_2, t \\ F_2(0,t) + g(x_1,0,t)u_2(x_1,0,t) - \bar{g}_2(x_1,0,t)\bar{u}_2(x_1,0,t) = 0 \\ \forall x_1, t \end{aligned} \tag{6.37}$$

♠♠

Notice that these conditions are sufficient only for transforming the system in $\mathcal{P}$, see remark 6.7.

**Proof** The proof of lemma 6.4 is straightforward. Inequalities (6.35) and (6.36) simply guarantee that $\frac{\partial V_i}{\partial x_i}(f_i + g_i u_i - \bar{g}_i \bar{u}_i) \leq -\gamma_i \beta_i(\| x_i \|)$, and (6.37) guarantees that $\dot{x}_i = f_i + g_i u_i - \bar{g}_i \bar{u}_i$ has $x_i = 0$ as equilibrium point. Thus $\tilde{\Sigma}_L$ is in $\mathcal{P}$. ♠♠

**Example 6.2** Consider the following system:

$$\begin{cases} \dot{x}_1 = F_1(x_1,t) + G_1(x_1,x_2,t)u_1 \\ y_1 = \frac{\partial V_1}{\partial x_1} = u_2 \end{cases} \tag{6.38}$$

$$\begin{cases} \dot{x}_2 = F_2(x_2,t) - G_1^T(x_1,x_2,t)u_2 \\ y_2 = \frac{\partial V_2}{\partial x_2} = u_1 \end{cases} \tag{6.39}$$

with $\frac{\partial V_i}{\partial t} + \frac{\partial V_i}{\partial x_i}^T f_i \leq -\gamma_i \beta_i(\| x_i \|)$, $\gamma_i \geq 0$, $f_i(0,t) = 0$ for all $t \geq t_0$, and $V$ satisfies (6.24) (6.25). Then we get along trajectories of (6.38) (6.39) , $\dot{V} = \dot{V}_1 + \dot{V}_2 \leq -\gamma_1\beta_1(\| x_1 \|) - \gamma_2\beta_2(\| x_2 \|)$. However the subsystems in (6.38) and (6.39) are not passive, as they do not verify the KYP property. The conditions (6.35) and (6.36) reduce to:

$$\frac{\partial V_1}{\partial x_1}^T G_1 \frac{\partial V_2}{\partial x_2} + \frac{\partial V_1}{\partial x_1}^T \bar{g}_1 \bar{g}_2^T \frac{\partial V_2}{\partial x_2} = 0 \tag{6.40}$$

as in this case $\frac{\partial V_1}{\partial x_1}^T G_1 h_2 = -\frac{\partial V_2}{\partial x_2}^T G_2 h_1$. Now choose $\bar{g}_1 = -G_1$, $\bar{g}_2 = 1$, $\bar{u}_1 = -\frac{\partial V_2}{\partial x_2}$, $\bar{u}_2 = -G_1^T \frac{\partial V_1}{\partial x_1}$: (6.40) is verified. ♠♠

In conclusion, the system in (6.38) (6.39) is not convenient because its outputs and inputs have not been properly chosen. By changing the definitions of the inputs and outputs of the subsystems in (6.38) (6.39), leaving the closed-loop system unchanged, we transform the system such that it belongs to $\mathcal{P}$. In most of the cases, the functions $g_i$, $h_i$ and $f_i$ are such that the only possibility for the equivalent systems in (6.33) (6.34) to be Lyapunov stable with Lyapunov functions $V_1$ and $V_2$ respectively is that $\bar{g}_i\bar{u}_i \equiv g_i u_i$, i.e. we only have to rearrange the inputs and the outputs to prove passivity.
From lemma 6.4 we can deduce the following result:

**Corollary 6.1** Consider the system in (6.30) (6.31). Assume (6.32) is satisfied, and that $\frac{\partial V_1}{\partial x_1}^T G_1 h_2 = -\frac{\partial V_2}{\partial x_2}^T G_2 h_1$ (let us denote this equality as the Cross Terms Cancellation Equality CTCE). Then **i)** If one of the subsystems in (6.30) or (6.31) is passive, the system in (6.30) (6.31) can be transformed into a system that belongs to $\mathcal{P}$. **ii)** If the system in (6.30) (6.31) is autonomous, it belongs to $\mathcal{P}$. ♠♠

**Proof**
Using the CTCE, one sees that inequalities in (6.35) (6.36) reduce either to:

$$\frac{\partial V_1}{\partial x_1}^T G_1 h_2 + \frac{\partial V_1}{\partial x_1}^T \bar{g}_1 \bar{g}_2^T \frac{\partial V_2}{\partial x_2} = 0 \tag{6.41}$$

or to

$$-\frac{\partial V_2}{\partial x_2}^T G_2 h_1 + \frac{\partial V_2}{\partial x_2}^T \bar{g}_2 \bar{g}_1^T \frac{\partial V_1}{\partial x_1} = 0 \qquad (6.42)$$

Suppose that the system in (6.31) is passive. Then $h_2 = G_2^T \frac{\partial V_2}{\partial x_2}$, thus it suffices to choose $\bar{g}_2 = G_2$, $\bar{g}_1 = -G_1$. If the system in (6.30) is passive, then $h_1 = G_1^T \frac{\partial V_1}{\partial x_1}$, and we can take $\bar{g}_2 = G_2$, $\bar{g}_1 = G_1$. The second part of the corollary follows from the fact that one has for all $x_1$ and $x_2$:

$$\frac{\partial V_1}{\partial x_1}^T G_1(x_1)h_2(x_2) = -\frac{\partial V_2}{\partial x_2}^T G_2(x_2)h_1(x_1) \qquad (6.43)$$

Then (6.30) (6.31) can be transformed into a system that belongs to $\mathcal{P}$. Necessarily $h_2 = G_2^T \frac{\partial V_2}{\partial x_2}$ and $h_1 = -G_1^T \frac{\partial V_1}{\partial x_1}$, or $h_2 = -G_2^T \frac{\partial V_2}{\partial x_2}$ and $h_1 = G_1^T \frac{\partial V_1}{\partial x_1}$, which correspond to solutions of (6.41) or (6.42) respectively. ♠♠

**Example 6.3** Throughout this chapter and chapter 7 we shall see several applications of lemmas 6.3 and 6.4. In particular it happens that the cancellation of cross terms in Lyapunov functions derivatives has been widely used for stabilization and almost systematically yields an interpretation *via* the passivity theorem. To illustrate those results let us reconsider the PD controller closed-loop dynamics in (6.6). Let us start from the knowledge of the Lyapunov function deduced from the storage function in (6.7). Letting $x_1 = (q, \dot{q})$ be the state of the rigid joint-rigid link dynamics and $x_2 = z_1$ be the state of the second subsystem in (6.12), one sees that the sum of the storage functions associated to each of these blocks forms a Lyapunov function that satisfies the conditions of lemma 6.3. Moreover the conditions of corollary 6.1 are satisfied as well, in particular the CTCE. Indeed from (6.13) we get (but the same could be done with the interconnection in (6.10))

$$\begin{aligned} \frac{\partial V_1}{\partial x_1}^T G_1(x_1)h_2(x_2) &= \left(g^T(q), \dot{q}^T M(q)\right) \begin{pmatrix} 0 \\ M^{-1}(q) \end{pmatrix} (-\lambda_2 \tilde{q}) \\ &= -\lambda_2 \dot{q}^T \tilde{q} \\ \frac{\partial V_2}{\partial x_2}^T G_2(x_2)h_1(x_1) &= \lambda_2 \tilde{q}^T \dot{q} \end{aligned} \qquad (6.44)$$

Hence the dynamics in (6.6) can effectively be interpreted as the negative feedback interconnection of two dissipative blocks. ♠♠

As another example consider theorem 4.14: notice that choosing the controller $u$ of the driving system as $u^T = -\left(L_{f_1} U(\zeta)\right)$ exactly corresponds to a CTCE. Hence the closed-loop system thereby constructed can be analyzed through the passivity theorem. This is the mechanism used in [112].

Such closed-loop interpretations of Lyapunov stable systems are not fundamental from a stability point of view, since the system is already known to be stable. However they have been widely used in the Systems and Control literature since they provide an elegant manner to analyze the closed-loop system (passivity is always thought of to shed some "physical" light on a stability analysis, and may help publishing a paper...). Moreover they may provide the designer with ideas linked to the properties of interconnections of passive systems. We shall illustrate again the application of lemmas 6.3 and 6.4 and corollary 6.1 in the sequel, see in particular sections 6.3, 6.5 and chapter 7.

### 6.2.5 Extensions of the PD controller for the tracking case

The tracking problem for the model in (5.108) can be easily solved using a linearizing feedback that renders the closed-loop system equivalent to a double integrator. Then all the classical machinery for linear systems can be applied. However we are not interested here in following this path. We would rather like to see how the PD control may be extended to the tracking case, i.e. how we can preserve and use the system dissipativity to derive a globally stable controller guaranteeing tracking of any sufficiently differentiable desired trajectory.

#### A first extension of the PD controller: the Paden and Panja scheme

The first idea is a direct extension of the PD structure, applying the control [151]:

$$\tau = M(q)\ddot{q}_d + C(q,\dot{q})\dot{q}_d + g(q) - \lambda_1\dot{\tilde{q}} - \lambda_2\tilde{q} \tag{6.45}$$

with $q_d \in C^2(I\!R^+)$. Setting $q_d$ constant one retrieves a PD controller with gravity compensation. The closed-loop system is given by:

$$M(q)\ddot{\tilde{q}} + C(q,\dot{q})\dot{\tilde{q}} + \lambda_1\dot{\tilde{q}} + \lambda_2\tilde{q} = 0 \tag{6.46}$$

This closed-loop dynamics resembles the one in (6.6). This motivates us to study its stability properties by splitting it into two subsystems as:

$$\left\{\begin{array}{l} M(q)\ddot{\tilde{q}} + C(q,\dot{q})\dot{\tilde{q}} = u_1 = -y_2 \\ \\ y_1 = \dot{\tilde{q}} = u_2 \end{array}\right. \tag{6.47}$$

and

$$\left\{\begin{array}{l} \dot{z}_1 = u_2 \\ \\ y_2 = \lambda_1 u_2 + \lambda_2 z_1 \\ \\ z_1(0) = q(0) - q_d(0) \end{array}\right. \tag{6.48}$$

Let us do the following (see lemma 5.4)

**Assumption 6.3** The matrix $C(q,\dot{q})$ is written in such a way that $\dot{M}(q,\dot{q}) - 2C(q,\dot{q})$ is skew-symmetric.

Then one computes that

$$\begin{aligned} \langle u_1, y_1 \rangle &= \int_0^t \dot{\tilde{q}}^T \left[ M(q)\ddot{\tilde{q}} + C(q,\dot{q})\dot{\tilde{q}} \right] ds \\ &= \frac{1}{2} \left[ \dot{\tilde{q}}^T M(q) \dot{\tilde{q}} \right]_0^t \\ &\geq -\frac{1}{2}\dot{\tilde{q}}(0)^T M(q(0))\dot{\tilde{q}}(0) \end{aligned} \tag{6.49}$$

and that

$$\begin{aligned} \langle u_2, y_2 \rangle &= \lambda_1 \int_0^t \dot{\tilde{q}}^T \dot{\tilde{q}} ds + \frac{1}{2} \left[ \tilde{q}(s)^T \tilde{q}(s) \right]_0^t \\ &\geq -\frac{1}{2}\tilde{q}(0)^T \tilde{q}(0) \end{aligned} \tag{6.50}$$

Notice that the second block is input strictly passive. Similarly to the PD controller analysis, one concludes that the dynamics in (6.46) can effectively be transformed into the interconnection of two passive blocks. We could also have deduced from lemma 6.4 that such an interconnection exists, checking that $\frac{1}{2}\dot{\tilde{q}}M(q)\dot{\tilde{q}} + \frac{1}{2}\lambda_2\tilde{q}^T\tilde{q}$ is a Lyapunov function for this system, whose derivative along the trajectories of (6.46) is semi-negative definite (i.e. $\gamma_1 = 0$ in lemma 6.3). However one cannot apply the Krasovskii-La Salle theorem to this system because it is not autonomous (the inertia and Coriolis matrices depend explicitly on time when the state is considered to be $(\tilde{q}, \dot{\tilde{q}})$). One has to resort to Matrosov's theorem to prove the asymptotic stability [151]. Equivalent representations (that are to be compared to the ones constructed for the PD control in subsection 6.2.1) are depicted in figures 6.3 and 6.4.

### The Slotine and Li controller

The above scheme has the advantage of being quite simple. However its extension to the adaptive case (when the inertia parameters are supposed to be unknown, one needs to introduce some on-line adaptation) is really not straightforward. One big challenge in the Robotics and Systems and Control fields during the eighties was to propose a feedback controller that guarantees tracking and which extends also to an adaptive version. Let us consider the following input [183] [168]:

$$\tau = M(q)\ddot{q}_r + C(q,\dot{q})\dot{q}_r + g(q) - \lambda_1 s \tag{6.51}$$

where $\dot{q}_r = \dot{q}_d - \lambda\tilde{q}$, $s = \dot{q} - \dot{q}_r = \dot{\tilde{q}} + \lambda\tilde{q}$. Introducing (6.51) into (5.108) one obtains

$$M(q)\dot{s} + C(q,\dot{q})s + \lambda_1 s = 0 \tag{6.52}$$

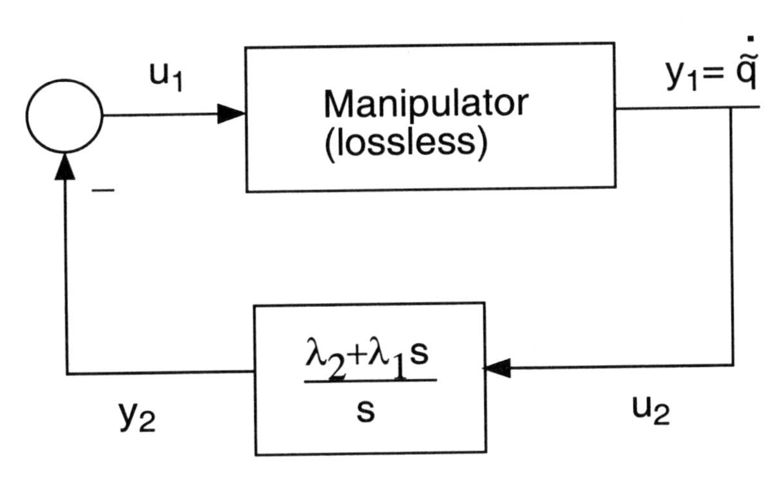


Figure 6.3: First interconnection: lossless manipulator dynamics.

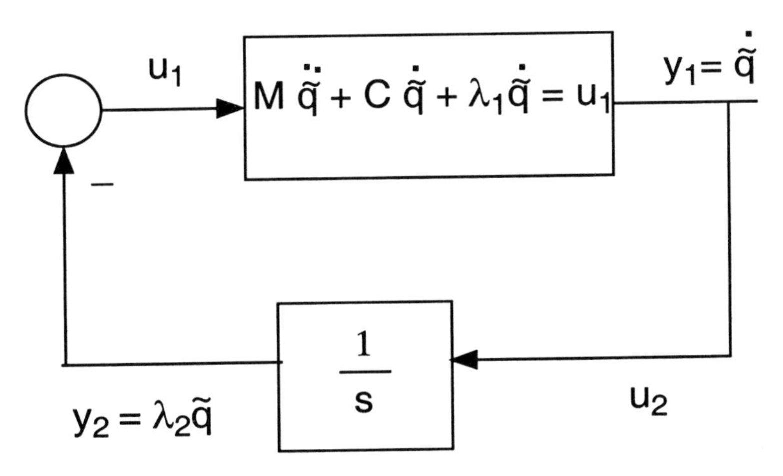


Figure 6.4: Second interconnection: OSP manipulator dynamics.

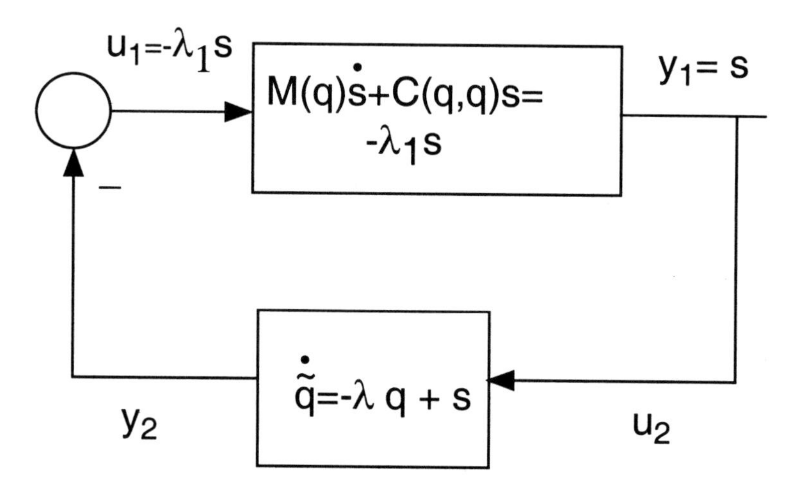


Figure 6.5: Closed-loop equivalent representation.

Notice that contrary to the scheme in (6.45), setting $q_d$ constant in (6.51) does not yield the PD controller. However the controller in (6.51) can be seen as a PD action ($\lambda s$) with additional nonlinear terms whose role is to assure some tracking properties. Before going on let us note that the whole closed-loop dynamics is not in (6.52) since this is an $n$th order system with state $s$, whereas the whole system is $2n$th order. To complete it one needs to add to (6.52):

$$\dot{\tilde{q}} = -\lambda \tilde{q} + s \tag{6.53}$$

It should clear from now all the foregoing developments that the subsystem in (6.52) defines a passive operator between $u_1 = -\lambda_1 s = -y_2$ and $y_1 = s = u_2$, with storage function $V_1(s) = \frac{1}{2} s^T M(q) s$ (which is a Lyapunov function for this subsystem which is zero-state observable). This is strongly based on assumption 6.3. The equivalent feedback interconnection of the closed-loop is shown in figure 6.5.

**Remark 6.5** The subsystem in (6.52) can be at once proved to define an asymptotically stable system since one can view it as the interconnection of a passive mapping $u \mapsto y = \dot{q}$ with zero-state detectable dynamics $M(q)\dot{s} + C(q,\dot{q})s = u$, with a static output feedback $u = -\lambda_1 y$. Hence theorem 4.9 applies and one concludes that $s \to 0$. ♠♠

The second subsystem obtained from (6.53) can be rewritten as

$$\begin{cases} \dot{z}_1 = -\lambda z_1 + u_2 \\ \\ y_2 = \lambda_1 u_2 \end{cases} \tag{6.54}$$

It therefore has a relative degree $n_2^* = 0$, and the state is not observable from the output $y_2$. However it is zero-state detectable since $\{y_2 = u_2 = 0\} \Rightarrow \lim_{t \to +\infty} z_1(t) = 0$. We also notice that this system is very strictly passive since

$$\begin{aligned} \langle u_2, y_2 \rangle &= \lambda_1 \int_0^t u_2^T u_2 ds \\ &= \tfrac{1}{\lambda_1} \int_0^t y_2^T y_2 ds \\ &= \tfrac{\lambda_1}{2} \int_0^t u_2^T u_2 ds + \tfrac{1}{2\lambda_1} \int_0^t y_2^T y_2 ds \end{aligned} \tag{6.55}$$

Let us compute storage functions for this system. Let us recall from (4.52) that for systems of the form $\dot{x} = f(x,t) + g(x,t)u, y = h(x,t) + j(x,t)u$ with $j(x,t) + j^T(x,t) = R$ full-rank, the storage functions are solutions of the differential inequation (that reduces to a Riccati inequation in the linear case)

$$\frac{\partial V}{\partial x}^T f(x,t) + \frac{\partial V}{\partial t} + \left(h^T - \frac{1}{2}\frac{\partial V}{\partial x}^T G\right) R^{-1} \left(h - \frac{1}{2} G \frac{\partial V}{\partial x}\right) \leq 0 \tag{6.56}$$

and that the available storage $V_a$ and the required supply $V_r$ (with $x(-t) = 0$) satisfy (6.56) as an equality. Thus from the fact that $n_2^* = 0$ we know that the storage functions $V(z_1)$ for the system in (6.54) are solutions of

$$-\lambda \frac{dV}{dz_1}^T z_1 + \frac{1}{2\lambda_1} \frac{dV}{dz_1}^T \frac{dV}{dz_1} \leq 0 \tag{6.57}$$

If we set the equality it follows that the two solutions satisfy

$$\begin{aligned} \tfrac{dV}{dz_1} &= 0 \\ \tfrac{dV}{dz_1} &= 2\lambda\lambda_1 z_1 \end{aligned} \tag{6.58}$$

from which one deduces that $V_a(z_1) = 0$ and $V_r(z_1) = \lambda\lambda_1 z_1^T z_1$, whereas any other storage function satisfies $0 = V_a(z_1) \leq V(z_1) \leq V_r(z_1)$. From the zero-state detectability property it follows that those functions are Lyapunov functions for the second subsystem, see lemma 4.6.

**Remark 6.6** Let us retrieve the available storage and the required supply from their variational formulations (notice that evidently the system in (6.54) is controllable so that the required supply can be defined):

$$V_a(z_1(0)) = \sup_{u_2:(0,z_1(0))\rightarrow} -\int_0^t \lambda_1 u_2^T u_2 ds = 0 \tag{6.59}$$

which means that the best strategy to recover energy from this system through the output $y_2$ is to leave it at rest (so as to recover nothing, actually!), and

$$\begin{aligned} V_r(z_1(0)) &= \inf_{u_2:(-t,0)\rightarrow(0,z_1(0))} \int_{-t}^0 u_2^T y_2 ds \\ &= \inf_{u_2:(-t,0)\rightarrow(0,z_1(0))} \lambda_1 \int_{-t}^0 \left\{(\dot{z}_1^T + \lambda z_1^T)(\dot{z}_1 + \lambda z_1)\right\} ds \\ &= \lambda_1 \lambda z_1^T(0) z_1(0) \end{aligned} \tag{6.60}$$

where the last step is performed by simple integration of the cross term and dropping the other two terms which are always positive, for any control strategy.
♠♠

We conclude that a suitable Lyapunov function for the closed-loop system in (6.52) (6.53) is given by the sum

$$V(s,\tilde{q},t) = \frac{1}{2} s^T M(\tilde{q},t) s + \lambda\lambda_1 \tilde{q}^T \tilde{q} \tag{6.61}$$

It is noteworthy that we have really deduced a Lyapunov function from the knowledge of some passivity properties of the equivalent interconnection form of the closed-loop system. Historically, the closed-loop system in (6.52) (6.53) has been studied first using the storage function of the first subsystem in (6.52) only, and then using additional arguments to prove the asymptotic convergence of the whole state towards zero [183]. It is only afterwards that the Lyapunov function for the whole closed-loop system has been proposed [189]. We have shown here that it is possible to construct it directly from passivity arguments. It must therefore be concluded on this example that the dissipativity properties allow one to directly find out the right Lyapunov function for this system.

**Remark 6.7** Lemmas 6.3 and 6.4 can in general be used if one starts from the knowledge of the Lyapunov function. However the cross-term-cancellation-equality (CTCE) is not satisfied since

$$\begin{aligned} \frac{\partial V_1}{\partial x_1}^T G_1(x_1) h_2(x_2) &= s^T M(q) M^{-1}(q) \lambda_1 s = \lambda_1 s^T s \\ \frac{\partial V_2}{\partial x_2}^T G_2(x_2) h_1(x_1) &= -\lambda\lambda_1 \tilde{q}^T s \end{aligned} \tag{6.62}$$

This comes from the fact that this time one has to add $\frac{\partial V_1}{\partial x_1}^T G_1(x_1)h_2(x_2) + \frac{\partial V_2}{\partial x_2}^T G_2(x_2)h_1(x_1) = -\lambda_1 s^T s + \lambda\lambda_1 \tilde{q}^T s$ to $\frac{\partial V_2}{\partial x_2}^T F_2(x_2) = -2\lambda^2\lambda_1 \tilde{q}^T \tilde{q}$ in order to get the inequality in (6.32). One may also check that the inequalities in (6.35) (6.36) can hardly be satisfied by any $\bar{g}_1$ and $\bar{g}_2$. Actually the conditions stated in lemma 6.4 and corollary 6.1 are sufficient only. For instance from (6.32) one can change the inequalities in (6.35) (6.36) to incorporate the terms $\frac{\partial V_1}{\partial x_1}^T F_1(x_1,t)$ and $\frac{\partial V_2}{\partial x_2}^T F_2(x_2,t)$ in the conditions required for the matrices $\bar{g}_1$ and $\bar{g}_2$. Actually lemmae 6.3 and 6.4 will be useful when we deal with adaptive control, see chapter 7, in which case the CTCE is generally satisfied.

**Remark 6.8** One major difference between what we can call *passivity-based* algorithms and feedback linearization schemes, is that the former more or less preserve the open-loop structure in the closed-loop dynamics, whereas the latter erase the original dynamics and ignore the open-loop physical properties. Although there is no rigorous proof of this, it has been widely argued that passivity-based schemes are more robust (with respect to which kind of uncertainties or neglected dynamics? This is not clear. But experimental results tend to prove that passivity-based schemes do possess interesting performance, see chapter 8).

**Remark 6.9** The main difference between the two schemes presented in subsection 6.2.5 is that the Paden and Panja controller reduces exactly to the PD controller+gravity compensation when $q_d$ is a constant signal, whereas the Slotine and Li scheme does not. This means that even for regulation, this scheme requires the knowledge of inertial parameters. However its stability analysis is much simpler since it does not rely upon advanced tools like Matrosov theorem.

### 6.2.6 Other types of state feedback controllers

#### Controllers using the skew-symmetry property

Let us replace the subsystem in (6.53) (whose input-output representation is given in (6.54)) by

$$\tilde{q}(p) = G(p)s(p) \tag{6.63}$$

where $p \in \mathbb{C}$ denotes the Laplace variable and $G(p)$ is a PR transfer function. In (6.53) one has $G(p) = \frac{1}{p+\lambda}$ (we do not use $s$ for the Laplace variable to avoid confusion with the signal $s(t)$ in (6.53)). Let us assume for the moment that the controller has been chosen such that the first closed-loop subsystem has the form in (6.52). Then a Lyapunov function for the whole system is given by the sum of $\frac{1}{2}s^T M(q)s$ and of any storage function of the dynamics associated to $G(p)$. Now it is easy to see that if the control signal $\dot{q}_r = \dot{q}_d - \lambda\tilde{q}$ is replaced by $\dot{q}_r = \dot{q}_d - K(p)\tilde{q}$ then $G(p) = [pI_n + K(p)]^{-1}$. One has therefore to choose $K(p)$ in a way such that $G(p)$ is PR.

**Remark 6.10** In fact the PR condition is too restrictive in this case. Indeed since it can be proved that $s \in \mathcal{L}_2(\mathbb{R}^+)$, it suffices that $G(p)$ be strictly proper and strictly stable to conclude that its output converges asymptotically towards zero, see theorem 4.1.

### Not relying on the skew-symmetry property

The use of the property in assumption 6.3 is not mandatory. Let us describe now a control scheme proposed in [76], that can be classified in the set of passivity-based control schemes, as will become clear after the analysis. Let us consider the following control input:

$$\begin{aligned} \tau = & -\tfrac{1}{2}\dot{M}(q,\dot{q})[\dot{\tilde{q}} + \lambda\tilde{q}] + M(q)[\ddot{q}_r - \lambda\dot{\tilde{q}}] + C(q,\dot{q})\dot{q} + g(q) \\ & - \left(\lambda_d + \tfrac{\lambda}{\lambda_1}\right)\dot{\tilde{q}} - \lambda\lambda_d\tilde{q} \end{aligned} \tag{6.64}$$

Introducing (6.64) into the dynamics (5.108) one obtains:

$$M(q)\dot{s} + \frac{1}{2}\dot{M}(q,\dot{q})s + \left(\lambda_d + \frac{\lambda}{\lambda_1}\right)\dot{\tilde{q}} + \lambda\lambda_d\tilde{q} = 0 \tag{6.65}$$

which we can rewrite equivalently as

$$M(q)\dot{s} + C(q,\dot{q})s + \left(\lambda_d + \frac{\lambda_1^2}{2}\right)\dot{\tilde{q}} + \frac{\lambda_d\lambda_2}{\lambda_1}\tilde{q} = -\frac{1}{2}\dot{M}(q,\dot{q})s + C(q,\dot{q})s \tag{6.66}$$

These two representations of the same closed-loop system are now analyzed from a "passivity theorem" point of view. Let us consider the following negative feedback interconnection:

$$\begin{cases} u_1 = -y_2 = -\tfrac{1}{2}\dot{M}(q,\dot{q})s + C(q,\dot{q})s \\ u_2 = y_1 = s \end{cases} \tag{6.67}$$

where the first subsystem has dynamics $M(q)\dot{s} + C(q,\dot{q})s + \left(\lambda_d + \frac{\lambda}{\lambda_1}\right)\dot{\tilde{q}} + \lambda_d\lambda\tilde{q} = u_1$ while the second one is a static operator between $u_2 = s$ and $y_2$ given by $u_2 = \frac{1}{2}\dot{M}(q,\dot{q})s - C(q,\dot{q})s$. It is easily checked that if assumption 6.3 is satisfied then

$$\langle u_2, y_2 \rangle = \frac{1}{2}\int_0^t s^T[\dot{M}(q,\dot{q}) - 2C(q,\dot{q})]s d\tau = 0 \tag{6.68}$$

and that the available storage of the second block is the zero function as well.

Concerning the first subsystem one has

$$\begin{aligned}\langle u_1, y_1\rangle &= \int_0^t s^T\left[M(q)\dot{s} + C(q,\dot{q})s + \left(\lambda_d + \tfrac{\lambda}{\lambda_1}\right)\dot{\tilde{q}} + \lambda_d\lambda\tilde{q}\right]d\tau \\ &= \tfrac{1}{2}[s^T M(q)s]_0^t + \tfrac{1}{2}\left(2\lambda\lambda_d + \tfrac{\lambda^2}{\lambda_1}\right)[\tilde{q}^T\tilde{q}]_0^t \\ &+ \int_0^t \left\{\left(\lambda_d + \tfrac{\lambda}{\lambda_1}\right)\dot{\tilde{q}}^T\dot{\tilde{q}} + \lambda^2\lambda_d\tilde{q}^T\tilde{q}\right\}d\tau \\ &\geq -\tfrac{1}{2}s(0)^T M(q(0))s(0) - \tfrac{1}{2}\left(2\lambda\lambda_d + \tfrac{\lambda^2}{\lambda_1}\right)\tilde{q}(0)^T\tilde{q}(0)\end{aligned} \tag{6.69}$$

which proves that it is passive with respect to the supply rate $u_1^T y_1$. It can also be calculated that the available storage function of this subsystem is given by:

$$\begin{aligned}V_a(\tilde{q}(0), s(0)) &= \sup_{u_1:[\tilde{q}(0),s(0)]\to} -\int_0^t s^T\{M(q)s + C(q,\dot{q})s \\ &+ \left(\lambda_d + \tfrac{\lambda}{\lambda_1}\right)\dot{\tilde{q}} + \lambda\lambda_d\tilde{q}\right\}d\tau \\ &= \tfrac{1}{2}s(0)^T M(q(0))s(0) + \left(\lambda\lambda_d + \tfrac{\lambda^2}{2\lambda_1}\right)\tilde{q}^T(0)\tilde{q}(0)\end{aligned} \tag{6.70}$$

Since this subsystem is zero-state detectable ($u_1 \equiv s \equiv 0 \Rightarrow \tilde{q} \to 0$ as $t \to +\infty$) one concludes that the available storage in (6.70) is actually a Lyapunov function for the corresponding unforced system, whose fixed point $(\tilde{q}, s) = (0,0)$ (or $(\tilde{q}, \dot{\tilde{q}}) = (0,0)$) is asymptotically stable. This also holds for the complete closed-loop system since the second block has storage functions equal to zero and the dynamics in (6.65) is zero-state detectable when one considers the input to be $u$ in the left-hand-side of (6.65) and $y = y_1 = s$ (set $u \equiv 0$ and $s \equiv 0$ and it follows from (6.65) that $\tilde{q} \to 0$ exponentially). Actually, the derivative of $V_a(\tilde{q}, s)$ in (6.70) along trajectories of the first subsystem is given by:

$$\dot{V}_a(\tilde{q}, s) = -\left(\lambda_d + \frac{\lambda}{\lambda_1}\right)\dot{\tilde{q}}^T\dot{\tilde{q}} - \lambda^2\lambda_d\tilde{q}^T\tilde{q} \leq 0 \tag{6.71}$$

It is noteworthy that the result in (6.71) can be obtained without using the skew-symmetry property in assumption 6.3 at all. But skew-symmetry was used to prove the dissipativity of each block in (6.67).

**Remark 6.11** Originally the closed-loop system in (6.65) has been proven to be Lyapunov stable using the Lyapunov function

$$V(\dot{\tilde{q}}, \tilde{q}) = \frac{1}{2}\dot{\tilde{q}}^T M(q)\dot{\tilde{q}} + \dot{\tilde{q}}^T M(q)\tilde{q} + \frac{1}{2}\tilde{q}^T[\lambda^2 M(q) + \lambda_1 I_n]\tilde{q} \tag{6.72}$$

which can be rearranged as

$$V(s, \tilde{q}) = \frac{1}{2}s^T M(q)s + \frac{1}{2}\lambda_1 \tilde{q}^T \tilde{q} \tag{6.73}$$

The derivative of $V$ in (6.72) or (6.73) along closed-loop trajectories is given by:

$$\dot{V} = -\dot{\tilde{q}}^T \left(\lambda_d + \frac{\lambda_1}{\lambda}\right) \dot{\tilde{q}} - 2\lambda_d \lambda \dot{\tilde{q}}^T \tilde{q} - \lambda^2 \lambda_d \tilde{q}^T \tilde{q} \tag{6.74}$$

Notice that $V_a$ in (6.70) and $V$ in (6.73) are not equal one to each other. One concludes that the passivity analysis of the closed-loop permits to discover a (simpler) Lyapunov function.

**Remark 6.12** The foregoing stability analysis does not use the CTCE of lemma 6.4. One concludes that the schemes that do not base on the skew-symmetry property in assumption 6.3 do not lend themselves very well to an analysis through the passivity theorem. We may however consider the controller in (6.64) to be passivity-based since it does not attempt at linearizing the system, similarly to the Slotine and Li scheme.

## 6.3 Rigid joint-rigid link: position feedback

The controllers we have analyzed until now use the measurements of both the position and the velocity of the systems. It is clear that since the dissipativity property of the mechanical structure holds between the generalized forces and the generalized velocity, the design of closed-loop systems that lend themselves well to an interconnection as two dissipative blocks is facilitated if the velocity is available in the control action. The PD, SSPR controls in subsections 6.2.1 and 6.2.3 use velocity feedback. However one should keep in mind that the studied interconnections are *only a way to interpret* the closed-loop dynamics. In other words, it is not because the interconnection considers $\dot{q}$ as the output of the mechanical dynamics and as the input of the control block, that $\dot{q}$ is necessarily used in the controller. On the other hand the PID in subsection 6.2.2 shows that it is not because the velocity is measured that the closed-loop nicely lends itself to an interconnection in two dissipative blocks. In view of this, it is interesting to investigate how dissipative controllers that use only output (position) feedback may be designed.
Usually most manipulators are equipped with position and velocity sensors, and controlled point-to-point with a PD. The tracking case requires more, as we saw. However the controllers structure becomes more complicated, hence less robust. It is of some interest to try to extend the separation principle for linear systems (a stable observer can be connected to a stabilizing controller without destroying the closed-loop stability), towards some classes of nonlinear systems. The rigid joint-rigid link manipulato case seems to constitute a good

candidate, due to its nice properties. At the same time such systems are nonlinear enough, so that the extension is not trivial. In the continuity of what has been done in the preceding sections, we shall investigate how the dissipativity properties of the Slotine and Li and of the Paden and Panja schemes can be used to derive (locally) stable controllers not using velocity feedback.
In the following we shall start by the regulation case (see subsection 6.3.1), and then analyze the tracking of trajectories (see subsections 6.3.2 and 6.3.3).

### 6.3.1 P + observer control

In this subsection we present the extension of the PD controller when the velocity is not available as done in [203] [16]. Basically the structure of output (position) feedback controllers is that of the original input where the velocity $\dot{q}$ is replaced by some estimated value. Let us consider the dynamics in (5.108) with the controller:

$$\begin{cases} \tau = g(q_d) - \lambda_1 \tilde{q} - \frac{1}{\lambda_2}(\tilde{q} - z) \\ \\ \dot{z} = \lambda_3(\tilde{q} - z) \end{cases} \tag{6.75}$$

so that the closed-loop dynamics is given by

$$\begin{cases} M(q)\ddot{q} + C(q,\dot{q})\dot{q} + g(q) - g(q_d) + \frac{1}{\lambda_2}(\tilde{q} - z) = -\lambda_1 \tilde{q} \\ \\ \dot{z} - \dot{q} = \lambda_3(\tilde{q} - z) - \dot{q} \end{cases} \tag{6.76}$$

Let us make now a direct application of corollary 6.1. Let us first rewrite (6.76) in a state-space form, with $x_1 = \begin{pmatrix} x_{11} \\ x_{12} \end{pmatrix} = \begin{pmatrix} \tilde{q} \\ \dot{q} \end{pmatrix}$ and $x_2 = \tilde{q} - z$. We obtain:

$$\begin{cases} \dot{x}_{11} & = x_{12} \\ \\ \dot{x}_{12} & = -M^{-1}(x_{11} + q_d)[C(x_{11} + q_d, x_{12})x_{12} + g(x_{11} + q_d) \\ \\ & -g(q_d) + \lambda_1 x_{11}] + M^{-1}(x_{11} + q_d)h_2 \\ \\ \dot{x}_2 & = -\lambda_3 x_2 + h_1 \\ \\ h_2 & = -\frac{1}{\lambda_2}x_2 \\ \\ h_1 & = x_{12} \end{cases} \tag{6.77}$$

The closed-loop scheme can be shown to be globally asymptotically Lyapunov stable with the Lyapunov function $V(x_{11}, x_{12}, x_2) = V_1(x_{11}, x_{12}) + V_2(x_2)$

defined as:

$$\begin{aligned} V_1(x_{11}, x_{12}) &= \lambda_2 \left[\tfrac{1}{2} x_{12}^T M(x_{11}+q_d) x_{12} + \tfrac{\lambda_1}{2} x_{11}^T x_{11} + U_g(x_{11}+q_d) \right. \\ &\left. -U_g(q_d) - x_{11}^T g(q_d)\right] \end{aligned} \tag{6.78}$$

and

$$V_2(x_2) = \frac{1}{2} x_2^T x_2 \tag{6.79}$$

It can be shown that $V_1$ is positive definite and has a global minimum at $(x_{11}, x_{12}) = (0,0)$ provided $\lambda_1 \geq \gamma$ where $\gamma$ is a Lipschitz constant for $g(\cdot)$. Differentiating $V$ along the trajectories of (6.76) or equivalently (6.77) one finds:

$$\dot{V} = -\lambda_3 x_2^T x_2 \tag{6.80}$$

where the CTCE is satisfied since $\frac{\partial V_1}{\partial x_1}^T G_1 h_2 = -x_{12}^T x_2 = -\frac{\partial V_2}{\partial x_2}^T G_2 h_1$. Since the system is autonomous, corollary 6.1 **ii)** applies. Now it is easy to see that the second subsystem with state vector $x_2$, input $u_2 = h_1$ and ouput $y_2 = -h_2$ is passive:

$$\begin{aligned} \langle u_2, y_2 \rangle &= \int_0^t \tfrac{1}{\lambda_2} x_2^T u_2 ds \\ &= \int_0^t \tfrac{1}{\lambda_2} x_2^T (\dot{x}_2 + \lambda_3 x_2) ds \\ &= \tfrac{1}{2\lambda_2} [x_2^T x_2]_0^t + \tfrac{\lambda_3}{\lambda_2} \int_0^t x_2^T x_2 ds \end{aligned} \tag{6.81}$$

and one recognizes a storage function equal to $\frac{1}{\lambda_2} V_2$ with $V_2$ in (6.79). Notice that the second subsystem (with state $x_2$) is strictly passive in the sense of lemma 4.4, but it is also output strictly passive. The other subsystem is defined with input $u_1 = -y_2$ and output $y_1 = u_2$ and is passive as one can check:

$$\begin{aligned} \langle u_1, y_1 \rangle &= \int_0^t x_{12}^T [M(x_{11}+q_d)\dot{x}_{12} + C(x_{11}+q_d, x_{12}) x_{12} \\ &+ g(x_{11}+q_d) - g(q_d) + \lambda_1 x_{11}] ds \\ &= V_1(t) - V_1(0) \end{aligned} \tag{6.82}$$

**Remark 6.13** .

- In connection with remark 6.4, let us note that this time the closed-loop scheme has an order strictly larger than the open-loop one.

- It is noteworthy that (6.82) does not define $V_1$ as it is given in (6.78): it just defines a storage function that is equal to $V_1$ plus a constant. If we had started by the passivity theorem, not *a priori* knowing that $V$ is a

Lyapunov function for the closed-loop system, we could state applying lemma 4.6 that

$$\begin{aligned} V^* &= \lambda_2 \left[\tfrac{1}{2} x_{12}^T M(x_{11}+q_d) x_{12}\right] + \tfrac{\lambda_1}{2} x_{11}^T x_{11} + U_g(x_{11}+q_d) \\ &-x_{11}^T g(q_d) + \tfrac{1}{2} x_2^T x_2 \end{aligned} \tag{6.83}$$

is a storage function, has a strict global minimum at $(q, x_{12}, x_2) = (q_d, 0, 0)$ so that $V = V^* - V^*(q_d, 0, 0)$ is a Lyapunov function for the system. Since $V^*(q_d, 0, 0) = U_g(q_d)$ one retrieves the definition of $V$ in (6.78) plus (6.79).

- The output strict passivity plus zero state detectability of the second block is important because it is precisely these properties that allow one to use the Krasovskii-La Salle theorem to prove the asymptotic stability.

**Remark 6.14** .

- In practice most of the time the velocity is deduced from position measurements *via* a crude differentiation. Actually this works well in most applications, see chapter 8, where it appears that the main problem is rather the measurement noise in the position signal. This does not preclude the observer design for nonlinear mechanical systems to be of significant theoretical interest, and it is worth presenting it in the body of passivity-based controllers.

- These observer-based schemes are non-trivial extensions of the state feedback controllers. We content ourselves here, as announced, to present the passivity interpretation of the closed-loop systems. It will become apparent that although their basic structure resembles that of the state feedback case, the position feedback controllers have storage functions which are not really evident to discover.

### 6.3.2 The Paden and Panja + observer controller

The material that follows is mainly taken from [17]. In fact it is to be expected that the separation principle does not extend completely to the nonlinear systems we deal with. Indeed the presented schemes assure local stability only (more exactly they assure semi-global stability, i.e. the region of attraction of the closed-loop fixed point can be arbitrarily increased by increasing some feedback gains). In what follows we shall not develop the whole stability proofs. We shall just focus on the passivity interpretation of the obtained closed-loop system, and in particular on the local stability that results from the fact that the storage function satisfies the dissipation inequality locally only.
The foregoing subsection was devoted to an extension of PD controllers and concerns global regulation around a fixed position only. It is of interest to

consider the tracking case which is, as one expects, much more involved due to the non-autonomy of the closed-loop scheme. Let us consider the following fixed parameter scheme (compare with the expression in (6.45)):

$$\textbf{Controller} \quad \left\{ \begin{array}{l} \tau = M(q)\ddot{q}_d + C(q,\dot{q}_0)\dot{q}_d + g(q) - \lambda_1(\dot{q}_0 - \dot{q}_r) \\ \\ \dot{q}_r = \dot{q}_d - \lambda_2 e \\ \\ \dot{q}_0 = \dot{\hat{q}} - \lambda_3 \tilde{q} \end{array} \right. \tag{6.84}$$

$$\textbf{Observer} \quad \left\{ \begin{array}{l} \dot{\hat{q}} = z + \lambda_4 \tilde{q} = z + (\lambda_6 + \lambda_3)\tilde{q} \\ \\ \dot{z} = \ddot{q}_d + \lambda_5 \tilde{q} = \ddot{q}_d + \lambda_6\lambda_3\tilde{q} \end{array} \right.$$

where $e = q - q_d(t)$ is the tracking error, $\tilde{q} = q - \hat{q}$ is the estimation error, $\lambda_i > 0$ for all $i = 1, \cdots, 6$. Let us denote $s_1 = \dot{q} - \dot{q}_r = \dot{e} + \lambda_2 e$ and $s_2 = \dot{q} - \dot{q}_0 = \dot{\tilde{q}} + \lambda_3 \tilde{q}$, so that $(\dot{q}_0 - \dot{q}_r) = s_1 - s_2$. Introducing (6.84) into (5.108) and using some properties of the matrix $C(q,\dot{q})$ (like the fact that $C(q,y)x = C(q,x)y$ and $C(q, z + \alpha x)y = C(q,z)y + \alpha C(q,x)y$ for all $x, y \in I\!R^n$ and $\alpha \in I\!R$) one gets the following error equation:

$$\left\{ \begin{array}{l} M(q)\ddot{e} + C(q,\dot{q})s_1 + \lambda_1 s_1 = \lambda_1 s_2 + C(q,\dot{q})\lambda_2 e - C(q,s_2)\dot{q}_d \\ \\ \dot{e} = -\lambda_2 e + s_1 \\ \\ M(q)\dot{s}_2 + C(q,\dot{q})s_2 + [\lambda_6 M - \lambda_1 I_n]s_2 = -\lambda_1 s_1 + C(q, s_2 - \dot{q})\dot{e} \\ \\ \dot{\tilde{q}} = -\lambda_3 \tilde{q} + s_2 \end{array} \right. \tag{6.85}$$

Define $K_1(q,e) = \lambda_2^2[2\frac{\lambda_1}{\lambda_2} - M(q)]$ and $K_2(q,\tilde{q}) = 2\lambda_3\lambda_1$. It can be shown using the positive definite function

$$\begin{array}{rl} V(e, s_1, \tilde{q}, s_2) & = \frac{1}{2}s_1^T M(q)s_1 + \frac{1}{2}e^T K_1(q,e)e + \frac{1}{2}s_2^T M(q)s_2 \\ & +\frac{1}{2}\tilde{q}^T K_2(q,\tilde{q})\tilde{q} \end{array} \tag{6.86}$$

that for a suitable choice of the initial data within a ball $B_r$ whose radius $r$ is directly related to the control gains, the closed-loop fixed point $(e, s_1, \tilde{q}, s_2) = (0,0,0,0)$ is (locally) exponentially stable. As pointed out above $r$ can actually be varied by varying $\lambda_6$ or $\lambda_1$, making the scheme semi-global. An intuitive decomposition of the closed-loop system in (6.85) is as follows, noting that $M(q)\ddot{e} = M(q)\dot{s}_1 - \lambda_2 M(q)e$:

$$\left\{ \begin{array}{lll} \bar{M}(q)\dot{s} + \bar{C}(q,\dot{q})s = u_1, & \dot{\tilde{q}} = -\lambda_2\tilde{q} + s_1, & \dot{e} = -\lambda_3 e + s_2 \\ \\ y_1 = s, \quad u_2 = y_1, & y_2 = -T(q,\dot{q},s) = -u_1 & \end{array} \right. \tag{6.87}$$

where

$$s = \begin{bmatrix} s_1 \\ s_2 \end{bmatrix} \tag{6.88}$$

$$T(q,\dot{q},s) = -\begin{bmatrix} \lambda_1 s_2 + \lambda_2 C(q,\dot{q})\dot{e} - C(q,\dot{q}_d)s_2 + \lambda_2 M(q)\dot{e} \\ \\ -\lambda_1 s_1 + C(q, s_2 - \dot{q})\dot{e} \end{bmatrix} \tag{6.89}$$

$$\bar{M}(q) = \text{diag}[M(q), M(q)] \tag{6.90}$$

$$\bar{C}(q,\dot{q}) = \text{diag}[C(q,\dot{q}), C(q,\dot{q})] \tag{6.91}$$

The first subsystem is clearly passive with respect to the supply rate $u_1^T y_1$. The second subsystem is a memoryless operator $u_2 \mapsto -T(q,\dot{q},u_2)$. If it can be shown that locally $-u_2^T T(q,\dot{q},u_2) \geq -\delta u_2^T u_2$, then the system with input $u = u_1 + y_2$ and output $y = y_1$ is output strictly passive. Indeed

$$\begin{aligned} \langle u, y\rangle = \langle u_1 + y_2, y\rangle &= \langle u_1, y_1\rangle + \langle y_2, u_2\rangle \\ &\geq -\tfrac{1}{2}s(0)^T \bar{M}(q(0))s(0) + \delta \int_0^t u_2^T u_2 ds \end{aligned} \tag{6.92}$$

for some $\delta > 0$. In other words the function in (6.86) satisfies the dissipation inequality along the closed-loop trajectories: $\frac{dV}{dx}^T [f(x) + g(x)u] \leq u^T h(x) - \delta h^T(x)h(x)$ for all $u$ and $x$ locally only, where $x^T = (e^T, s_1^T, \tilde{q}^T, s_2^T)$ and $y = h(x)$. Then under suitable zero-state detectability properties, any storage function which is positive definite with respect to the closed-loop fixed point is a strict (local) Lyapunov function. Notice that the total closed-loop system is zero-state detectable since $y_1 = s \equiv 0$ and $u \equiv 0$ implies that $y_2 \equiv 0$, hence $u_1 \equiv 0$ and $e \to 0$ and $\tilde{q} \to 0$ as $t \to +\infty$.

### 6.3.3 The Slotine and Li + observer controller

Let us consider the following fixed parameter scheme:

$$\textbf{Controller} \quad \begin{cases} \tau = M(q)\ddot{q}_r + C(q,\dot{q}_0)\dot{q}_r + g(q) - \lambda_1(\dot{q}_0 - \dot{q}_r) \\ \quad -\lambda_2 e \\ \dot{q}_r = \dot{q}_d - \lambda(\hat{q} - q_d) \\ \dot{q}_0 = \dot{\hat{q}} - \lambda(q - \hat{q}) \end{cases} \tag{6.93}$$

$$\textbf{Observer} \quad \begin{cases} \dot{\hat{q}} = z + \lambda_3(q - \hat{q}) \\ \dot{z} = \ddot{q}_r + \lambda_4(q - \hat{q}) + \lambda_2 M^{-1}(q)[q_d - \hat{q}] \end{cases}$$

Introducing (6.93) into (5.108) one obtains the closed-loop error equation:

$$\left\{\begin{array}{l} M(q)\dot{s}_1 + C(q,\dot{q})s_1 + \lambda_1 s_1 + \lambda_2 e = \lambda_1 s_2 - C(q,s_2)\dot{q}_r \\ \\ \dot{e} = -\lambda(e-\tilde{q}) + s_1, \quad \dot{\tilde{q}} = -\lambda\tilde{q} + s_2 \\ \\ M(q)\dot{s}_2 + C(q,\dot{q})s_2 + (\lambda_6 M(q) - \lambda_1 I_n)s_2 + \lambda_2\tilde{q} = \\ \\ \qquad -\lambda_2 s_1 + C(q,s_1)[s_2 - \dot{q}] \end{array}\right. \tag{6.94}$$

with $\lambda_3 = \lambda_6 + \lambda$, $\lambda_4 = \lambda_6\lambda$. Again a natural decomposition of the closed-loop scheme is similarly done as in the previous case, i.e.

$$\left\{\begin{array}{ll} \bar{M}(q)\dot{s} + \bar{C}(q,\dot{q})s = u_1, & \dot{e} = -\lambda(e-\tilde{q}) + s_1, \dot{\tilde{q}} = -\lambda\tilde{q} + s_2 \\ \\ y_1 = s, u_2 = y_1, & y_2 = -T(q,\dot{q},s) = -u_1 \end{array}\right. \tag{6.95}$$

where this time

$$T(q,\dot{q},s) = \left[\begin{array}{c} \lambda_1 s_1 - [\lambda_1 + C(q, s_1 - \dot{q})]s_2 \\ \\ \lambda_1 - C(q, s_2 - \dot{q})s_1 + [\lambda_6 M(q) - \lambda_1 I_n]s_2 \end{array}\right] \tag{6.96}$$

It can be shown that locally $T(q,\dot{q},s) > 0$ so that $\langle u_2, y_2\rangle \geq \delta \int_0^t u_2(s)^T u_2(s)ds$ for some $\delta > 0$. The same conclusions as above follow about semi-global asymptotic Lyapunov stability of the closed-loop fixed point.

## 6.4 Flexible joint-rigid link: state feedback

### 6.4.1 A passivity-based controller

In section 5.4 we saw how the dissipativity properties derived for the rigid joint-rigid link manipulator case extend to the flexible joint-rigid link case, and we presented what we called *passivity-based* schemes. Considering the Lyapunov function in (6.61) let us try the following [29] [113] [111]:

$$\begin{array}{rl} V(\tilde{q}_1,\tilde{q}_2,s_1,s_2) & = \frac{1}{2}s_1^T M(q_1)s_1 + \frac{1}{2}s_2^T J s_2 + \lambda\lambda_1\tilde{q}_1^T\tilde{q}_1 + \lambda\lambda_1\tilde{q}_2\tilde{q}_2 \\ & +\frac{1}{2}\left(\tilde{q}_1 - \tilde{q}_2\right)^T K\left(\tilde{q}_1 - \tilde{q}_2\right) \end{array} \tag{6.97}$$

The various signals have the same definition as in the rigid case. One sees that similarly to (6.61) this positive definite function mimics the total energy function of the open-loop unforced system. In order to make it a Lyapunov function for the closed-loop system, one can classically compute its derivative along the trajectories of (5.115) and try to find out a $u$ that makes its derivative negative definite. Since we already have analyzed the rigid joint-rigid link case, we can intuitively guess that one goal is to get a closed-loop system of the form

$$\left\{\begin{array}{l} M(q_1)\dot{s}_1 + C(q_1,\dot{q}_1)s_1 + \lambda_1 s_1 = f_1(s_1,s_2,\tilde{q}_1,\tilde{q}_2) \\ \\ J\dot{s}_2 + \lambda_1 s_2 = f_2(s_1,s_2,\tilde{q}_1,\tilde{q}_2) \end{array}\right. \tag{6.98}$$

For the moment we do not fix the functions $f_1(\cdot)$ and $f_2(\cdot)$. Since the Lyapunov function candidate preserves the form of the system's total energy, it is also to be strongly expected that the potential energy terms appear in the closed-loop dynamics. Moreover we desire that the closed-loop system consists of two passive blocks in negative feedback. Obviously $V$ in (6.97) contains the ingredients for lemmas 6.3 and 6.4 to apply. The first block may be chosen with state vector $x_1 = \begin{pmatrix} \tilde{q}_1 \\ s_1 \\ \tilde{q}_2 \\ s_2 \end{pmatrix}$. We know it is passive with respect to the supply rate $u_1^T y_1$ with input $u_1 = (s_1 s_2)$ and output $y_2 = \begin{pmatrix} K(\tilde{q}_1 - \tilde{q}_2) \\ -K(\tilde{q}_1 - \tilde{q}_2) \end{pmatrix}$. One storage function for this subsystem is

$$V_1(x_1,t) = \frac{1}{2}s_1^T M(q_1)s_1 + \frac{1}{2}s_2^T J s_2 + \lambda\lambda_1 \tilde{q}_1^T \tilde{q}_1 + \lambda\lambda_1 \tilde{q}_2 \tilde{q}_2 \tag{6.99}$$

However notice that we have not fixed the input and output of this subsystem, since we leave for the moment $f_1(\cdot)$ and $f_2(\cdot)$ free. Now the second subsystem must have a storage function equal to:

$$V_2(x_2,t) = \frac{1}{2}(\tilde{q}_1 - \tilde{q}_2)^T K (\tilde{q}_1 - \tilde{q}_2) \tag{6.100}$$

which we know also is passive with respect to the supply rate $u_2^T y_2$, with an input $u_2 = y_1$ and an output $y_2 = -u_1$, and from (6.100) a state vector $x_1 = K(\tilde{q}_1 - \tilde{q}_2)$. Its dynamics is consequently given by

$$\dot{x}_2 = -\lambda x_2 + K(s_2 - s_1) \tag{6.101}$$

In order for lemmas 6.3 and 6.4 to apply we also require the CTCE to be satisfied, i.e. $\frac{\partial V_1}{\partial x_1}^T G_1 h_2 = -\frac{\partial V_2}{\partial x_2}^T G_2 h_1$, where we get from (6.98)

$$s_1^T f_1 + s_2^T f_2 = -(\tilde{q}_2 - \tilde{q}_1)^T K(s_2 - s_1) \tag{6.102}$$

from which one deduces that $f_2(s_1,s_2,\tilde{q}_1,\tilde{q}_2) = K(\tilde{q}_1 - \tilde{q}_2)$ and $f_1(s_1,s_2,\tilde{q}_1,\tilde{q}_2) = K(\tilde{q}_2 - \tilde{q}_1)$. Thus since we have fixed the input and output of the second subsystem so as to make it a passive block, we can deduce from lemma 6.4 that the closed-loop system that consists of the feedback interconnection of the dynamics in (6.98) and in (6.101) can be analyzed through the passivity theorem.

Notice however that we have not yet checked whether a state feedback exists that assures this closed-loop form. This is what we develop now. Let us consider the following controller:

$$\begin{cases} u = J\ddot{q}_{2r} + K(q_{2d} - q_{1d}) - \lambda_1 s_2 \\ \\ q_{2d} = K^{-1}u_r + q_{1d} \end{cases} \tag{6.103}$$

where $\dot{q}_{2r} = \dot{q}_{2d} - \lambda\tilde{q}_2$ and $u_r$ is given by the rigid joint-rigid link controller in (6.51), i.e.

$$u_r = M(q_1)\ddot{q}_r + C(q_1, \dot{q}_1)\dot{q}_r + g(q_1) - \lambda_1 s_1 \tag{6.104}$$

It is noteworthy that the controller is thus formed of two controllers similar to the one in (6.51): one for the first "rigid link" subsystem and the other for the motorschaft dynamics. The particular form of the interconnection between them makes it possible to pass from the first dynamics to the second one easily. It should be noted that the form in (6.103) (6.104) depends on the state $(\tilde{q}_1, s_1, \tilde{q}_2, s_2)$ only, and not on any acceleration nor jerk terms.

**Remark 6.15** .

- It is possible to replace the potential energy terms in (6.97) by

$$\left(\int_0^t [s_1 - s_2]d\tau\right)^T K \left(\int_0^t [s_1 - s_2]d\tau\right) \tag{6.105}$$

  This does not modify significantly the structure of the scheme, apart from the fact that this introduces a dynamic state feedback term in the control loop. Actually as shown in [29] the static state feedback scheme has the advantage over the dynamic one of not constraining the initial conditions on the open-loop state vector and on $q_{1d}(0)$, $\dot{q}_{1d}(0)$ and $\ddot{q}_{1d}(0)$. Note that the function $V(t, \tilde{q}_1, s_1, \tilde{q}_2, s_2)$ in (6.97) with (6.105) is not a positive definite function of the vector $(\tilde{q}_1, s_1, \tilde{q}_2, s_2)$: indeed $V(t, 0, 0, 0, 0)$ is not necessarily zero for all $t \geq 0$, but it can take any positive value depending on the evolution of $s_1(t) - s_2(t)$. However it is a Lyapunov function for the extended state $(\tilde{q}_1, s_1, \tilde{q}_2, s_2, z)$ with $z = \int_0^t [s_1(\tau) - s_2(\tau)]d\tau$.

- A strong property of the controller in (6.103) (6.104) in closed-loop with the dynamics in (5.115), with the Lyapunov function in (6.97), is that they converge towards the closed-loop system in (6.52) (6.53) when $K \to +\infty$ (all the entries diverge). Indeed one notices that $K(q_{2d} - q_{1d}) = u_r$ for all $K$ and that $q_{2d} \to q_{1d}$ as $K \to \infty$. Noting that all the closed-loop signals remain uniformly bounded for any $K$ and introducing these results into $u$ in (6.103) one sees that $u = J\ddot{q}_r + u_r - \lambda_1 s_1$ which is exactly the controller in (6.51) applied to the system in (5.115), letting $q_1 \equiv q_2$ and adding both subsystems. We therefore have constructed a real family of controllers that share some fundamental features of the plant dynamics.

### 6.4.2 Other globally tracking feedback controllers

#### A recursive method for control design

As pointed out one may also view the passivity-based controller in (6.103) as the result of a procedure that consists of stabilizing first the rigid part of the dynamics, using the signal $q_{2d}(t)$ as a fictitious intermediate input, and then looking at the rest of the dynamics. However instead of looking at the rest as a whole and considering it as a passive second order subsystem, one may treat it step by step: this is the core of a popular method known under the name of *backstepping*. Let us develop it now for the flexible joint-rigid link manipulators.

- **Step 1:**

  Any type of globally stabilizing controller can be used. Let us still use $u_r$ in (6.104), i.e. let us set:

$$q_{2d} = K^{-1}u_r + q_1 \tag{6.106}$$

  so that we get

$$M(q_1)\dot{s}_1 + C(q_1, \dot{q}_1)s_1 + \lambda_1 s_1 = K\tilde{q}_2 \tag{6.107}$$

  The system in (6.107) with $\tilde{q}_2 \equiv 0$ thus defines a globally uniformly asymptotically stable system with Lyapunov function $V_1(\tilde{q}_1, s_1) = \frac{1}{2}s_1^T M(q_1)s_1 + \lambda\lambda_1\tilde{q}_1^T\tilde{q}_1$. The interconnection term is therefore quite simple (as long as the stiffness matrix is known!). Let us take its derivative to obtain

$$\dot{\tilde{q}}_2 = \dot{q}_2 - \dot{q}_{2d} = \dot{q}_2 + f_1(q_1, \dot{q}_1 q_2) \tag{6.108}$$

  where $f_1(\cdot)$ can be computed using the dynamics (actually $\dot{q}_{2d}$ is a function of the acceleration $\ddot{q}_1$ which can be expressed in terms of $q_1$, $\dot{q}_1$ and $q_2$ by simply inverting the first dynamical equation in (5.115)).

- **Step 2:**

  Now if $\dot{q}_2$ was the input we would set $\dot{q}_2 = -f_1(q_1, \dot{q}_1 q_2) - \lambda_2\tilde{q}_2 - Ks_1$ so that the function $V_2 = V_1 + \frac{1}{2}\tilde{q}_2^T\tilde{q}_2$ has a negative definite derivative along the partial closed-loop system in (6.107) and

$$\dot{\tilde{q}}_2 = -\lambda_2\tilde{q}_2 - Ks_1 \tag{6.109}$$

However $\dot{q}_2$ is not an input, so that we shall rather define a new error signal as $e_2 = \dot{q}_2 - e_{2d}$, with $e_{2d} = -f_1(q_1, \dot{q}_1 q_2) - \lambda_2 \tilde{q}_2 - K s_1$. One obtains:

$$\begin{aligned} \dot{e}_2 &= \ddot{q}_2 - \dot{e}_{2d} = \ddot{q}_2 + f_2(q_1, \dot{q}_1, q_2, \dot{q}_2) \\ &= J^{-1}[K(q_1 - q_2) + u] + f_2(q_1, \dot{q}_1, q_2, \dot{q}_2) \end{aligned} \tag{6.110}$$

- **Step 3:** Since the real control input appears in (6.110) this is the last step. Let us choose

$$u = K(q_2 - q_1) + J[-f_2(q_1, \dot{q}_1, q_2, \dot{q}_2) - e_2 - \tilde{q}_2] \tag{6.111}$$

so that we get:

$$\dot{e}_2 = -\lambda_3 e_2 - \tilde{q}_2 \tag{6.112}$$

where the term $-\tilde{q}_2$ has been chosen to satisfy the CTCE (see lemma 6.4) when the function $V_2$ is augmented to

$$V_3(\tilde{q}_1, s_1, \tilde{q}_2, e_2) = V_2 + \frac{1}{2} e_2^T e_2 \tag{6.113}$$

Then along the closed-loop trajectories of the system in (6.107) (6.98) (6.112) one gets:

$$\dot{V}_3(\tilde{q}_1, s_1, \tilde{q}_2, e_2) = -\lambda_1 \dot{\tilde{q}}_1^T \dot{\tilde{q}}_1 - \lambda^2 \lambda_1 \tilde{q}_1^T \tilde{q}_1 - \tilde{q}_2^T \tilde{q}_2 - e_2^T e_2 \tag{6.114}$$

which shows that this closed-loop system is globally uniformly exponentially stable.

**Remark 6.16** .

- It is noteworthy that $e_2$ is not the time derivative of $q_2$. Therefore the backstepping method hinges upon a state variable transformation which actually depends on the system dynamics in the preceding steps.
- The control law in (6.111) can be computed from the definition of $q_{2d}$ in (6.106) and $\dot{q}_{2d}$ as well as $\ddot{q}_{2d}$ are to be calculated using the dynamics to express the acceleration $\ddot{q}_1$ and the jerk $q_1^{(3)}$ as functions of positions and velocities only (take the first dynamical equation in (5.115) and invert it to get the acceleration. Differentiate it again and introduce the expression obtained for the acceleration to express the jerk). Clearly $u$ is a complicated nonlinear function of the state, but it is a static state feedback. This apparent complexity is shared by all the nonlinear controllers described in section 6.4. Notice however that after all, it is only a matter of additions and multiplications, nothing else!

- We noticed in remark 6.15 that the passivity-based controller tends towards the Slotine and Li input when the joint stiffness tends to infinity. This is no longer the case with the backstepping controller derived here. Even more, after some manipulations, it can be shown [34] that the controller in (6.111) can be equivalently rewritten as:

$$\left\{ \begin{array}{l} u = J[\ddot{q}_{2d} - (\lambda_2 + \lambda_3)\dot{\tilde{q}}_2 - (1 + \lambda_2\lambda_3)\tilde{q}_2 - K(\dot{s}_1 + s_1)] \\ \\ q_{2d} = K^{-1}u_r + q_1 \end{array} \right. \tag{6.115}$$

where it immediately appears that the term $K(\dot{s}_1 + s_1)$ is not bounded as $K$ grows without bound. Here comes into play the "flexibility" of the backstepping method: let us modify the function $V_2$ above to $V_2 = V_1 + \frac{1}{2}\tilde{q}_2^T K \tilde{q}_2$. Then in step 2 it is sufficient to choose $\dot{q}_2 = -f_1(q_1, \dot{q}_1 q_2) - \lambda_2\tilde{q}_2 - s_1$, so that the final input becomes

$$\left\{ \begin{array}{l} u = J[\ddot{q}_{2d} - (\lambda_2 + \lambda_3)\dot{\tilde{q}}_2 - (1 + \lambda_2\lambda_3)\tilde{q}_2 - (\dot{s}_1 + s_1)] \\ \\ q_{2d} = K^{-1}u_r + q_1 \end{array} \right. \tag{6.116}$$

Such a modification may appear at first sight quite innocent, easy to do, and very slight: it is not! The experimental results presented in chapter 8 demonstrate it. Actually the term $K(\dot{s}_1 + s_1)$ introduces a high-gain in the loop that may have disastrous effects. This may be seen through simulations, see [34]. It is noteworthy that even with quite flexible systems (some of the reported experiments were led with a system whose stiffness is $k = 3.5$ Nm/rad) this term makes the control law in (6.111) behave less satisfactorily than the one in (6.116). More details can be found in chapter 8.

- This recursive design method applies to all systems that possess a triangular structure [119].

- Compare (6.115) and (6.116) to (6.103). Although these controllers have the same degree of complexity and can be considered as similar, they have significant discrepencies as explained above. For instance in (6.103) one has $K(q_{2d} - q_{1d}) = u_r$ while in (6.115) (6.116), $K(q_{2d} - q_{1d}) = u_r + \tilde{q}_1$.

### A passivity theorem interpretation

As we pointed out in the foregoing subsubsection, the procedure relies on the CTCE at each step. Since the first subsystem in (6.107) is output strictly passive with respect to the supply rate $u_1^T y_1$ with $u_1 = K\tilde{q}_2$ and $y_1 = s_1$, we are tempted to apply the result of lemmas 6.3 and 6.4 to interpret the closed-loop scheme in (6.107) (6.98) (6.112) as an interconnection of passive blocks. From the developments concerning the rigid joint-rigid link case we know that the first subsystem can be seen as the interconnection of two passive blocks in (6.52) and (6.54). However now the first subsystem is passive when the input is changed to $u_1 = K\tilde{q}_2 - \lambda_1 s_1$. We shall therefore define four subsystems as follows:

$$
\begin{cases}
(H1)\begin{cases}
(H11): & M(q_1)\dot{s}_1 + C(q_1,\dot{q}_1)s_1 = K\tilde{q}_2 - \lambda_1 s_1 \\
& u_{11} = K\tilde{q}_2 - \lambda_1 s_1, y_{11} = s_1, \text{state } s_1 \\
(H12): & \dot{\tilde{q}}_1 = -\lambda_1\tilde{q}_1 + s_1 \\
& u_{12} = s_1, y_{12} = \lambda_1 s_1, \text{state } \tilde{q}_1
\end{cases} \\
(H2)\begin{cases}
(H21): & \dot{\tilde{q}}_2 = -\lambda_2\tilde{q}_2 + e_2 - Ks_1 \\
& u_{21} = e_2 - Ks_1, y_{21} = \tilde{q}_2, \text{state } \tilde{q}_2 \\
(H22): & \dot{e}_2 = -\lambda_3 e_2 - \tilde{q}_2 \\
& u_{22} = \tilde{q}_2, y_{22} = -e_2, \text{state } e_2
\end{cases}
\end{cases}
\tag{6.117}
$$

Then the closed-loop system can be viewed as the negative feedback interconnection of the block $(H1)$ with $u_1 = u_{11} + y_{12} = K\tilde{q}_2$, $y_1 = y_{11}$, with the block $(H2)$ with input $-K^{-1}u_2 = s_1 = y_1$ and output $-Ky_2 = -K\tilde{q}_2 = -u_1$. This is depicted in figure 6.6.

**Remark 6.17** Consequently the backstepping procedure also yields a closed-loop system that can be analyzed through the passivity theorem. However the major difference with the passivity-based method is that the block $(H2)$ is not related to any physical relevant energetical term. In a sense this is similar to what one would get by linearizing the rigid joint-rigid link dynamics, applying a new linear feedback so as to impose some second order linear dynamics which may define an "artificial" passive system.

# 6.5 Flexible joint-rigid link: output feedback

## 6.5.1 PD control

We have seen in subsection 6.2.1 that a PD controller stabilizes globally and asymptotically rigid joint-rigid link manipulators. It is a combination of passivity and detectability properties that makes such a result hold: the former is a guide for the choice of a Lyapunov function, while the latter allows the Krasovskii-La Salle invariance principle to apply. More precisely, the output strict passivity property is crucial, because output strict passivity together

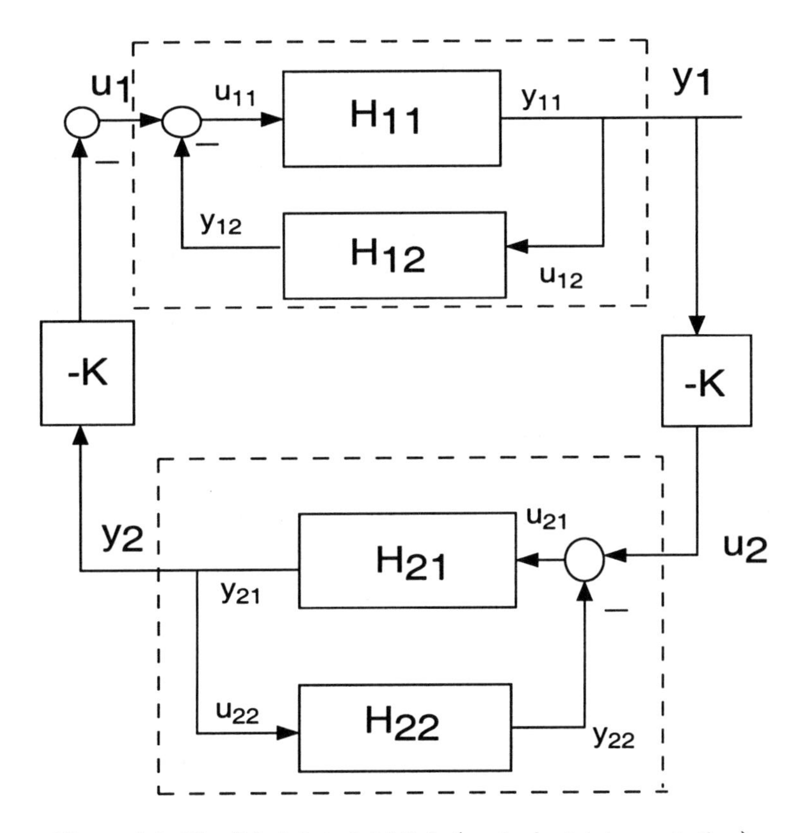


Figure 6.6: Flexible joint-rigid link (equivalent interpretation).

with zero-state detectability of a system, imply its asymptotic stability in the sense of Lyapunov (see corollary 4.4). Let us consider the dynamics in (5.115) and the following controller:

$$u = -\lambda_1 \dot{q}_2 - \lambda_2 (q_2 - q_d) \tag{6.118}$$

with $q_d$ a constant signal, so that the closed-loop system is given by

$$\begin{cases} M(q_1)\ddot{q}_1 + C(q_1, \dot{q}_1)\dot{q}_1 + g(q_1) = K(q_2 - q_1) \\ J\ddot{q}_2 + \lambda_1 \dot{\tilde{q}}_2 + \lambda_2 (q_2 - q_d) = K(q_1 - q_2) \end{cases} \tag{6.119}$$

Let us proceed as for the rigid joint-rigid link case, i.e. let us first "guess" a Lyapunov function candidate from the available storage function, and then show how the application of the passivity theorem applies equally well.

**Remark 6.18** When we deal with state feedback tracking control, we are really looking for a controller. We therefore cannot use the same strategy as here, i.e. propose a controller and then propose to show the closed-loop stability afterwards. What we are doing with the PD control is a sort of verification of its closed-loop stability.

### The closed-loop available storage

Similarly as for the rigid joint-rigid link case, one may guess that a PD controller alone will not enable one to stabilize any fixed point. The closed-loop fixed point is given by

$$\begin{aligned} g(q_1) &= K(q_2 - q_1) \\ \lambda_2 (q_2 - q_d) &= K(q_1 - q_2) \end{aligned} \tag{6.120}$$

and we may assume for simplicity that this set of nonlinear equations (which are not in general algebraic but transcendental) possesses a unique root $(q_1, q_2) = (q_{10}, q_{20})$. We aim at showing the stability of this point. One has

$$\begin{aligned} & V_a(\tilde{q}_1, \dot{q}_1, \tilde{q}_2, \dot{q}_2) \\ & = \sup_{u:(0,\tilde{q}_1(0),\dot{q}_1(0),\tilde{q}_2(0),\dot{q}_2(0))\rightarrow} - \int_0^t \dot{q}_2^T u ds \\ & = \sup_{u:(0,\tilde{q}_1(0),\dot{q}_1(0),\tilde{q}_2(0),\dot{q}_2(0))\rightarrow} - \int_0^t u^T [J\ddot{q}_2 + K(q_2 - q_1) + \lambda_1 \dot{q}_2 + \lambda_2 \tilde{q}_2] ds \\ & = \tfrac{1}{2}\dot{q}_1(0)^T M(q_1(0))\dot{q}_1(0) + U_g(q_1(0)) + \tfrac{1}{2}\dot{q}_2(0)^T J \dot{q}_2(0) \\ & + \tfrac{1}{2}\lambda_2 \tilde{q}_2^T(0)\tilde{q}_2 + \tfrac{1}{2}(q_2 - q_1)^T K(q_2 - q_1) \end{aligned} \tag{6.121}$$

where $\tilde{q}_i = q_i - q_{i0}$, $i = 1, 2$. Now the supply rate satisfies $w(0, \dot{q}_2) \leq 0$ for all $\dot{q}_2$, and obviously $(\tilde{q}_1, \dot{q}_1, \tilde{q}_2, \dot{q}_2) = (0, 0, 0, 0)$ is a strict (global) minimum of $V_a$ in (6.121) provided $U_g(q_1)$ has a strict minimum at $q_{10}$. Notice that $\tilde{q}_2 = 0 \Rightarrow (q_1 - q_2) = 0 \Rightarrow g(q_1) = 0$ so that $q_1 = q_{10}$ is a critical point for $U_g(q_1)$ (that we might assume to be strictly globally convex, but this is only sufficient). Hence from lemmae 4.6 and 4.7 one deduces that the closed-loop system in (6.119) is Lyapunov stable. To show asymptotic stability, one has to resort to Krasovskii-La Salle invariance principle.

### Closed-loop feedback interconnections

Motivated by the rigid joint-rigid link case let us look for an equivalent feedback interconnection such that the overall system is strictly output passive and zero-state detectable. To this end let us consider the following two blocks

$$\begin{cases} u_1 = K(q_1 - q_2), \quad y_1 = \dot{q}_2 \\ \\ u_2 = y_1, \qquad\qquad\quad y_2 = -u_1 \end{cases} \tag{6.122}$$

where the first block has the dynamics $J\ddot{q}_2 + \lambda_1 \dot{q}_2 + \lambda_2(q_2 - q_d) = K(q_1 - q_2)$, while the second one has the dynamics $M(q_1)\ddot{q}_1 + C(q_1, \dot{q}_1)\dot{q}_1 + g(q_1) = K(q_2 - q_1)$. It is easy to calculate the following:

$$\begin{aligned} \langle u_1, y_1 \rangle \ & \geq -\tfrac{1}{2}\dot{q}_2(0)^T J \dot{q}_2(0) - \lambda_2 (q_2(0) - q_d)^T (q_2(0) - q_d) \\ & +\lambda_1 \int_0^t \dot{q}_2(s)^T \dot{q}_2(s) ds \\ \langle u_2, y_2 \rangle \ & \geq -\tfrac{1}{2}[q_1(0) - q_2(0)]^T K [q_1(0) - q_2(0)] \\ & -\tfrac{1}{2}\dot{q}_1(0)^T M(q_1(0)) \dot{q}_1(0) - U_g(q_1(0)) \end{aligned} \tag{6.123}$$

from which one deduces that the first block is output strictly passive (actually if we added Rayleigh dissipation in the first dynamics, the second block would not be output strictly passive with the proposed decomposition). Each block possesses its own storage functions which are Lyapunov functions for them. The concatenation of these two Lyapunov functions forms the available storage in (6.121). Let us now consider the overall system with input $u = u_1 + y_2$ and output $y = y_1$. Setting $u \equiv y \equiv 0$ implies $\tilde{q}_2 \equiv 0$ and $\dot{q}_1 \to 0$, $\tilde{q}_1 \to 0$ asymptotically. The system is zero-state detectable. Hence by lemmae 4.6 and 4.7 its fixed point is globally asymptotically Lyapunov stable.

**Remark 6.19 (Collocation)** The *collocation* of the sensors and the actuators is an important feature for closed-loop stability, as we saw in section 2.17. It is clear here that if the PD control is changed to:

$$u = -\lambda_1 \dot{\tilde{q}}_1 - \lambda_2 \tilde{q}_1 \tag{6.124}$$

then the above analysis no longer holds. It can even be shown that there are some gains for which the closed-loop system is unstable [190]. One choice for the location of the sensors may be guided by the passivity property between their output and the actuators torque (in case the actuator dynamics is neglected). ♠♠

### 6.5.2 Motor position feedback

A position feedback controller similar to the one in subsection 6.3.1 can be derived for flexible joint-rigid link manipulators [15]. It may be seen as a PD controller with the velocity feedback replaced by an observer feedback. It is given by:

$$\begin{cases} u = g(q_d) - \lambda_1 \tilde{q}_2 - \frac{1}{\lambda_2}(\tilde{q}_2 - z) \\ \\ \dot{z} = \lambda_3(\tilde{q}_2 - z) \end{cases} \tag{6.125}$$

with $\tilde{q}_2 = q_2 - q_d + K^{-1}g(q_d)$, and $q_d$ is the desired position for $q_1$. The analysis is quite close to the one done for the rigid joint-rigid link case. Due to the autonomy of the closed-loop ($q_d$ is constant) corollary 6.1 is likely to apply. The stability proof bases on the following global Lyapunov function:

$$\begin{aligned} V(\tilde{q}_1, \dot{q}_1 \quad \tilde{q}_2, \dot{q}_2) = \lambda_2 \big( & \tfrac{1}{2}\dot{q}_1^T M(q_1)\dot{q}_1 + \tfrac{1}{2}\dot{q}_2^T J \dot{q}_2 + \tfrac{1}{2}\tilde{q}_1^T K \tilde{q}_1 \\ & + \tfrac{1}{2}\tilde{q}_2^T (K + \lambda_1 I_n)\tilde{q}_2 \big) - 2\lambda_2 \tilde{q}_1^T K \tilde{q}_2 + \tfrac{1}{2}(\tilde{q}_2 - z)^T(\tilde{q}_2 - z) \end{aligned} \tag{6.126}$$

Compare with $V = V_1 + V_2$ in (6.78) (6.79): the structure of $V$ in (6.126) is quite similar. It is a positive definite function provided $K + \frac{dg(q)}{dq}(q_d) > 0$ and $\lambda_1 I_n + K - K\left(K + \frac{dg(q)}{dq}(q_d)\right)^{-1} > 0$, for all $q_d$. This implies that $K$ and $\lambda_1$ are sufficiently large. The decomposition into two subsystems as in (6.77) can be performed, choosing $x_2 = \tilde{q}_2 - z$ and $x_1^T = (\tilde{q}_1^T, \dot{q}_1^T, \tilde{q}_2^T, \dot{q}_2^T) = (x_{11}^T, x_{12}^T, x_{13}^T, x_{14}^T)$. The closed-loop scheme is given by:

$$\begin{cases} \dot{x}_{11} = x_{12} \\ \\ \dot{x}_{12} = -M(x_{11} + q_d)[C(x_{11} + q_d, x_{12})x_{12} + K(x_{11} - x_{12}) \\ \\ \quad +g(x_{11} + q_d) - g(q_d)] \\ \\ \dot{x}_{13} = x_{14} \\ \\ \dot{x}_{14} = J^{-1}[K(x_{11} - x_{13}) - g(q_d) - \lambda_1 x_{13} - \frac{1}{\lambda_2} x_2] \\ \\ \dot{x}_2 = -\lambda_3 x_2 + x_{14} \end{cases} \tag{6.127}$$

Define $h_2 = \frac{1}{\lambda_2}x_2$ and $h_1 = x_{14}$ it follows that the CTCE is satisfied since $\frac{\partial V_1}{\partial x_1}^T G_1 h_2 = -x_{14}^T x_2 = -\frac{\partial V_2}{\partial x_2}^T G_2 h_1$. Indeed one may calculate that $G_1^T = (0, 0, 0, J^{-1}) \in \mathbb{R}^{n \times 4n}$ whereas $G_2 = I_n$. Hence once again corollary 6.1 applies and the closed-loop system can be interpreted *via* the passivity theorem.

**Remark 6.20** .

- Battilotti et al [16] have presented a result that allows one to recast the dynamic position feedback controllers presented in this subsection and in subsections 6.3 into the same general framework. It bases on passifiability and detectability properties. The interpretation of the P + observer schemes in subsections 6.3.1 and 6.5.2 *via* corollary 6.1 is however original.

- It is also possible to derive a globally stable P + observer controller using only the measurement of $q_1$ [16]. Its structure is however more complex than the above one. This is easily understandable since in this case the actuators and sensors are non-collocated.

# 6.6 Including actuator dynamics

## 6.6.1 Armature-controlled DC motors

We have seen in section 5.6 that the available storage of the interconnection between the rigid joint-rigid link manipulator model and the armature-controlled DC motor is given by

$$V_a(q, \dot{q}, I) = \frac{1}{2}I^T L I + \frac{1}{2}\dot{q}^T M(q)\dot{q} + U_g(q) \tag{6.128}$$

Motivated by the method employed for the design of stable controllers for rigid joint-rigid link and flexible joint-rigid link manipulators, let us consider the following positive definite function:

$$V(\tilde{q}, s, \tilde{I}) = \frac{1}{2}\tilde{I}^T L \tilde{I} + \frac{1}{2}s^T M(q)s + +2\lambda\lambda_1\tilde{q}^T\tilde{q} \tag{6.129}$$

where $s = \dot{\tilde{q}} + \lambda\tilde{q}$. Let us consider the dynamics in (5.148) which we recall here for convenience

$$\begin{cases} M(q)\ddot{q} + C(q, \dot{q})\dot{q} + g(q) = \tau = K_t I \\ RI + L\frac{dI}{dt} + K_t\dot{q}dt = u \end{cases} \tag{6.130}$$

Let us set

$$I_d = K_t^{-1}[M(q)\ddot{q}_r + C(q, \dot{q})\dot{q}_r + g(q) - \lambda_1 s] \tag{6.131}$$

where $s = \dot{q} - \dot{q}_r$, so that the manipulator dynamics in (5.148) become

$$M(q)\dot{s} + C(q,\dot{q})s + \lambda_1 s = K_t \tilde{I} \tag{6.132}$$

where $\tilde{I} = I - I_d$. Then it is easy to see that the control input

$$u = RI - k_v \dot{q} + L^{-1}\dot{I}_d - L^{-1}K_t s - \tilde{I} \tag{6.133}$$

(which is a state feedback) leads to

$$\dot{\tilde{I}} = -\tilde{I} + L^{-1}K_t s \tag{6.134}$$

Taking the derivative of $V(\tilde{q}, s, \tilde{I})$ in (6.129) along closed-loop trajectories in (6.132) (6.134) one gets:

$$\dot{V}(\tilde{q}, s, \tilde{I}) = -\tilde{I}^T L \tilde{I} - \lambda_1 \dot{\tilde{q}}^T \dot{\tilde{q}} - \lambda^2 \lambda_1 \tilde{q}^T \tilde{q} \tag{6.135}$$

showing that the closed-loop fixed point $(\tilde{q}, s, \tilde{I}) = (0,0,0)$ is globally asymptotically uniformly stable in the sense of Lyapunov.

**Remark 6.21 (Regulation of cascade systems)** Consider the system in (6.130) with Rayleigh dissipation in the manipulator dynamics. Let us write the second subsystem in (6.130) as

$$\dot{I} = -L^{-1}RI - L^{-1}K_t\dot{q} + L^{-1}u \tag{6.136}$$

Let $L^{-1}u = L^{-1}K_v\dot{q} + \mathbf{u}$ so that we obtain the cascade system

$$\begin{cases} M(q)\ddot{q} + C(q,\dot{q})\dot{q} + g(q) + \frac{dR}{d\dot{q}} = K_t y \\ \dot{I} = -L^{-1}RI + \mathbf{u} \\ y = I \end{cases} \tag{6.137}$$

The terms corresponding to 4.72 can be easily identified by inspection. One sees that the conditions of theorem 4.14 are satisfied (provided the potential energy $U(q)$ satisfies the requirements of assumption 6.1), so that this (partially) closed-loop system is feedback equivalent to a strictly passive system. In other words there exists a feedback input $\mathbf{u} = \alpha(I,q,\dot{q}) + \mathbf{v}$ such that there exists a positive definite function $V(I,q,\dot{q})$ of the fixed point $(I,q,\dot{q}) = (0,0,0)$ and a positive definite function $S(I,q,\dot{q})$ such that

$$V(t) - V(0) = \int_0^t \mathbf{v}^T y ds - \int_0^t S(I,q,\dot{q}) ds \tag{6.138}$$

Thus the unforced system (i.e. take $\mathbf{v} = 0$) has a globally asymptotically stable fixed point (in the sense of Lyapunov).
A similar analysis for the field-controlled DC motor case can be led. The dissipativity properties of the driven and the driving subsystems allow the designer to construct a globally stabilizing feedback law.

**Remark 6.22 (Nested passive structure)** The computation of $\dot{V}$ relies on a CTCE as required in lemma 6.4 and corollary 6.1. Thus if we had started from the *a priori* knowledge of the function $V$ we could have deduced that the closed-loop system can be analyzed as the negative feedback interconnection of two passive blocks, one with input $u_1 = K_t\tilde{I}$ and output $y_1 = s$ and dynamics in (6.132), the second one with dynamics in (6.134) and $u_2 = y_1$, $y_2 = -u_1$. Recall from section 6.2.5 that the first subsystem can be in turn decomposed as a negative feedback interconnection of two passive blocks given in (6.52) and (6.53): the overall system therefore possesses a structure of nested negative feedback interconnections of passive systems.

### 6.6.2 Field-controlled DC motors

Let us recall the model of rigid joint-rigid link manipulators in cascade with a field-controlled DC motor:

$$\begin{cases} L_1\frac{dI_1}{dt} + R_1 I_1 = u_1 \\ L_2\frac{dI_2}{dt} + R_2 I_2 + K_t(I_1)\dot{q} = u_2 \\ M(q)\ddot{q} + C(q,\dot{q})\dot{q} + g(q) + K_{vt}\dot{q} = \tau = K_t(I_1)I_2 \end{cases} \tag{6.139}$$

The regulation problem around the constant fixed points $(q,\dot{q},I_1,I_2) = (q_0, 0, I_{1d}, 0)$ or $(q_0, 0, 0, I_{2d})$ is solvable, where $q_0$ is as in assumption 6.1. Indeed the subsystem can be seen as a cascade system as in 4.72 that satisfies the requirements of theorem 4.14. Hence it is feedback equivalent to a strictly passive system (in the sense of theorem 4.12), whose unforced version is Lyapunov globally asymptotically stable. One remarks that the tracking control problem is quite similar to that of the flexible joint-rigid link manipulators with torque input. However this time the matrix that premultiplies $I_2$ is no longer constant invertible. Actually $K_t(I_1)$ may pass through singular values each time $I_{1i} = 0$ for some $i \in \{1, \cdots, n\}$. The extension of the regulation case is therefore not trivial. Nevertheless if the goal is to track a reference trajectory for $(q,\dot{q})$ only, then one may keep $I_1$ constant such that $K_t(I_1)$ remains full-rank, through a suitable $u_{21}$, so that the armature-controlled DC motor case is recovered.

**Remark 6.23** All the preceding developments apply to flexible joint-rigid link manipulators. Notice also that induction motors have the same complexity as field-controlled DC motors for control since the generated torque for each motor is given by $\tau = L_{sr}(I_2 I_3 - I_1 I_4)$ , see remark 5.20 for details.

## 6.7 Constrained mechanical systems

In real robotic tasks, the manipulators seldom evolve in a space free of obstacles. A general task may be thought as involving free-motion as well as constrained motion phases. In this section we shall focus on the case when the system is assumed to be in a permanent contact with some environment. In other words the constraint between the controlled system and the obstacle is supposed to be bilateral. In all the sequel we assume that the potential energy of the controlled system $U_g(z)$ and of the passive environment $U_{g_e}(z_1)$ each have a unique strict minimum, and to simplify further that they are positive (they have been chosen so).

### 6.7.1 Regulation with a position PD controller

Before going on with particular environment dynamics, let us analyze the regulation problem for the system in (5.194). To this end let us define the PD control

$$\bar{\tau} = -\lambda_2 \tilde{z} - \lambda_1 \dot{z} \tag{6.140}$$

where $\tilde{z} = z(t) - z_d$, $z_d$ a constant signal. Since we have assumed that the constraints are bilateral, we do not have to restrict $z_d$ to a particular domain of the state space (i.e. we do not care about the sign of the interaction force). Let us "invent" a Lyapunov function candidate by mimicing the available storage in (5.195), i.e.

$$\begin{aligned} V(\tilde{z}, \dot{z}, z_1) &= \tfrac{1}{2}\dot{z}^T M(z)\dot{z} + \tfrac{1}{2}\dot{z}_1 M_e(z_1)\dot{z}_1 \\ &\quad + \tfrac{1}{2}\lambda_2 \tilde{z}^T \tilde{z} + U_g(z) + U_{g_e}(z_1) + \tfrac{1}{2} z_1^T K_e z_1 \end{aligned} \tag{6.141}$$

Instead of computing the derivative of this function along the closed-loop system (5.194) (6.140), let us decompose the overall system into two blocks. The first block contains the controlled subsystem dynamics, and has input $u_1 = F_z = \begin{pmatrix} \lambda_z \\ 0 \end{pmatrix}$, output $y_1 = \dot{z}$. The second block has the dynamics of the environment, output $u_2 = -\lambda_z$ and input $u_2 = \dot{z}$. These two subsystems are passive since

$$\begin{aligned} \langle u_1, y_1 \rangle &= \int_0^t \dot{z}^T \left[\bar{M}(z)\ddot{z} + \bar{C}(z,\dot{z})\dot{z} + \bar{g}(z) + \lambda_2 \tilde{z} + \lambda_2 \dot{z}\right] ds \\ &\geq -\tfrac{1}{2} z(0)^T \bar{M}(z(0)) z(0) - U_g(z(0)) - \tfrac{1}{2}\lambda_2 z(0)^T z(0) \end{aligned} \tag{6.142}$$

and

$$\begin{aligned}
&\langle u_2, y_2 \rangle \\
&= \int_0^t \dot{z}_1^T \left[ M_e(z_1)\ddot{z}_1 + C_e(z_1, \dot{z}_1)\dot{z}_1 + \tfrac{dR_e}{d\dot{z}_1} + K_e z_1 + g_e(z_1) \right] ds \qquad (6.143) \\
&\geq -\tfrac{1}{2}\dot{z}_1(0)^T M_e(z_1(0))\dot{z}_1(0) - \tfrac{1}{2} z_1(0)^T K_e z_1(0) - U_{g_e}(z_1(0))
\end{aligned}$$

Now the inputs and outputs have been properly chosen so that the two subsystems are already in the required form for the application of the passivity theorem. Notice that they are both controllable and zero-state detectable from the chosen inputs and outputs. Therefore the storage functions that appear in the right-hand-sides of (6.142) and (6.143) are Lyapunov functions (see lemmae 4.6 and 4.7) and their concatenation is the Lyapunov function candidate in (6.141) which is a Lyapunov function. The asymptotic stability of the closed-loop system fixed point can be shown using the Krasovskii-La Salle theorem, similarly to the case of rigid joint-rigid link manipulators controlled by a PD feedback. Notice that similarly to (6.8) the fixed points are given as solutions of the following equation (obtained by summing the dynamics of the two subsystems)

$$\begin{pmatrix} K_e z_1 + \bar{g}_1(z) + g_e(z_1) + \lambda_2 \tilde{z}_1 \\ \\ \lambda_2 \tilde{z}_2 + \bar{g}_2(z) \end{pmatrix} = \begin{pmatrix} 0_{m\times 1} \\ \\ 0_{(n-m)\times 1} \end{pmatrix} \qquad (6.144)$$

We may assume that this equation has only one root $z = z_i$ so that the fixed point $(z, \dot{z}) = (z_i, 0)$ is globally asymptotically stable.

**Remark 6.24** It is noteworthy that this interpretation works well because the interconnection between the two subsystems satisfies Newton's principle of mutual actions. The open-loop system is therefore "ready" for a decomposition through the passivity theorem.

**Remark 6.25** Let us notice that there is no measurement of the environment state in (6.140). The coordinate change presented in section 5.8.2 just allows one to express the generalized coordinates for the controlled subsystem in a frame that coincides with a "natural" frame associated to the obstacle. It is clear however that the transformation relies on the exact knowledge of the obstacle geometry.

## 6.7.2 Holonomic constraints

Let us now analyze the case when $M_e(z_1)\ddot{z}_1 = 0$ and the contact stiffness $K_e$ and damping $R_e(\dot{z}_1)$ tend to infinity, in which case the controlled subsystem is subject to a bilateral holonomic constraint $\phi(q) = 0$ [2]. In the transformed coordinates $(z_1, z_2)$ the dynamics is given in (5.188), see subsection 5.8.1. We saw that the open-loop properties of the unforced system transport from the free-motion to the reduced constrained motion systems. Similarly, it is clear that any feedback controller that applies to the dynamics in (5.108) apllies equally well to the reduced order dynamics $(z_2, \dot{z}_2)$ in (5.188). The real problem now (which has important practical consequences) is to design a controller such that the contact force tracks some desired signal. Let us investigate the extension of the Slotine and Li scheme in this framework. The controller in (6.51) is slightly transformed into:

$$\begin{cases} \bar{\tau}_1 = \bar{M}_{12}\ddot{z}_{2r} + \bar{C}_{12}(z_2, \dot{z}_2)\dot{z}_{2r} + \bar{g}_1 - \lambda_2\lambda_{zd} \\ \\ \bar{\tau}_2 = \bar{M}_{22}\ddot{z}_{2r} + \bar{C}_{22}(z_2, \dot{z}_2)\dot{z}_{2r} + \bar{g}_2 - \lambda_2 s_2 \end{cases} \tag{6.145}$$

where all the terms keep the same definition as for (6.51). $\lambda_d$ is some desired value for the contact force $\lambda_{z_1}$. The closed-loop system is therefore given by:

$$\begin{cases} \bar{M}_{12}(z_2)\dot{s}_2 + \bar{C}(z_2, \dot{z}_2)s_2 = \lambda_2(\lambda_{z_1} - \lambda_d) \\ \\ \bar{M}_{22}(z_2)\dot{s}_2 + C(z_2, \dot{z}_2)s_2 + \lambda_1 s_2 = 0 \\ \\ \dot{\tilde{z}}_2 = -\lambda\tilde{z}_2 + s_2 \end{cases} \tag{6.146}$$

The dissipativity properties of the free-motion closed-loop system are similar to those of (6.52) and (6.53). Notice that due to the asymptotic stability properties of the fixed point $(\tilde{z}_2, s_2)$ one gets $\lambda_{z_1}(t) \to \lambda_d(t)$ as $t \to +\infty$.

**Remark 6.26 (Nonsmooth mechanical systems)** In practice one often has to face *unilateral* or *inequality* constraints where (5.187) is replaced by $\phi(q) \geq 0$, which model the fact that contact may be lost or established with obstacles. We do not enter into the details of the dynamics of systems subject to unilateral constraints, which is a complicated matter. Let us just point out that this yields to *nonsmooth* systems containing impact rules (or state reinitializations) and so-called *complementarity* relationships between $\lambda_{z_1}$ and $z_1$, of the form

$$\lambda_{z_1} \geq 0, \quad z_1 \geq 0, \quad \lambda_{z_1}^T z_1 = 0 \tag{6.147}$$

Such systems form an important class of *hybrid* dynamical systems whose modeling, well-posedness, numerical analysis, controllability and stabilizability properties are still the object of many research activities. To keep contact one has to assure that $\lambda_{z_1} > 0$. The scheme in (6.145) may therefore not be

[2] Actually the way these coefficients tend to infinity is important to pass from the compliant case to the rigid body limit. This is analyzed for instance in [117] through a singular perturbation approach.

suitable except if the feedback gain $\lambda_2$ is taken large enough and if $\lambda_d > 0$. This may constitute a drawback of passivity-based schemes for constrained systems with respect to feedback linearization for which it can be assured that $\lambda_{z_1} \equiv \lambda_d$ for all $t \geq 0$ (evidently assuming that $z_1 \equiv 0$ for all $t \geq 0$). More details on nonsmooth mechanical systems dynamics and control can be found in [12] [27] [33] [32] [58].

# Chapter 7

# Adaptive Control

This chapter is dedicated to present so-called *direct adaptive* controllers applied to mechanical and to linear invariant systems. We have already studied some applications of dissipativity theory in the stability of adaptive schemes in chapters 1, 2, 3 and 4. Direct adaptation means that one has been able to rewrite the fixed parameter input $u$ in a form that is linear with respect to some unknown parameters, usually written as a vector $\theta \in \mathbb{R}^p$, i.e. $u = \phi(x,t)\theta$, where $\phi(x,t)$ is a known matrix (called the regressor) function of measurable [1] terms. The parameters $\theta_i$, $i \in \{1, \cdots, p\}$, are generally nonlinear combinations of the physical parameters (for instance in the case of mechanical systems, they will be nonlinear combinations of moments of inertia, masses). When the parameters are unknown, one cannot use them in the input. Therefore one replaces $\theta$ in $u$ by an estimate, that we shall denote $\hat{\theta}$ in the sequel. In other words, $u = \phi(x,t)\theta$ is replaced by $u = \phi(x,t)\hat{\theta}$ at the input of the system, and $\hat{\theta}$ is estimated on-line with a suitable identification algorithm. As a consequence, one easily imagines that the closed-loop system stability analysis will become more complex. However through the passivity theorem (or the application of lemma 6.4) the complexity reduces to adding a passive block to the closed-loop system that corresponds to the estimation algorithm dynamics. The rest of the chapter is composed of several examples that show how this analysis mechanism work. It is always assumed that the parameter vector is constant: the case of time-varying parameters, although closer to the reality, is not treated here due to the difficulties in deriving stable adaptive controllers in this case. This is a topic in itself in adaptive control theory and is clearly outside the scope of this book.

---

[1] In the technological sense, not in the mathematical one.

# 7.1 Lagrangian systems

## 7.1.1 Rigid joint-rigid link manipulators

In this subsection we first examine the case of a PD controller with an adaptive gravity compensation. Indeed it has been proved in subsection 6.2.1 that gravity hampers asymptotic stability of the desired fixed point, since the closed-loop system possesses an equilibrium that is different from the desired one. Then we pass to the case of tracking control of $n$ degree-of-freedom manipulators.

### PD + adaptive gravity compensation

**A first simple extension** Let us consider the following controller + estimation algorithm:

$$\begin{cases} \tau = -\lambda_1 \dot{q} - \lambda_2 \tilde{q} + Y_g(q)\hat{\theta}_g \\ \\ \dot{\hat{\theta}}_g(t) = \lambda_3 Y_g^T \dot{q} \end{cases} \tag{7.1}$$

where we suppose that the gravity generalized torque $g(q) = Y_g(q)\theta_g$ for some known matrix $Y_g(q) \in I\!R^{n \times p}$ and unknown vector $\theta_g$, and $\tilde{\theta}_g = \theta_g - \hat{\theta}_g$. The estimation algorithm is of the gradient type, and we know from subsection 4.3.1 that such an estimation law defines a passive operator $\dot{q} \mapsto \tilde{\theta}_g^T Y_g(q)$, with storage function $V_2(\tilde{\theta}_g) = \frac{1}{2}\tilde{\theta}^T\tilde{\theta}$. This strongly suggests to decompose the closed-loop system obtained by introducing (7.1) into (5.108) into two blocks as follows:

$$\begin{cases} M(q)\ddot{q} + C(q,\dot{q})\dot{q} + \lambda_1 \dot{q} + \lambda_2 \tilde{q} = -Y_g(q)\tilde{\theta} \\ \\ \dot{\tilde{\theta}}_g = \lambda_3 Y_g^T \dot{q} \end{cases} \tag{7.2}$$

Obviously the first block with the rigid joint-rigid link dynamics and input $u_1 = -Y_g(q)\tilde{\theta}(= -y_2)$ and output $y_1 = \dot{q}(= u_2)$ defines an output strictly passive operator with storage function $V_1(\tilde{q},\dot{q}) = \frac{1}{2}\dot{q}^T M(q)\dot{q} + \frac{\lambda_2}{2}\tilde{q}^T\tilde{q}$, see subsection 6.2.1. One is tempted to conclude about the asymptotic stability with a Lyapunov function $V(\tilde{q},\dot{q},\tilde{\theta}) = V_1(\tilde{q},\dot{q}) + V_2(\tilde{\theta}_g)$. However notice that the overall system with input $u = u_1 + y_2$ and output $y = y_1$, although output strictly passive, is not zero-state detectable. Indeed $u \equiv y \equiv 0$ implies $\lambda_2\tilde{q} = Y_g(q)\tilde{\theta}_g$ and $\dot{\tilde{\theta}}_g = 0$, nothing more. Hence very little has been gained by adding an estimation of the gravity, despite the passivity theorem applies well.

**How to get asymptotic stability?** The lack of zero-state detectability of the system in (7.2) is an obstacle to the asymptotic stability of the closed-loop scheme. The problem is therefore to keep the negative feedback interconnection structure of the two highlighted blocks, while introducing some detectability property in the loop. However the whole state is now $(q, \dot{q}, \tilde{\theta}_g)$ and it is known in identification and adaptive control theory that the estimated parameters converge to the real ones (i.e. $\tilde{\theta}_g(t) \to 0$) only if some persistent excitation conditions are fullfilled. Those conditions are related to the spectrum of the signals entering the regressor matrix $Y_g(q)$. Such a result is hopeless here since we are dealing with regulation. Hence the best one may expect to obtain is convergence of $(\tilde{q}, \dot{q})$ towards zero. We may however hope that there exists a feedback adaptive controller that can be analyzed through the passivity theorem and such that the underlying storage function can be used as a Lyapunov function with Krasovskii-La Salle theorem to prove asymptotic convergence. Let us consider the estimation algorithm proposed in [203]:

$$\dot{\tilde{\theta}}_g = \lambda_3 Y_g^T \left( \lambda_4 \dot{q} + \frac{2\tilde{q}}{1 + 2\tilde{q}^T \tilde{q}} \right) \tag{7.3}$$

Notice that this is still a gradient update law. It defines a passive operator $\left(\lambda_4 \dot{q} + \frac{2\tilde{q}}{1+2\tilde{q}^T\tilde{q}}\right) \mapsto Y_g(q)\tilde{\theta}_g$, not $\dot{q} \mapsto Y_g(q)\tilde{\theta}_g$. We therefore have to look at the dissipativity properties of the subsystem with dynamics $M(q)\ddot{q} + C(q, \dot{q})\dot{q} + \lambda_1 \dot{q} + \lambda_2 \tilde{q} = u_1$, $y_1 = \left(\lambda_4 \dot{q} + \frac{2\tilde{q}}{1+2\tilde{q}^T\tilde{q}}\right)$: this is new compared to what we have seen from the begining! Let us analyze it in detail:

$$\begin{aligned}
&\langle u_1, y_1 \rangle \\
&= \int_0^t \left(\lambda_4 \dot{q} + \tfrac{2\tilde{q}}{1+2\tilde{q}^T\tilde{q}}\right)^T [M(q)\ddot{q} + C(q,\dot{q})\dot{q} + \lambda_1 \dot{q} + \lambda_2 \tilde{q}] ds \\
&= q \int_0^t \left\{ \lambda_4 \dot{q}^T (\lambda_2 \tilde{q} + \lambda_1 \dot{q}) + \tfrac{d}{ds}\left( \tfrac{\lambda_4}{2} \dot{q}^T M(q) \dot{q} + \tfrac{2\dot{q}^T M(q)\tilde{q}}{1+2\tilde{q}^T\tilde{q}} \right) \right\} ds \\
&+ \int_0^t \left\{ -\tfrac{2\dot{q}^T M(q)\dot{q} + 2\dot{q}^T C(q,\dot{q})\tilde{q}}{1+2\tilde{q}^T\tilde{q}} + \tfrac{8\dot{q}^T M(q)\tilde{q}\dot{q}^T\tilde{q}}{1+2\tilde{q}^T\tilde{q}} 2 \tfrac{\tilde{q}}{1+2\tilde{q}^T\tilde{q}} (\lambda_2 \tilde{q} + \lambda_1 \dot{q}) \right\} ds \\
&\geq \tfrac{\lambda_4 \lambda_2}{2} \left[\tilde{q}^T \tilde{q}\right]_0^t + \left[ \tfrac{\lambda_4}{2} \dot{q}^T M(q) \dot{q} + \tfrac{2\dot{q}^T M(q)\tilde{q}}{1+2\tilde{q}^T\tilde{q}} \right]_0^t + \lambda_4 \lambda_1 \int_0^t \dot{q}^T \dot{q} ds \\
&+ \int_0^t \left\{ 2\lambda_2 \tfrac{\tilde{q}^T\tilde{q}}{1+2\tilde{q}^T\tilde{q}} - \left(4\lambda_M + \tfrac{k_c}{\sqrt{2}}\right) \dot{q}^T \dot{q} - 2 \tfrac{\lambda_1 ||\dot{q}|| \cdot ||\tilde{q}||}{1+2\tilde{q}^T\tilde{q}} \right\} ds \\
&\geq -\tfrac{\lambda_4 \lambda_2}{2} \tilde{q}(0)^T \tilde{q}(0) - \tfrac{\lambda_4}{2} \dot{q}(0)^T M(q(0)) \dot{q}(0) + \tfrac{2\dot{q}(0)^T M(q(0))\tilde{q}(0)}{1+2\tilde{q}(0)^T\tilde{q}(0)} \\
&+ \lambda_4 \lambda_1 \int_0^t \dot{q}^T \dot{q} ds
\end{aligned} \tag{7.4}$$

where we have used the fact that due to the skew-symmetry of $\dot{M}(q) - 2C(q, \dot{q})$ we have $\dot{M}(q) = C(q, \dot{q}) + C^T(q, \dot{q})$, and where $\lambda_4 >$

$\max\left\{\frac{1}{\lambda_1}\left(\frac{\lambda_1^2}{2\lambda_2}+4\lambda_M+\frac{k_c}{\sqrt{2}}\right),\frac{2\lambda_M}{\sqrt{\lambda_m\lambda_2}}\right\}$, with $\lambda_m I_n \leq M(q) \leq \lambda_M I_n$, $||C(q,\dot{q})|| \leq k_c||\dot{q}||$ for any compatible matrix and vector norms. Under these gain conditions, one sees from (7.4) that the first subsystem is passive with respect to the supply rate $u_1^T y_1$, and a storage function is given by

$$V_1(\tilde{q},\dot{q}) = \frac{\lambda_4\lambda_2}{2}\tilde{q}^T\tilde{q} + \frac{\lambda_4}{2}\dot{q}^T M(q)\dot{q} + \frac{2\dot{q}^T M(q)\tilde{q}}{1+2\tilde{q}^T\tilde{q}} \tag{7.5}$$

The first subsystem even possesses some strict passivity property, see (7.4). Finally a complete storage function is provided by the sum $V(\tilde{q},\dot{q},\tilde{\theta}_g) = V_1(\tilde{q},\dot{q}) + V_2(\tilde{\theta}_g)$, and it can be shown that its derivative is semi-negative definite and that the largest invariant set contained in the set $\dot{V}\equiv 0$ is contained in the set $(q,\dot{q}) = (0,0)$ which ends the proof.

**Remark 7.1** .

- The storage function associated to the first subsystem is quite original. It looks like the available storage of the closed-loop system when a PD controller is applied, but the added term comes from "nowhere"! Our analysis has been done easily because we knew beforehand that such a storage function was a good one. The intuition behind it is not evident. It was first discovered in [89] and then used in [203].
- This is typically an instance where the passivity theorem and the dissipativity properties alone are not sufficient to show asymptotic stability.
- Concerning the application of lemma 6.4, one sees that the majoration that occurs in the passivity proof in (7.4) does not concern the interconnection terms since the CTCE is satisfied. It just concerns the first subsystem. This is in contrast with the following example:

$$\left\{\begin{array}{l}\dot{x}_1 = -x_1^5 - x_1 + x_1^2 x_2\\ \\ \dot{x}_2 = -x_2^3 - x_2 + x_2 x_1\end{array}\right. \tag{7.6}$$

  where the subsystems considered as independent systems are trivially stable with Lyapunov functions $V_1(x_1) = \frac{1}{2}x_1^2$ and $V_2(x_2) = \frac{1}{2}x_2^2$. It is possible to show that $V = V_1 + V_2$ is a Lyapunov function for the interconnected system, but this relies on majorations of the cross-terms. The inequalities (6.35) and (6.36) reduce to finding two functions $\bar{g}_1(x_1,x_2)$ and $\bar{g}_2(x_1,x_2)$ such that $x_1x_2(x_1^2+\bar{g}_1\bar{g}_2) \leq 0$ and $x_1x_2(x_2-\bar{g}_1\bar{g}_2) \leq 0$, which seems a difficult task, since this is equivalent to solving $[h(x_1,x_2)+g(x_1,x_2)]x_1x_2 = -x_1^2 - x_2$, $h(x_1,x_2) \geq 0$, $g(x_1,x_2) \geq 0$.

### The adaptive Slotine and Li controller

Let us now pass to the controller presented in subsection 6.2.5 in (6.51). It turns out that this scheme yields a much more simple stability analysis than the PD with adaptive gravity compensation: this is due to the fact that as pointed out earlier, it uses the inertia matrix explicitly even for regulation.

**Gradient estimation law** Consider the following controller:

$$\begin{cases} \tau &= \hat{M}(q)\ddot{q}_r + \hat{C}(q,\dot{q})\dot{q}_r + \hat{g}(q) - \lambda_1 s = Y(q,\dot{q},t)\hat{\theta} \\ &= M(q)\ddot{q}_r + C(q,\dot{q})\dot{q}_r + g(q) - \lambda_1 s - Y(q,\dot{q},t)\tilde{\theta} \\ \dot{\tilde{\theta}} &= \lambda_2 Y^T(q,\dot{q},t)s \end{cases} \tag{7.7}$$

where we used the fact that the fixed parameter controller can be rewritten under the required linear form $Y(q,\dot{q},t)\hat{\theta}$, where $\theta$ is a vector of unknown inertia parameters. Actually one has $M(q)\ddot{q} + C(q,\dot{q})\dot{q} + g(q) = Y(q,\dot{q},\ddot{q})\theta$. The closed-loop system is therefore given by:

$$\begin{cases} M(q)\dot{s} + C(q,\dot{q})s + \lambda_1 s = Y(q,\dot{q},t)\tilde{\theta} \\ \dot{\tilde{q}} = -\lambda\tilde{q} + s \\ \dot{\tilde{\theta}} = \lambda_2 Y^T(q,\dot{q},t)s \end{cases} \tag{7.8}$$

The interpretation through the passivity theorem is obvious: the update law in (7.7) is a gradient that defines a passive operator $s \mapsto Y(q,\dot{q},t)\tilde{\theta}$ and the first subsystem has state $(\tilde{q}, s)$. From the developments in subsection 6.2.5 one therefore sees that the adaptive version of the Slotine and Li controller just yields a closed-loop system that is identical to the one in (6.52) (6.53) with an additional passive block interconnected to the two previous ones in (6.52) and (6.54), see figure 7.1 and compare with figure 6.5. The storage function follows immediately. Similarly to the PD with adaptive compensation scheme, one cannot expect to get asymptotic stability of the whole state because of the parameter estimates that generally do not converge towards the real ones. The passivity theorem allows however one to conclude that $s \in \mathcal{L}_2(\mathbb{R}^+)$, from which it follows using a classical result that $\tilde{q}$ converges towards zero (the output of a strictly proper and strictly stable transfer function tends to zero when the input is $\mathcal{L}_2$-bounded).

**Remark 7.2** Historically the passivity interpretation of the Slotine and Li scheme has been deduced from lemma 6.4, see [28] [31], where most of the adaptive schemes (including e.g. [169]) designed for rigid manipulators have been analyzed through the passivity theorem. Indeed this is based on a CTCE

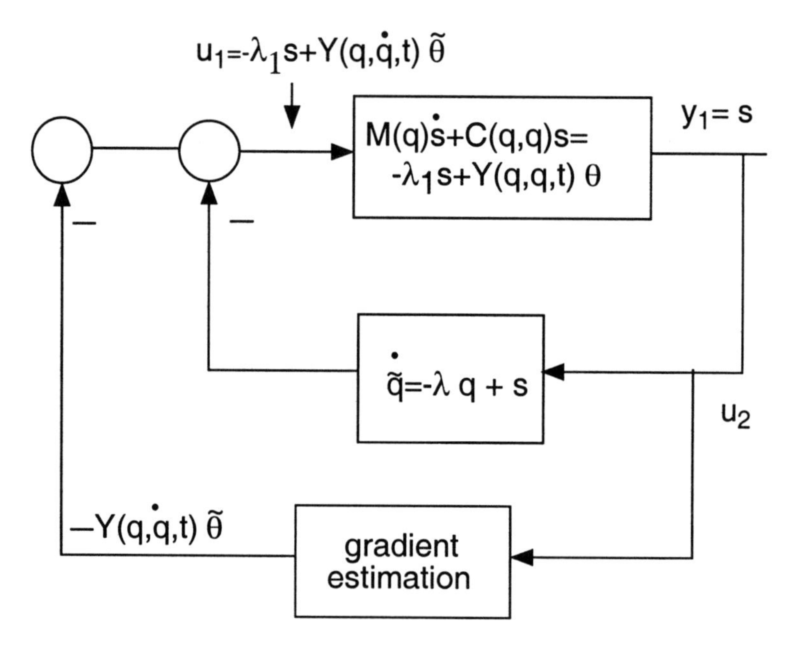


Figure 7.1: Closed-loop equivalent representation (adaptive case).

as defined in lemma 6.4. Actually the first subsystem in (7.8) with state $s$ has relative degree one between its input $u_1 = Y(q, q, t)\tilde{\theta}$ and its output $y_1 = s$. As we shall remark when we have presented the adaptive control of linear invariant systems with relative degree one, the CTCE is ubiquitous in direct adaptive control.

**Least-squares estimation law** Until now we have presented only gradient-type update laws. It is clear that the estimation block can be replaced by any system that has $\tilde{\theta}$ inside the state and is passive with respect to the same supply rate. The classical recursive least-squares estimation algorithm does not satisfy such requirements. However it can be "passified" as explained now. First of all let us recall the form of the classical least-squares algorithm:

$$\begin{cases} \dot{\hat{\theta}}_{ls} = P(t)Y^T(q, \dot{q}, t)s \\ \dot{P} = -PY(q, \dot{q}, t)Y^T(q, \dot{q}, t)P, \quad P(0) > 0 \end{cases} \tag{7.9}$$

The required passivity property is between $s$ and $-Y(q,\dot{q},t)\tilde{\theta}$ (recall we have defined $\tilde{\theta} = \theta - \hat{\theta}$). Let us compute the available storage of this system:

$$
\begin{aligned}
V_a(\tilde{\theta}, P) &= \sup_{s:(0,\tilde{\theta}(0),P(0))\rightarrow} - \int_0^t s^T Y \tilde{\theta} ds \\
&= \sup_{s:(0,\tilde{\theta}(0),P(0))\rightarrow} - \frac{1}{2}\left[\tilde{\theta}^T P^{-1} \tilde{\theta}\right]_0^t + \frac{1}{2}\int_0^t \tilde{\theta}^T \dot{P}^{-1} \tilde{\theta} ds \qquad (7.10) \\
&= \sup_{s:(0,\tilde{\theta}(0),P(0))\rightarrow} - \frac{1}{2}\left[\tilde{\theta}^T P^{-1} \tilde{\theta}\right]_0^t + \frac{1}{2}\int_0^t \tilde{\theta}^T Y Y^T \tilde{\theta} ds
\end{aligned}
$$

where we used the fact that $\dot{P}^{-1} = YY^T$. One remarks that the available storage in (7.10) is not "far" from being bounded: it would suffice that $Y^T\tilde{\theta}$ be $L^2$-bounded. However it seems difficult to prove this. Consequently let us propose the following modified least-squares estimation algorithm [2]:

$$
\left\{
\begin{array}{l}
\hat{\theta} = \hat{\theta}_{ls} + \mathcal{S} \\
\dot{P} = \alpha(t)\left[-P\left(\frac{Y^TY}{1+\mathrm{tr}(Y^TY)} + \lambda R\right)P + \lambda P\right] \\
\alpha(t) = \frac{s^T Y Y^T s}{(1+s^T s)(1+\mathrm{tr}(Y^TY))} \\
A = \frac{Y^TY}{1+\mathrm{tr}(Y^TY)} + \lambda R \\
\mathcal{S} = \frac{Y^T}{1+\mathrm{tr}(Y^TY)} \frac{s}{1+s^Ts}\left(\hat{\theta}_{ls}^T A \hat{\theta}_{ls} + M(1+\lambda\lambda_{\max}(R))\right) \\
\lambda \geq 0, \quad R > 0 \\
\lambda_{\min}(R) I_n \leq P^{-1}(0) \leq \left(\lambda_{\max}(R) + \frac{1}{\lambda}\right) I_n \\
M \geq \theta^T \theta
\end{array}
\right. \qquad (7.11)
$$

Then the following is true [114] [30]:

**Lemma 7.1** .

- $\lambda_{\min}(R) \leq \lambda_i(P^{-1}) \leq \lambda_{\max}(R) + \frac{1}{\lambda}$, where $\lambda_i(P^{-1})$ denotes the eigenvalues of $P^{-1}$.

[2] Let us note that the denomination "least-squares" somewhat loses its original meaning here, since it is not clear that the proposed scheme minimizes any quadratic criterion. However the name least-squares is kept for obvious reasons.

- $\int_0^t -s^T Y\tilde{\theta} ds = \frac{1}{2}\left[\tilde{\theta}_{ls}^T P^{-1}\tilde{\theta}_{ls}\right]_0^t - \frac{1}{2}\int_0^t \tilde{\theta}_{ls}^T \dot{P}^{-1}\tilde{\theta}_{ls} d\tau - \int_0^t s^T Y S d\tau$, where $\tilde{\theta}_{ls} = \theta - \hat{\theta}_{ls}$, where $\hat{\theta}_{ls}$ is the classical least-squares estimate $\dot{\hat{\theta}}_{ls} = PY^T s$.
- $-\frac{1}{2}\int_0^t \tilde{\theta}_{ls}^T \dot{P}^{-1}\tilde{\theta}_{ls} d\tau - \int_0^t s^T Y S d\tau \geq 0$ ♠♠

It follows that the mapping $s \mapsto -Y\tilde{\theta}$ is passive with storage function $\frac{1}{2}\tilde{\theta}_{ls}^T P^{-1}\tilde{\theta}_{ls}$. The proof of lemma 7.1 is not given here for the sake of briefness and also because despite its originality, it has not been proved that such passified least-square yields better closed-loop performance than the simple gradient update law (for instance in terms of parameter convergence speed and of robustness). It is therefore to be seen more like a theoretical exercise (find out how to passify the classical least-squares) rather than something motivated by applications. The interest for us here is to illustrate the modularity provided by passivity-based controllers. As we shall see further, it applies equally well to adaptive control of relative degree one and two linear invariant systems.

**Remark 7.3** Other schemes using the same basic structure as the Slotine and Li one, but with different update laws or control structure, have been derived in the control and robotics literature. Most of them have been shown to fit within the passivity theorem through the implicit use of lemma 6.4, see [28]. It is not worth developing all of them here. Just let us note that they were originally all designed through the choice of a Lyapunov function, and then reinterpreted with the passivity point of view.

### 7.1.2 Flexible joint-rigid link manipulators

In the whole section we will assume that $K$ is a known matrix (we refer the reader to a remark below for more precisions about this).

#### The backstepping algorithm

We have to solve two main problems in order to extend the fixed parameter scheme presented in subsection 6.4.2 towards an adaptive version:

1) The input $u$ in must be LP (Linear in some set of Parameters).

2) The signals $\tilde{q}_2$ and $e_2$ have to be available on line because they will be used in the update laws.

To solve 1), we can use the idea which consists of adding the determinant of the inertia matrix $\det(M(q_1))$ in the Lyapunov function $V_1$ (see subsection 6.4.2). The trick is that the nonlinearity in the unknown parameters comes from the terms containing the inverse of the inertia matrix $M^{-1}(q_1)$. Premultiplying by $\det(M)$ allows us to retrieve LP terms, as $\det(M)M^{-1}$ is indeed LP (the price to pay is an overparametrization of the controller). Moreover 2) implies that $q_{2d}$ (see (6.106)) and $e_{2d}$ (see after (6.109)) are available on line, and thus do not depend on unknown parameters. We can proceed as follows:

- **Step 1**

  The right-hand-side of (5.115) can be written as $Y_1(q_1, \dot{q}_1, t)\theta_1^*$. Thus we choose $q_{2d}$ in (6.106) as

$$Kq_{2d} = Y_1(q_1, \dot{q}_1, t)\hat{\theta}_1 \tag{7.12}$$

  where $\hat{\theta}_1$ stands for an estimate of $\theta_1^*$. Thus

$$\tilde{q}_2 = q_2 - K^{-1}Y_1(q_1, \dot{q}_1, t)\hat{\theta}_1 \tag{7.13}$$

  Adding $\pm Y_1(\cdot)\theta_1^*$ to the right-hand-side of the first equation in (5.115) and differentiating (7.13), one obtains:

$$\begin{aligned} &M(q_1)\dot{s}_1 + C(q_1, \dot{q}_1)s_1 + \lambda_1 s_1 = K\tilde{q}_2 - Y_1\theta_1^* \\ &\dot{\tilde{q}}_2 = \dot{q}_2 - K^{-1}\tfrac{d}{dt}(Y_1\theta_1^*) \end{aligned} \tag{7.14}$$

- **Step 2**

  Now consider $e_{2d}$ defined after (6.109). The first two terms are available but the third term is a function of unknown parameters and it is not LP (it contains $M^{-1}$). Assume now that $V_2$ is replaced by

$$V_{2a} = V_r(\tilde{q}_1, s_1, t) + \frac{1}{2}\tilde{\theta}_1^T\tilde{\theta}_1 + \frac{1}{2}\det M\tilde{q}_2^T\tilde{q}_2 \tag{7.15}$$

  Setting $\dot{q}_2 = e_{2d} + e_2$, i.e. $\dot{\tilde{q}}_2 = e_{2d} + e_2 - K^{-1}\frac{d}{dt}(Y_1\hat{\theta}_1)$, we get along trajectories of (7.14):

$$\begin{aligned} \dot{V}_{2a} \leq \; & -\lambda_1\dot{\tilde{q}}_1^T\dot{\tilde{q}}_1 - \lambda^2\lambda_1\tilde{q}_1^T\tilde{q}_1 - s_1^TY_1\tilde{\theta}_1 + \dot{\hat{\theta}}_1^T\tilde{\theta}_1 + \tilde{q}_2^TKs_1 \\ & \tilde{q}_2^T\det(M)e_2 + \tilde{q}_2^T\det(M)[e_{2d} - \dot{q}_{2d}] + \tilde{q}_2^T\tfrac{d}{dt}\{\tfrac{(\det M)}{2}\}\tilde{q}_2 \end{aligned} \tag{7.16}$$

  Let us denote $\det(M) = Y_2(q_1)\theta_2^*$, and choose

$$Y_2\hat{\theta}_2e_{2d} = -Y_3(q_1, \dot{q}_1, q_2, t)\hat{\theta}_3 - \tilde{q}_2 \tag{7.17}$$

where

$$Y_3(q_1, \dot{q}_1, q_2, t)\theta_3^* = \frac{d}{dt}\left\{\frac{\det(M)}{2}\right\}\tilde{q}_2 - \det(M)\dot{q}_{2d} + Ks_1 \tag{7.18}$$

Choose also

$$\dot{\hat{\theta}}_1 = Y_1^T(q_1, \dot{q}_1, t)s_1 \tag{7.19}$$

Thus we obtain

$$\dot{V}_{2a} \leq -\lambda_1 \dot{\tilde{q}}_1^T \dot{\tilde{q}}_1 - \lambda^2 \lambda_1 \tilde{q}_1^T \tilde{q}_1 + \tilde{q}_2^T \det(M) e_2 + \tilde{q}_2^T [Y_2 \theta_2^{*T} e_{2d} + Y_3 \theta_3^{*T}] \tag{7.20}$$

(We drop the arguments for convenience). Introducing $\pm \tilde{q}_2^T Y_2 \hat{\theta}_2 e_{2d}$ we obtain

$$\begin{aligned} \dot{V}_{2a} &\leq -\lambda_1 \dot{\tilde{q}}_1^T \dot{\tilde{q}}_1 - \lambda^2 \lambda_1 \tilde{q}_1^T \tilde{q}_1 + \tilde{q}_2^T \det(M) e_2 \\ &-\tilde{q}_2^T e_{2d} Y_2 \tilde{\theta}_2 + \tilde{q}_2^T Y_3 \tilde{\theta}_3 - \tilde{q}_2^T \tilde{q}_2 \end{aligned} \tag{7.21}$$

Defining $V_{3a} = V_{2a} + \frac{1}{2}\tilde{\theta}_2^T \tilde{\theta}_2 + \frac{1}{2}\tilde{\theta}_3^T \tilde{\theta}_3$ and setting

$$\dot{\hat{\theta}}_3 = -Y_3^T \tilde{q}_2 \tag{7.22}$$

$$\dot{\hat{\theta}}_2 = -Y_2^T e_{2d}^T \tilde{q}_2 \tag{7.23}$$

We therefore obtain

$$\dot{V}_{3a} \leq -\lambda_1 \dot{\tilde{q}}_1^T \dot{\tilde{q}}_1 - \lambda^2 \lambda_1 \tilde{q}_1^T \tilde{q}_1 + \tilde{q}_2^T \det(M) e_2 - \tilde{q}_2^T \tilde{q}_2 \tag{7.24}$$

**Remark 7.4** In order to avoid any singularity in the control input, the update law in (7.23) has to be modified using a projection algorithm, assuming that $\theta_2^*$ belongs to a known convex domain. We refer the reader to [113] for details about how this domain may be calculated, and the stability analysis related to the projection. For the sake of clarity of this presentation, we do not introduce this modification here, although we know it is necessary.

- **Step 3**

  At this stage our goal is partially reached, as we have defined signals $\tilde{q}_2$ and $e_2$ available on line.
  Now consider the function

$$V_{4a} = V_{3a} + \frac{1}{2}\det(M) e_2^T e_2 \tag{7.25}$$

We obtain

$$\begin{aligned}\dot{V}_{4a} \quad &\leq -\lambda_1 \dot{\tilde{q}}_1^T \dot{\tilde{q}}_1 - \lambda^2 \lambda_1 \tilde{q}_1^T \tilde{q}_1 + \tilde{q}_2^T \det(M) e_2 \\ &-\tilde{q}_2^T \tilde{q}_2 + e_2^T [v - \dot{e}_{2d}] + e_2^T \frac{d}{dt} \left\{ \frac{\det(M)}{2} \right\} e_2\end{aligned} \tag{7.26}$$

Notice that

$$-\det(M)\dot{e}_{2d} + \frac{d}{dt}\left\{\frac{\det(M)}{2}\right\} e_2 = Y_4(q_1, \dot{q}_1, q_2, \dot{q}_2)\theta_4^* \tag{7.27}$$

for some $Y_4$ and $\theta_4^*$ matrices of suitable dimensions. Let us denote this time $\det(M) = Y_2(q_1)\theta_5^*$ (This is strictly equal to $Y_2(q_1)\theta_2^*$ defined above, but we choose a different notation because the estimate of $\theta_5^*$ is going to be chosen differently). Let us choose $v = -\tilde{q}_2 + w$ and

$$Y_2 \hat{\theta}_5 w = -Y_4 \hat{\theta}_4 - e_2 \tag{7.28}$$

We obtain

$$\dot{V}_{4a} \leq -\lambda_1 \dot{\tilde{q}}_1^T \dot{\tilde{q}}_1 - \lambda^2 \lambda_1 \tilde{q}_1^T \tilde{q}_1 - \tilde{q}_2^T \tilde{q}_2 - e_2^T w Y_2 \tilde{\theta}_5 + e_2^T Y_4 \tilde{\theta}_4 - e_2^T e_2 \tag{7.29}$$

Finally we choose as a Lyapunov function for the whole closed-loop system

$$V = V_{4a} + \frac{1}{2}\tilde{\theta}_4^T \tilde{\theta}_4 + \frac{1}{2}\tilde{\theta}_5^T \tilde{\theta}_5 \tag{7.30}$$

and the following update laws

$$\dot{\hat{\theta}}_4 = -Y_4^T e_2 \tag{7.31}$$

$$\dot{\hat{\theta}}_5 = -Y_2^T w^T e_2 \tag{7.32}$$

(a projection algorithm has to be applied to $\hat{\theta}_5$ as well, see remark 7.4 above). We obtain

$$\dot{V} \leq -\lambda_1 \dot{\tilde{q}}_1^T \dot{\tilde{q}}_1 - \lambda^2 \lambda_1 \tilde{q}_1^T \tilde{q}_1 - \tilde{q}_2^T \tilde{q}_2 - e_2^T e_2 \tag{7.33}$$

We therefore conclude that $\hat{\theta} \in \mathcal{L}_\infty$, $\tilde{q}_2$, $e_2$, $\tilde{q}_1, s_1 \in \mathcal{L}_2 \cap \mathcal{L}_\infty$, $q_2 \in \mathcal{L}_\infty$, (see (7.13)) , $\dot{q}_2 \in \mathcal{L}_\infty$. Finally from the definition of $s_1$ and theorem 4.1 we conclude that $\tilde{q}_1 \in \mathcal{L}_2 \cap \mathcal{L}_\infty$, $\dot{\tilde{q}}_1 \in \mathcal{L}_2$ and $\tilde{q}_1 \to 0$ as $t \to +\infty$.

**Remark 7.5** .

- It is noteworthy that the above procedure can be seen as a modified (because of the linearity in the parameters problem, plus the *a priori* knowledge of $K$) version of the work in [82]. We retrieve the fact that the estimates are functions of the "previous" estimates (see (7.17) (7.31) (7.32) (7.22) (7.23) (7.13)), due to the backstepping procedure.

- We have considered for simplicity $K$ as a known matrix. However in practice it may be interesting to relax this assumption. The difficulty arises from the fact that the "intermediate" input $q_{2d}$ feeds the rigid dynamics through $K$, i.e. if $K$ is unknown, then the interconnection term between the two subsystems is unknown. This problem has been given a solution in [113] by replacing $K$ by an estimate $\hat{K}$, such that $\hat{K}$ remains full-rank and is twice differentiable.

- If $J$ is unknown, one can slightly modify $\frac{1}{2}\det(M)e_2^T e_2$ by $\frac{1}{2}\det(M)e_2^T J e_2$ in $V$ so that it can be incorporated in $\theta_4^\star$ (see (7.27) and (7.26)) .

- Although CTCEs have been used all along the procedure, the interpretation of this scheme is not a trivial addition of a passive block (containing the adaptation algorithms) to the block corresponding to the fixed parameter closed-loop scheme (see subsection 6.4.2). We leave it to the reader to explore the equivalent closed-loop interpretation of the scheme in (7.12)-(7.32) through the passivity theorem!

### The passivity-based schemes

We now turn our attention to the scheme developed in subsection 6.4.1. We will not give here a detailed analysis of the adaptive version of the passivity-based schemes (it can be found in [113]), but we rather focus on a particular problem, i.e. the choice of the "first stage" input $q_{2d}$ in (6.103). It is worth noting that $q_{2d}$ in (7.12) could have been chosen else, just as what can be done in the known parameters case. In particular, the ideas in [18] can be used to enhance the robustness of the scheme with respect to velocity noise. On the contrary, the design of $q_{2d}$ is much less obvious when we design the adaptive pasivty-based scheme of subsection 6.4.1. Roughly speaking, this is due to the fact that both $\tilde{q}_2$ and $\dot{q}_2$ have to be available on line to update some estimated parameters. This significantly reduces the possible choices for $\tilde{q}_2$ (and consequently for $q_{2d}$). It has been shown in [113] that the algorithm presented in [169] can successfully be used to design $q_{2d}$ so that $\dot{\tilde{q}}_2$ can be calculated without acceleration measurements. We refer the reader to [113] for details concerning the adaptive algorithm. It is a major conclusion that contrarily to the rigid joint-rigid link case, the extension of passivity-based schemes towards adaptive control is not trivial at all for rigid link-flexible joint manipulators.

## 7.2 Linear invariant systems

The problem of adaptive control of linear invariant systems has been a very active field of research since the beginning of the sixties. Two paths have been followed: the indirect approach which consists of estimating the process parameters, and using those estimated values into the control input, and the direct approach that we described in the introduction of this chapter. The direct approach has many attractive features, among them the nice passivity properties of the closed-loop system, which actually is a direct consequence of lemma 6.4. This is what we develop now.

### 7.2.1 A scalar example

Before passing to more general classes of systems, let us reconsider the following first order system similar to the one presented in subsection 1.4:

$$\dot{x} = a^* x + b^* u \tag{7.34}$$

where $x \in I\!R$, $a^*$ and $b^*$ are constant parameters, and $u \in I\!R$ is the input signal (we do not enter here into the problem of defining the input space; it is assumed that the basic conditions for existence and uniqueness of solutions are satisfied). The control objective is to make the state $x(\cdot)$ track some desired signal $x_d(\cdot) \in I\!R$ defined as follows:

$$\dot{x}_d = -x_d + r(t) \tag{7.35}$$

where $r(\cdot)$ is some time function. Let us assume first that $a^*$ and $b^*$ are known to the designer and define the tracking error as $e = x - x_d$. Then it is easy to see that the input

$$u = \frac{1}{b^*}(r - (a^* + 1)x) \tag{7.36}$$

forces the closed-loop to behave like $\dot{e} = -e$ so that $e \to 0$ as $t \to +\infty$.
Let us assume now that $a^*$ and $b^*$ are unknown to the designer, but that it is known that $b^* > 0$. Let us rewrite the input in (7.36) as $u = \theta^{*T}\phi$, where $\theta^{*T} = (-\frac{a^*+1}{b^*}, \frac{1}{b^*})$ and $\phi^T = (x, r)$ are the vector of unknown parameters and the regressor, respectively. Clearly it is possible to rewrite the error dynamics as

$$\dot{e} = -e + b^* \left(-\theta^{*T}\phi + u\right) \tag{7.37}$$

Since the parameters are unknown, let us choose (following the so-called *certainty equivalence principle*, which is not a principle but mainly a heuristic method) the control as

$$u = \hat{\theta}^T \phi \tag{7.38}$$

where $\hat{\theta}^T = (\hat{\theta}_1, \hat{\theta}_2)$ is a vector of control parameters to be estimated online. Notice that we intentionally do not impose any structure on these parameters, since they are not meant to represent the plant parameters, but the control parameters: this is what is called *direct* adaptive control. An *indirect* adaptive scheme would aim at estimating the plant parameters and then introducing these estimates in the control input: this is not the case in what we shall describe in this part of the book. Introducing (7.38) into (7.37) we obtain

$$\dot{e} = -e + b\tilde{\theta}^T \phi \tag{7.39}$$

where $\tilde{\theta} = \hat{\theta} - \theta^*$. The reader may have a look now at (7.2) and (7.8) to guess what will follow. The dynamics in (7.39) may be rewritten as $[e](s) = \frac{1}{1+s} b[\tilde{\theta}^T \phi](s)$, where $[\cdot](s)$ denotes the Laplace transform and $s \in \mathbb{C}$. Consequently a gradient estimation algorithm should suffice to enable one to analyze the closed-loop scheme with the passivity theorem, since $\frac{b}{1+s}$ is SPR. Let us choose

$$\dot{\hat{\theta}} = -\phi e \tag{7.40}$$

As shown in subsection 4.3.1, this defines a passive operator $e \mapsto -\tilde{\theta}^T \phi$. The rest of the stability analysis follows as usual (except that since we deal here with a time-varying system, one has to resort to Barbalat's lemma to prove the asymptotic convergence of $e$ towards 0. The zero state detectability property plus Krasovskii-La Salle invariance lemma do not suffice so that the various results exposed in section 4.5 cannot be directly applied).

**Remark 7.6** .

- The system in (7.35) is called a model of reference, and this adaptive technique approach is called the Model Reference Adaptive Control MRAC [95].

- One can easily deduce the storage functions associated to each subsystem and form a Lyapunov candidate function for the overall closed-loop scheme.

- One may also proceed with a Lyapunov function analysis, and then retrieve the passivity interpretation using the results in subsection 6.2.4.

- We have supposed that $b^* > 0$. Clearly we could have supposed $b^* < 0$. However when the sign of $b^*$ is not known, then things complicate a lot. A solution consists of an indirect adaptive scheme with a modified estimation algorithm [106]. The above passivity design is lost in such schemes.

### 7.2.2 Systems with relative degree $n^* = 1$

Let us consider the following controllable and observable system

$$\begin{aligned} \dot{x} &= Ax + Bu \\ y &= C^T x \end{aligned} \tag{7.41}$$

with $u \in I\!R$, $y \in I\!R$, $x \in I\!R^n$, whose transfer function is given by

$$H(s) = k\frac{B(s)}{A(s)} = C^T(sI_n - A)^{-1}B \tag{7.42}$$

where $s$ is the Laplace variable. The constant $k$ is the high-frequency gain of the system, and we assume in the following that:

- $k > 0$.
- $A(s)$ and $B(s)$ are monic polynomials, and $B(s)$ is Hurwitz (the system has strictly stable zero dynamics), with known order $m = n - 1$.

The problem is basically that of cancelling the dynamics of the process with a suitable dynamic output feedback in order to get a closed-loop system whose dynamics matches that of a given *reference model* with input $r(t)$ and output $y_m(t)$. The reference model transfer function is given by

$$H_m(s) = k_m \frac{B_m(s)}{A_m(s)} \tag{7.43}$$

where $H_m(s)$ is chosen as a SPR transfer function.
The control problem is that of output tracking, i.e. one desires to find out a differentiator-free dynamic output feedback such that all closed-loop signals remain bounded, and such that $\lim_{t \to +\infty} |y(t) - y_m(t)| = 0$. It is clear that one chooses $r(t)$ bounded so that $y_m(t)$ is. Due to the fact that the parameters of the polynomials $A(s)$ and $B(s)$ as well as $k$ are unknown, the exact cancellation procedure cannot be achieved. Actually the problem can be seen as follows: in the ideal case when the process parameters are known, one is able to find out a dynamic output controller of the following form

$$\begin{cases} u = \theta^T \phi(r, \omega_1^T, y, \omega_2^T) \\ \dot{\omega}_1 = \Lambda \omega_1 + bu, \qquad \dot{\omega}_2 = \Lambda \omega_2 + by \\ \phi^T = [r, \omega_1^T, y, \omega_2^T], \qquad \theta^T = [k_c, \theta_1, \theta_0, \theta_2] \end{cases} \tag{7.44}$$

with $\omega_1(t)$, $\theta_1$, $\theta_2$ and $\omega_2(t) \in I\!R^{n-1}$, $\theta_0 \in I\!R$, and $(\Lambda, b)$ is controllable. One sees immediately that $u$ in (7.44) is a dynamic output feedback controller with a feedforward term. The set of gains $[k, \theta_1, \theta_0, \theta_2]$ can be properly chosen such that the closed-loop transfer function

$$H_0(s) = \frac{k_c k B(s)\lambda(s)}{(\lambda(s) - C(s))A(s) - kB(s)D(s)} = H_m(s) \tag{7.45}$$

where the transfer function of the feedforward term is given by $\frac{\lambda(s)}{\lambda(s)-C(s)}$ while that of the feedback term is given by $\frac{D(s)}{\lambda(s)}$. $C(s)$ has order $n-2$ and $D(s)$ has order $n-1$. Notice that $\lambda(s)$ is just the characteristic polynomial of the matrix $\Lambda$, i.e. $\lambda(s) = (sI_{n-1} - \Lambda)^{-1}$ and is therefore Hurwitz. We do not develop further the model matching equations here (see e.g. [144] or [172] for details). Let us just denote the set of "ideal" controller parameters such that (7.45) holds as $\theta^*$. In general those gains will be combinations of the process parameters. Let us now write down the state space equations of the whole system. Notice that we have

$$\dot{z} \triangleq \begin{bmatrix} \dot{x} \\ \dot{\omega}_1 \\ \dot{\omega}_2 \end{bmatrix} = \begin{bmatrix} A & 0 & 0 \\ 0 & \Lambda & 0 \\ bC^T x & 0 & \Lambda \end{bmatrix} z + \begin{bmatrix} B \\ b \\ 0 \end{bmatrix} u \tag{7.46}$$

from which one deduces using (7.44) that

$$\dot{z} = \begin{bmatrix} A + B\theta_0^{*T} C^T & B\theta_1^{*T} & B\theta_2^{*T} \\ b\theta_0^* C^T & \Lambda + b\theta_1^{*T} & b\theta_2^{*T} \\ bC^T & 0 & \Lambda \end{bmatrix} z + \begin{bmatrix} Bk^* \\ bk^* \\ 0 \end{bmatrix} r(t) \tag{7.47}$$

Now since the process parameters are unknown, so is $\theta^*$. The controller in (7.44) is thus replaced by its estimated counterpart, i.e. $u = \hat{\theta}\phi$. This gives rise to exactly the same closed-loop structure as in (7.47), except that $\theta^*$ is replaced by $\hat{\theta}$. Notice that the system in (7.47) is not controllable nor observable, but it is stabilizable and detectable. Also its transfer function is exactly equal to $H_0(s)$ when the input is $r(t)$ and the output is $y$. This is therefore a SPR transfer function.

Now we have seen in the manipulator adaptive control case that the classical way to proceed is to add and substract $\theta^{*T}\phi$ in the right-hand-side of (7.47) in order to get (see (7.46) and (7.47)) a system of the form

$$\dot{z} = A_m z + B_m \tilde{\theta}^T \phi + B_m k^* r(t) \tag{7.48}$$

where $A_m$ is given in the right-hand-side of (7.47) while $B_m$ is in the right-hand-side of (7.46) (actually in (7.47) the input matrix is given by $B_m k^*$).

We are now ready to set the last step of the analysis: to this end notice that we can define the same type of dynamical structure for the reference model as the one that has been developed for the process. One can define filters of the input $r(t)$ and of the output $y_m(t)$ similarly to the ones in (7.44). Let us denote their state as $\omega_{1m}$ and $\omega_{2m}$, whereas the total reference model state will be denoted as $z_m$. In other words one is able to write

$$\dot{z}_m = A_m z_m + B_m k^* r(t) \tag{7.49}$$

Defining $e = z - z_m$ and introducing (7.49) into (7.48) one gets the following error equation:

$$\dot{e} = A_m e + B_m \tilde{\theta}^T \phi \tag{7.50}$$

This needs to be compared with (7.8) and (7.2). Let us define the signal $e_1 = C_m^T e = C^T(x - x_m)$: clearly the transfer function $C_m^T(sI_{3n-2} - A_m)^{-1}B_m$ is equal to $H_m(s)$ which is SPR by definition. Hence the subsystem in (7.50) is strictly passive with input $\tilde{\theta}^T\phi$ and output $e_1$ (in the sense of lemma 4.4) and is also output strictly passive since it has relative degree $n^* = 1$ (see example 4.4). A gradient estimation algorithm of the form:

$$\dot{\tilde{\theta}} = -\lambda_1 \phi e_1 \tag{7.51}$$

$\lambda_1 > 0$, is passive with respect to the supply rate $u_2 y_2$ with $y_2 = -\tilde{\theta}^T\phi$ and $u_2 = e_1$. Due to the stabilizability properties of the first block in (7.50), it follows from the Meyer-Kalman-Yakubovic lemma that the overall system is asymptotically stable (take $\gamma = 0$). Indeed there exists a storage function $V_1(e) = e^T P e$ associated to the first block, and such that $V(e, \tilde{\theta}) = V_1(x) + \frac{1}{2}\tilde{\theta}^T\tilde{\theta}$ is a Lyapunov function for the system in (7.50) (7.51), i.e. one gets $\dot{V} = -e^T q q^T e \le 0$. Notice that in general the closed-loop system is not autonomous, hence Krasovskii-La Salle theorem does not apply. One has to resort to Barbalat's lemma (see the appendix) to prove the asymptotic convergence of the tracking error $e$ towards 0. Notice also that the form of $\dot{V}$ follows from a CTCE so that lemma 6.4 directly applies.

### 7.2.3 Systems with relative degree $n^* = 2$

Let us now concentrate on the case when the plant in (7.41) and (7.42) has relative degree two. Let us pass over the algebraic developments that allow one to show that there is a controller such that when the process parameters are known, then the closed-loop system has the same transfer function as the model reference. Such a controller is a dynamic output feedback of the form $u = \theta^{*T}\phi$. It is clear that one can repeat exactly the above relative degree one procedure to get a system as in (7.50) (7.51). However this time $H_m(s)$ cannot be chosen as a SPR transfer function, since it has relative degree two! Thus the interconnection interpretation through the passivity theorem no longer works.

The basic idea is to modify the input $u$ so that the transfer function between the estimator output and the first block output $e_1$ is no longer $H_m(s)$ but $(s+a)H_m(s)$ for some $a>0$ such that $(s+a)H_m(s)$ is SPR. To this end let us define a filtered regressor $\bar{\phi} = \frac{1}{s+a}[\phi]$, i.e. $\dot{\bar{\phi}} + a\bar{\phi} = \phi$. Since we aim at obtaining a closed-loop system such that $e_1 = H_m(s)(s+a)\tilde{\theta}^T\bar{\phi}$, let us look for an input that realizes this goal:

$$
\begin{aligned}
e_1 &= H_m(s)(s+a)\tilde{\theta}^T\bar{\phi} \\
&= H_m(s)[\dot{\tilde{\theta}}^T\bar{\phi} + \tilde{\theta}^T\dot{\bar{\phi}} + a\tilde{\theta}^T\bar{\phi}] \\
&= H_m(s)[\dot{\tilde{\theta}}^T\bar{\phi} + \tilde{\theta}^T(\phi - a\bar{\phi}) + a\tilde{\theta}^T\bar{\phi}] \\
&= H_m(s)[\dot{\hat{\theta}}^T\bar{\phi} + \tilde{\theta}^T\phi] \\
&= H_m(s)[u - \theta^{*T}\phi]
\end{aligned}
\tag{7.52}
$$

It follows that a controller of the form

$$
u = \dot{\hat{\theta}}^T\bar{\phi} + \hat{\theta}^T\phi \tag{7.53}
$$

will be suitable. Indeed one can proceed as for the relative degree one case, i.e. add and substract $\theta^{*T}\phi$ to $u$ in order to get $\dot{z} = A_m z + B_m(\dot{\tilde{\theta}}^T\bar{\phi} + \tilde{\theta}^T\phi)$ such that the transfer function between $\tilde{\theta}^T\bar{\phi}$ and $e_1$ is $H_m(s)(s+a)$. Then the update law can be logically chosen as

$$
\dot{\tilde{\theta}} = -\lambda_1\bar{\phi}e_1 \tag{7.54}
$$

(compare with (7.51)), and the rest of the proof follows.

### 7.2.4 Systems with relative degree $n^* \geq 3$

The controller in (7.53) is implementable without differentiation of the plant output $y$ because the derivative $\dot{\hat{\theta}}$ is available. The extension of the underlying idea towards the case $n^* \geq 3$ would imply it is possible to have at one's disposal an estimation algorithm that provides the higher order derivatives of the estimates: this is not the case of a simple gradient update law. The relative degree problem has been for a long time a major obstacle in direct adaptive control theory. The next two paragraphs briefly present two solutions: the first one uses the backstepping method that we already used in subsection 6.4.2 to derive a globally stable tracking controller for the flexible joint-rigid link manipulators. It was presented in [91]. The second method is due to

Morse [139] can be considered as an extension of the controllers in subsections 7.2.2 and 7.2.3. It bases on the design of update laws which provide $\hat{\theta}$ as well as its derivatives up to the order $n^* - 1$. In the following we shall restrict ourselves to the presentation of the closed-loop error equations: the whole developments would take us too far.

### The backstepping approach

Given a plant defined as in (7.41), $n^* = n - m$, it is possible to design $u(t)$ such that the closed-loop system becomes :

$$\left\{\begin{aligned} \dot{z} &= A(z,t,\Gamma)z + b(z,t,\Gamma)(\omega^T\tilde{\theta} + \varepsilon_2) \\ \dot{\tilde{\theta}} &= -\Gamma\omega b^T(z,t,\Gamma)z \\ \dot{\varepsilon} &= A_0\varepsilon \\ \dot{\tilde{\eta}} &= A_0\tilde{\eta} + e_n z_1 \\ \dot{\tilde{\zeta}} &= A_b\tilde{\zeta} + \bar{b}z_1 \end{aligned}\right. \tag{7.55}$$

where $\tilde{\theta} \in I\!R^{(m+n)\times 1}$, $\omega \in I\!R^{(m+n)\times 1}$, $\bar{b} \in I\!R^{m\times 1}$, $b \in I\!R^{n^*\times 1}$, $z \in I\!R^{n^*\times 1}$, $e_n \in I\!R^{n\times 1}$ and is the $n^{th}$ coordinate vector in $I\!R^n$, , $\tilde{\eta} \in I\!R^{n\times 1}$, $\tilde{\zeta} \in I\!R^{m\times 1}$. $z_1$ is the first component of $z$ and $z_1 = y - y_r$ is the tracking error, $y_r$ is the reference signal, all other terms come from filtered values of the input $u(t)$ and the output $y(t)$. $A_b$ and $A_0$ are stable matrices. What is important in the context of our study is that the closed-loop system in (7.55) can be shown to be stable using the function

$$V = V_z(z) + V_\varepsilon(\varepsilon) + V_{\tilde{\theta}}(\tilde{\theta}) + V_{\tilde{\eta}}(\tilde{\eta}) + V_{\tilde{\zeta}}(\tilde{\zeta}) \tag{7.56}$$

whose time derivative along trajectories of (7.55) is

$$\dot{V} \le -\sum_{i=1}^{n^*} \lambda_i z_i^2 - \lambda_\varepsilon \varepsilon^T \varepsilon - \lambda_{\tilde{\eta}} \tilde{\eta}^T \tilde{\eta} - \lambda_{\tilde{\zeta}} \tilde{\zeta}^T \tilde{\zeta} \tag{7.57}$$

with $V_z$,$V_\varepsilon$,$V_{\tilde{\theta}}$,$V_{\tilde{\eta}}$,$V_{\tilde{\zeta}}$ positive definite functions, $\lambda_i > 0$, $1 \le i \le$, $\lambda_\varepsilon > 0$, $\lambda_{\tilde{\eta}} > 0$, $\lambda_{\tilde{\zeta}} > 0$.

Now let us have a look at equations in (7.55): Notice that we can rewrite the closed-loop system similarly as in (6.30) (6.31) as follows ($\bar{e}_1$ is the first component vector in $\mathbb{R}^{n^*}$):

$$\begin{pmatrix} \dot{z} \\ \dot{\tilde{\eta}} \\ \dot{\tilde{\zeta}} \end{pmatrix} = \begin{pmatrix} A & 0 & 0 \\ e_n\bar{e}_1^T & A_0 & 0 \\ \bar{b}\bar{e}_1^T & 0 & A_b \end{pmatrix} \begin{pmatrix} z \\ \tilde{\eta} \\ \tilde{\zeta} \end{pmatrix} + \begin{pmatrix} b\omega^T\tilde{\theta} \\ 0 \\ 0 \end{pmatrix} + \begin{pmatrix} b\varepsilon_2 \\ 0 \\ 0 \end{pmatrix} \tag{7.58}$$

$$\dot{\tilde{\theta}} = -\Gamma\omega b^T z \tag{7.59}$$

We can thus directly conclude from lemma 6.4 that the closed-loop system can be transformed into a system in $\mathcal{P}$ [3]. With the notations of the preceding section, we get $V_1{=}V_{\tilde{\theta}} = \frac{1}{2}\tilde{\theta}^T\Gamma^{-1}\tilde{\theta}$, $V_2{=}V_z + V_{\tilde{\eta}} + V_{\tilde{\zeta}}$, $y_2{=}-u_1{=}z$, $y_1{=}u_2{=}b\omega^T\tilde{\theta}$, $g_1{=}\Gamma\omega b^T$ $(\frac{\partial V_{\tilde{\theta}}}{\partial \theta} {=}\Gamma^{-1}\tilde{\theta})$, $g_2{=}\begin{pmatrix} I_{n^*} \\ 0_{n\times n^*} \\ 0_{m\times n^*} \end{pmatrix}$ $(\frac{\partial V_2}{\partial x_2}{=}\begin{pmatrix} \frac{\partial V_z}{\partial z} \\ \frac{\partial V_{\tilde{\eta}}}{\partial \tilde{\eta}} \\ \frac{\partial V_{\tilde{\zeta}}}{\partial \tilde{\zeta}} \end{pmatrix})$. The CTCE is verified as $\frac{\partial V_1}{\partial x_1}^T g_1 u_1{=}-\frac{\partial V_2}{\partial x_2}^T g_2 u_2{=}-z^T b\omega^T\tilde{\theta}$. $I_{n^*}$ is the $n^*$-dimensional identity matrix, $0_{i\times j}$ is the $i \times j$ zero matrix.

### Morse's high order tuners

Similarly to the preceding case, we only present here the closed-loop equations without entering into details on how the different terms are obtained. The interested reader can consult the original paper [139] and [149] for a comprehensive study of high order tuners. The closed-loop equations are the following

$$\dot{e} = -\lambda e + q_0\tilde{\theta}^T\omega + q_0\sum_{i=1}^{m}\omega_i\bar{c}z_i + \varepsilon \tag{7.60}$$

$$\dot{z}_i = \bar{A}z_i(1+\mu\omega_i^2) - \text{sign}(q_0)\bar{A}^{-1}\bar{b}\omega_i e, i \in \mathbf{m} \tag{7.61}$$

$$k_i - h_i = \bar{c}z_i, i \in \mathbf{m} \tag{7.62}$$

$$\dot{\tilde{\theta}} = -\text{sign}(q_0)\omega e \tag{7.63}$$

where $\mathbf{m} = \{1,...,m\}$, $e$ is the scalar tracking error, $\lambda > 0$, $q_0$ is the high frequency gain of the open-loop system, $|q_0| \leq \bar{q}$, $k \in \mathbb{R}^m$ is the vector of estimated parameters to be tuned, $h$ is an internal signal of the high order tuner, $\tilde{\theta} = h - q_P$, $q_P \in \mathbb{R}^m$ is a vector of unknown parameters, $(\bar{c}, \bar{A}, \bar{b})$ is the minimal realization of a stable transfer function, $\omega \in \mathbb{R}^m$ is a regressor, and $\varepsilon$ is an exponentially decaying term due to non-zero initial conditions. $k_i$ and

[3] $\varepsilon_2$ can be seen as a $\mathcal{L}_2$-bounded disturbance and is therefore not important in our study.

$h_i$ denote the $i^{th}$ component of $k$ and $h$ respectively, whereas $\mu$ is a constant satisfying $\mu > \frac{2m\bar{q}\|\bar{c}^T\|.\|P\bar{A}^{-1}\bar{b}\|}{\lambda}$.
It is proved in [139] that the system in (7.60) through (7.63) is stable using the function

$$V = e^2 + |q_0|\tilde{\theta}^T\tilde{\theta} + \delta\sum_{i=1}^{m} z_i^T P z_i \tag{7.64}$$

where $\bar{A}^T P + P\bar{A} = -1_m$ ($1_m$ is the m-dimensional identity matrix), $\delta = \frac{\bar{q}\|\bar{c}^T\|}{\|P\bar{A}^{-1}\bar{b}\|}$. The time derivative of $V$ along trajectories of (7.60) through (7.63) is given by

$$\dot{V} \leq -\lambda^\star e^2 + \frac{1}{\lambda^\star}\varepsilon^2 \tag{7.65}$$

with $\lambda^\star = \lambda - \frac{2m\bar{q}\|\bar{c}^T\|.\|P\bar{A}^{-1}\bar{b}\|}{\mu}$.
Now let us rewrite the system in (7.60) through (7.63) as follows:

$$\begin{pmatrix} \dot{e} \\ \dot{z}_1 \\ \dot{z}_2 \\ \cdots \\ \dot{z}_m \end{pmatrix} =$$

$$\begin{pmatrix} -\lambda & q_0\omega_1\bar{c} & \cdots & \cdots & \cdots & q_0\omega_m\bar{c} \\ -sgn(q_0)\bar{A}^{-1}\bar{b}\omega_1 & \bar{A}(1+\mu\omega_1^2) & 0 & \cdots & \cdots & 0 \\ -sgn(q_0)\bar{A}^{-1}\bar{b}\omega_2 & 0 & \bar{A}(1+\mu\omega_2^2) & 0 & \cdots & 0 \\ \cdots & \cdots & \cdots & \cdots & \cdots & \cdots \\ -sgn(q_0)\bar{A}^{-1}\bar{b}\omega_m & 0 & \cdots & \cdots & 0 & \bar{A}(1+\mu\omega_m^2) \end{pmatrix}.$$

$$\times \begin{pmatrix} e \\ z_1 \\ z_2 \\ \cdots \\ z_m \end{pmatrix} + \begin{pmatrix} q_0\tilde{\theta}^T\omega \\ 0 \\ 0 \\ \cdots \\ 0 \end{pmatrix} + \begin{pmatrix} \varepsilon \\ 0 \\ 0 \\ \cdots \\ 0 \end{pmatrix} \tag{7.66}$$

$$\dot{\tilde{\theta}} = -\text{sgn}(q_0)\omega e \tag{7.67}$$

We conclude from Corollary 4 that the system in (7.66) (7.67) belongs to $\mathcal{P}$, with $V_1 = |q_0|\tilde{\theta}^T\tilde{\theta}$, $V_2 = e^2 + \delta\sum_{i=1}^m z_i^T P z_i$, $g_1 = -q_0\omega$, $g_2 = \begin{pmatrix} 1 \\ 0 \\ 0 \\ \cdots \\ 0 \end{pmatrix}$, $u_1 = -y_2 = -e$, $u_2 = y_1 = q_0\omega^T\tilde{\theta}$. (We can neglect $\varepsilon$ in the analysis or consider it as a $L_2$-bounded disturbance).

**Remark 7.7** .

- Comparing (7.58)(7.59) and (7.66) (7.67) we conclude that the closed-loop error equations in both cases are very much similar. However, this similarity is limited to the closed-loop system stability analysis. Firstly, the basic philosophies of each scheme are very different: Roughly speaking, the *high order tuners* philosophy aims at rendering the operator between the tracking error and the estimates strictly passive (using a control input that is the extension of "classical" certainty equivalent control laws), while preserving stability of the overall system with an appropriate update law; On the contrary, the *backstepping* method is based on the use of a very simple "classical" update law (a passive gradient), and the difficulty is to design a control input (quite different in essence from the certainty equivalent control laws) which guarantees stability. Secondly, notice that $\tilde{\theta}$ in (7.59) truly represents the unknown parameters estimates, while $\tilde{\theta}$ in (7.67) is the difference between the vector of unknown plant parameters and a signal $h$ internal to the high order update law (the control input being computed with the estimates $k$ and their derivatives up to the plant relative degree minus one). Thirdly, the tracking error in the *backstepping* scheme is part of a $n^*$-dimensional differential equation (see the first equation in (7.55)), while it is the solution of a first order equation in the *high order tuner* method (see (7.60)).
- In [149], it is proved that the high order tuner that leads to the error equations in (7.60) through (7.63) defines a passive operator between the tracking error $e$ and $(k - q_P)^T \omega$, and that this leads to nice properties of the closed-loop system, such as guaranteed speed of convergence of the tracking error towards zero. In [92], it has been shown that the backstepping method also possesses interesting transient performances. Such results tend to prove that the schemes that base on passivity properties possess nice closed-loop properties.

## 7.3 Conclusions and comments

Let us propose a tentative conclusion of chapters 5, 6 and 7. The general tendency that emerges from these developments is that for tracking control of Lagrangian systems, one just tries to mimic the storage function of the open-loop system and hopes (prays?) that a state feedback controller can be designed this way. In this sense passivity-based methods differ from those methods that rely on a step-by-step construction of a Lyapunov function, since the Lyapunov function candidate is fixed *a priori*. Roughly, the major ingredients for showing asymptotic stability are the passivity theorem, the strict passivity of the operator associated to the closed-loop system, and Krasovskii-La Salle invariance principle in case one gets only strict *output* passivity. In this

setting lack of zero-state detectability is closely related to the loss of asymptotic stability. It does not seem that any general method for stabilization of open-loop dissipative systems may ever be settled. The flexible joint-rigid link manipulator case (see remark 5.11), as well as the case when actuator dynamics are taken into account, show that an apparently slight modification of the drift term of the dynamics or of the interconnection terms, may result in great difficulty for the realization of this passivity-based stabilization program. The adaptive gravity compensation is also a good example of such difficulty, where the storage function is quite original, similarly to the adaptive extension of passivity-based schemes for flexible joint-rigid link manipulators. The position feedback schemes also prove that the extension of state feedback passivity-based controllers towards output feedback is far from being trivial. But such drawback is shared by any more or less general method for the stabilization of nonlinear systems, anyway, and the numerous presented schemes show that passivity-based controllers apply to a wide range of physical systems, despite their limitations. It may be argued that dissipativity tools do not bring much compared with Lyapunov functions. Let us say that this is at the same time true and false. On one hand, it is clear that the fundamental stability properties of dissipative systems coincide with Lyapunov stability: all dissipative systems (under some restrictions) are Lyapunov stable, but the contrary is not true. But on the other hand, dissipativity can help in designing stable controllers, because it may guide the designer towards solutions that would not be seen just looking at a Lyapunov function without any physical intuition. Recall also the nice property of interconnections of dissipative blocks in negative feedback interconnections, or of the sum of dissipative blocks. Altogether these results have often allowed the discovering of new controllers.
Let us note that the control of flexible joint-rigid link manipulators was at the beginning of the nineties a big challenge in the Robotics and Systems and Control communities. In particular the extension of passivity-based control for rigid joint-rigid link manipulators towards flexible joint ones, together with a globally stable adaptive version (for any unknown joint stiffness) was first presented in [111] [113] [29]. From a historical perspective, [113] was published almost simultaneously with [82] where the so-called *backstepping* method is developed for a class of triangular nonlinear systems (see [25] for a survey of backstepping techniques applied to flexible joint manipulators). Actually the study in [113] really hinged upon passivity arguments, from the choice of the positive definite function $V$ in (6.97), and $q_{2d}(t)$ was just thought to be an intermediate input whose value could be chosen so as to obtain a negative definite $\dot{V}$. It may therefore be argued that the backstepping-based scheme in subsection 6.4.2 and the passivity-based one in subsection 6.4.1 are quite similar and designed with the same method. Notice however that our point of view in the presented analysis bases on the interconnection of two passive blocks, and not on a particular state space representation structure of the overall system. Moreover recall that one difficulty in the design of backstepping-

based controllers is in the choice of the Lyapunov functions at each step. The passivity-oriented method possesses the advantage of providing a controller in one-shot. Moreover experimental results (see chapter 8) confirm that they do possess interesting closed-loop performance. It is also worth recalling that the adaptive control of flexible joint-rigid link manipulators with the model in (5.115) when the joint stiffness is unknown is not a trivial application of the general results of backstepping methods. Indeed the interconnection term is then unknown and has to be estimated. A smooth kind of projection is used in [113] to cope with this problem. Moreover the fixed parameter controller is nonlinear in the unknown inertial parameters, due to the term $M^{-1}(q_1)$ which appears when one replaces $\ddot{q}_1$ from the first dynamical equation. Hence another trick has to be used as described above. Clearly the application of dissipative system theory in the field of feedback control is far from being closed. On the one hand it is expected that dissipativity properties of more complex (e.g. nonsmooth [27] or infinite dimensional systems [210] as in sections 2.12 2.13 2.17 4.11) than the ones studied in this book, will be needed. On the other hand (according to the experimental results exposed in chapter 8, see also [136] [137]) the existing results are likely to be applied to other classes of systems whose control problem will be raised by industrial or academic applications. It is not possible in such an introductory monograph to exhaustively present all the applications of dissipativity in control. The interested reader who desires to go further is therefore kindly referred to the cited literature.

# Chapter 8

# Experimental Results

## 8.1 Flexible joint manipulators

### 8.1.1 Introduction

In this chapter we present experimental results on two experimental mechanical systems. They illustrate the applicability of the methodologies exposed in the foregoing chapters. The first concerns flexible-joint manipulators, whose dynamics and control have been thoroughly explained. The second one focuses on an underactuated system, the inverted pendulum, which does not fall into the classes of mechanical systems presented so far. The state feedback control problem of flexible joint manipulators has constituted an interesting challenge in the Systems and Control and in the Robotics scientific communities. It was motivated by practical problems encountered for instance in industrial robots equipped with harmonic drives, that may decrease the tracking performances, or even sometimes destabilize the closed-loop system. Moreover as we pointed out in the previous chapter, it represented a pure academic problem, due to the particular structure of the model. From a historical point of view, the main directions that have been followed to solve the tracking and adaptive control problems have been: Singular Perturbation techniques (the stability results then require a high enough stiffness value at the joints so that the stability theoretical results make sense in practice) [195] [196], and nonlinear global tracking controllers derived from design tools such as the backstepping or the passivity-based techniques. We have described these last two families of schemes in the previous chapter, see sections 6.4, 6.4.2. In this section we aim at illustrating on two laboratory processes how these schemes work in practice and whether they bring significant performance improvement with respect to PD and the Slotine and Li controllers (which both can be cast into the passivity-based schemes, but do not a priori incorporate flexibility effects in their design). What follows is taken from [35] and [36]. More generally the goal of this section is to present experimental results for passivity-based controllers with increasing complexity, starting from the PD input. Let us stress that the reliability of the presented experimental works is increased by the fact that theoretical and numerical investigations predicted reasonably well the obtained behaviours of the real closed-loop plants, see [34]. The experimental results that follow should not be considered as a definitive answer to the question: "What is the best controller?". Indeed the answer to such a question may be very difficult, possibly impossible to give. Our goal is only to show that the concepts that were presented in the previous chapters may provide nice results in practice.

### 8.1.2 Controller design

In this work the model as introduced in [188] is used, see (5.115). As we saw in section 5.4 this model possesses nice passivity properties that make it quite attractive for control design, see sections 6.4, 6.4.2 and 6.5.1. Only fixed parameter controllers are considered here. As shown in [34] (see (6.103) and (6.116)), the three nonlinear controllers for flexible joint manipulators which are tested can be written shortly as follows:
*Controller 1*

$$\begin{cases} u = & J[\ddot{q}_{2d} - 2\dot{\tilde{q}}_2 - 2\tilde{q}_2 - K(\dot{s}_1 + s_1)] + K(q_2 - q_1) \\ q_{2d} = & K^{-1}u_R + q_1 \end{cases} \tag{8.1}$$

*Controller 2*

$$\begin{cases} u = & J[\ddot{q}_{2d} - 2\dot{\tilde{q}}_2 - 2\tilde{q}_2 - (\dot{s}_1 + s_1)] + K(q_2 - q_1) \\ q_{2d} = & K^{-1}u_R + q_1 \end{cases} \tag{8.2}$$

*Controller 3*

$$\begin{cases} u = & J\ddot{q}_{2r} + K(q_{2d} - q_{1d}) - B_2 s_2 \\ q_{2d} = & K^{-1}u_R + q_{1d} \end{cases} \tag{8.3}$$

where $u_R = M(q_1)\ddot{q}_{1r} + C(q_1, \dot{q}_1)\dot{q}_{1r} + g(q_1) - \lambda_1 s_1$ is as in (6.104). $\dot{q}_{1r} = \dot{q}_{1d} - \lambda\tilde{q}_1$, $s_1 = \dot{\tilde{q}}_1 + \lambda\tilde{q}_1$ are the classical signals used in the design of this controller (the same definitions apply with subscript 2). Let us reiterate that the expressions in (8.1), (8.2) and (8.3) are equivalent closed-loop representations. In particular no acceleration measurement is needed for the implementation, despite $\dot{s}_1$ may appear in the equivalent form of $u$.
As pointed out in remark 6.15, the last controller is in fact an improved version (in the sense that it is a static state feedback) of the dynamic state feedback proposed in [113] [29], that can be written as:

$$\begin{cases} u = & J\ddot{q}_{2r} - K[q_{1d} - q_{2d} - \int_0^t(\lambda_1\tilde{q}_1 - \lambda_2\tilde{q}_2)d\tau] - \lambda_2 s_2 \\ q_{2d} = & p[pI + \lambda_2]^{-1}\left\{K^{-1}u_R + q_{1d} - \int_0^t(\lambda_1\tilde{q}_1 - \lambda_2 q_2)d\tau\right\} \end{cases} \tag{8.4}$$

This controller has not been considered in the experiments, because it is logically expected not to provide better results than its simplified counterpart: It is more complex, but based on the same idea. Controllers 1 and 2 are designed following a backstepping approach. Let us recall that the backstepping method has been invented to apply to certain classes of triangular systems [119], where one constructs the controller recursively step by step, and then proves the stability with a suitable Lyapunov function of a new transformed state vector. Then convergence and boundedness of the original state follows. On the contrary for the passivity-based method, one chooses a positive-definite function whose form is inspired by storage functions of the process, and then designs the controller so that it becomes a Lyapunov function. The two backstepping controllers differ from the fact that in controller 2, the joint stiffness $K$ no longer appears before $\dot{s}_1 + s_1$ in the right-hand-side of the $u$-equation. This modification is expected to decrease significantly the input magnitude when $K$ is large. This will indeed be confirmed experimentally.
In [34] these controllers have been commented and discussed from several points of views. Most importantly it was shown that when the joint stiffness grows unbounded (i.e. the rigid manipulator model is retrieved), then the only controller that converges to the rigid Slotine and Li control law is the passivity-based controller 3 in (8.3). In this sense, it can be concluded that controller 3 is the extension of the rigid case to the flexible joint case, which cannot be stated for the other two control laws. We believe that this elegant physical property plays a major role in the closed-loop behaviour of the plant. It is however clear that there is no (analytical) rigorous proof of this claim. As shown in section 6.4.2 the backstepping schemes presented here *do* possess some closed-loop passivity properties. However they are related to transformed coordinates, as the reader may see in section 6.4.2. On the contrary, the passivity-based schemes possess this property in the original generalized coordinates $\tilde{q}$: consequently they are closer to the physical system than the other schemes. This is to be considered as an intuitive explanation of the good experimental results obtained with passivity-based schemes (PD, Slotine and Li, and controller 3).

### 8.1.3 The experimental devices

This subsection is devoted to present the two experimental devices in detail: a planar two degree-of-freedom (dof) manipulator, and a planar system of two pulleys with one actuator. They are shown on photographs 8.1.5 and 8.1.5 respectively. We shall concentrate on two points: the mechanical structure and the real time computer connected to the process. Actually we focus essentially in this description on the first plant, that is a two dof planar manipulator of the Laboratoire d'Automatique de Grenoble, named Capri. The second process is much more simple and is depicted in figure 8.1. It can be considered as an equivalent one dof flexible joint manipulator. Its dynamics is linear. Its physical parameters are given by: $I_1 = 0.0085$ kg.m$^2$, $I_2 = 0.0078$ kg.m$^2$, $K = 3.4$ Nm/rad.

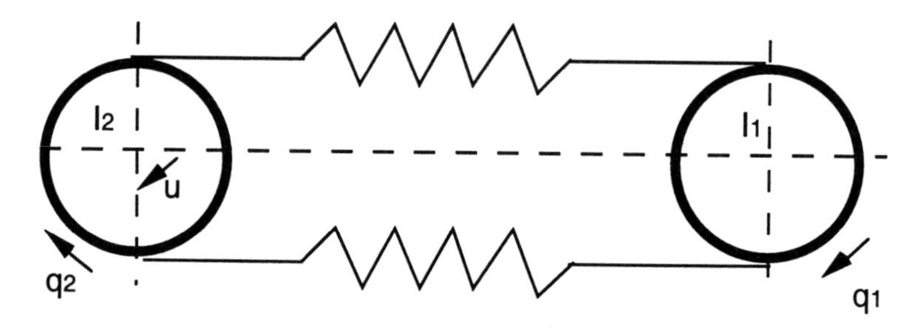


Figure 8.1: A one dof flexible joint manipulator.

### Mechanical structure of the Capri robot

The Capri robot is a planar mechanism constituted by two links, of respective lengths 0.16 m and 0.27 m, connected by two hubs. The first link is an aluminium AU4G, U-frame to improve stiffness, with respect to the forearm which can be designed less rigid. The second link has a more peculiar structure because it supports the applied forces: It is designed as a pipe of diameter 0.05 m, and it is equipped with force piezo-electric sensors. The force magnitude, point of application and orientation can be measured and calculated. The sides of the forearm with Kistler quartz load washers can measure extension and compression forces, and the half-spherical extremity possesses a Kistler 3 components force transducer (only two of them are used) from which it is possible to calculate the magnitude and the orientation of the applied force. In this work these force measurement devices are not needed, since we are concerned by motion control only.
The robot arm is actuated by two DC motors located at the underside of the basement table (therefore the Capri robot is a parallel-drive manipulator for which the model in (5.115) is the "exact" one, see remark 5.11). They are coupled to the links by reducers (gears and notched belts), each of them with ratio 1/50. The first motor (Infranor MX 10) delivers a continuous torque of 30 N.cm and a peak torque of 220 N.cm for a total weight of 0.85 kg. The second motor (Movinor MR 08) provides a continuous torque of 19 N.cm and a peak torque of 200 N.cm, for a weight of 0.65 kg. The drive arrangement

is such that the weight is not boarded on the links, to increase speed. Both motors are equipped with a 500 pulses/turn incremental encoder and a DC tachometer making joint position $q_2$ and velocity $\dot{q}_2$ available for feedback. The position $q_1$ is measured by a potentiometer mounted on the last link. In the experiments the velocity $\dot{q}_1$ has been obtained by differentiating the position signal (a filtering action has been incorporated by calculating the derivative from one measurement every four only, i.e. every 4 sampling times).
The effective working area of the robot arm is bounded by sensors: an inductive sensor prevents the first arm from doing more than one turn, i.e. $q_{11} \in [-\frac{\pi}{2}, \frac{\pi}{2}]$ (see figure 8.2 for the definition of the angles). Two microswitches prevents the second arm from overlapping on the first one. They both inhibit the inverters (Infranor MSM 1207) controlling the DC motors.

**Remark 8.1** The Capri robot has been modeled as a parallel-drive rigid-link robot, with the second joint elastic. It is clear that such a model is only a crude approximation of the real device. Some approximations may be quite justified, like the rigidity of the first joint and of the links. Some others are much more inaccurate.

**i)** The belt that couples the second actuator and the second joint is modeled as a spring with constant stiffness, which means that only the first mode of its dynamic response is considered.

**ii)** There is some clearance in the mechanical transmission (especially at the joints, due to the belts and the pulleys), and a serious amount of dry friction.

**iii)** The frequency inverters that deliver the current to the motors possess a nonsymmetric dead zone. Therefore, different amounts of current are necessary to start motion in one direction or the other.

**iv)** The value of $\dot{q}_1$ used in the algorithm and obtained by differentiating a potentiometer signal is noisy, despite a filtering action.

**v)** The inertial parameters have been calculated by simply measuring and weightening the mechanical elements of the arms. The second joint stiffness has been measured statically off-line. It has been found to be 50 Nm/rad. This value has been used in the experiments without any further identification procedure.

**vi)** Some saturation on the actuators currents has been imposed by software, for obvious safety reasons. Since nothing ***a priori*** guarantees stability when the inputs are saturated, the feedback gains have to be chosen so that the control input remains inside these limits.

Some of these approximations stem from the process to be controlled, and cannot be avoided (points **i, ii, iii**): this would imply modifying the mechanical structure. The measurement noise effects in **iv** could perhaps be avoided via the use of observers or of position dynamic feedbacks. However on one hand the robustness improvement is not guaranteed and would deserve a deep analytical study. On the other hand the structure of the obtained schemes would be significantly modified (compare for instance the schemes in sections 6.2.5 and 6.3 respectively). A much more simple solution consists of replacing the potentiometer by an optical encoder. The saturation in **vi** is necessary to protect the motors, and has been chosen in accordance with the manufacturer recommendations and our own experience on their natural "robustness". The crude identification procedure in **v** has been judged sufficient, because the aim of the work was not to make a controller perform as well as possible in view of an industrial application, but rather to compare several controllers and to show that nonlinear control schemes behave well. In view of this the most important fact is that they be all tested with the same (acceptable) parameters values, i.e. if one controller proves to behave correctly with these set of parameters, do the others behave as well or not? Another problem is that of the choice of the control parameters, i.e. feedback gains. We will come back on this important point later.

### Real-time computer

A real-time computer is connected to both processes in the workshop of the Laboratoire d'Automatique de Grenoble. It consists of a set of DSpace boards and a host PC. The PC is a HP Vectra running at 66 MHz with 8 Mo of RAM and a hard disk of 240 Mo. The DSpace system is made of:

- A DS 1002 floating-point processor board built around the Texas Instruments TMS/320C30 digital signal processor. This processor allows 32 bits floating point computation at 33 MFlops. A static memory of 128 K words of 32 bits is available on this board. A 2 K words dual-port RAM is used simultaneously by the host PC and the DSP.

- A DS 2002 multi-channel ADC board with 2 A/D 16 bits resolution converters (5 $\mu$s conversion time) and a 16 channel multiplexer for each converter.

- A DS 2001 D/A converter board comprising 5 parallel analog output channels with 12 bits DAC (3 $\mu$s conversion time)

- A DS 3001 incremental encoder board with 5 parallel input channels. A fourfold pulse multiplication, a digital noise filter and a 24 bits width counter are used for each channel.

- A DS 4001 digitak I/O and timer board with 32 digital I/O lines configurable as inputs or outputs in groups of 8 lines.

All these boards are attached together by the 32 bits PHS-Bus at a 16 MB/sec transfer speed. They are located in a separate rack connected to the host PC by a cable between two adaptation boards.
The PC is used for developments and supervision of the application. Several softwares are available for the DSpace system:

- SED30 and MON30 are used to configure the hardware.
- C30 is the Texas Instruments Compiler for the TMS320C30.
- TRACE30W is a graphical real-time software which permits to display the selected variables of the application.

The application itself is made of two parts: The control algorithm runing on the DSP, sampled at 1 ms in our case, and the dialogue interface running on the PC which allows the operator to supervise the execution of the control through the dual port memory.
To guarantee repeatability of the experiments, there is an initialization procedure that is to be activated each time the origins have been lost, or at the begining of the experiment.

### 8.1.4 Experimental results

In this section we present the experimental results obtained by implementing the three controllers described above on each plant. A PD controller as in (6.118), and the scheme in (6.51) have also been implemented, as if the manipulator had both joints rigid (i.e. one replaces $q$ in (6.51) by $q_2$). This allows to clearly dissociate the effects of the nonlinearities (the reference trajectories have been chosen fast enough so that Coriolis and centrifugal effects are effective), from the effects of the flexibility (once the "rigid" controllers are implemented, one can see how the "flexible" ones improve the closed-loop behaviour, if they do). In the case of the linear system in figure 8.1, the scheme in (6.51) reduces to a PD control.
In order to perform the experiments, three different desired trajectories have been implemented for the Capri robot (see figure 8.2 for the definition of the angles, due to the fact that the Capri robot is a parallel-drive manipulator):

- Desired trajectory 1: $q_{1d} = \begin{pmatrix} q_{11d} \\ q_{12d} \end{pmatrix} = \begin{pmatrix} 0.8\sin(ft) \\ -0.8\sin(ft) \end{pmatrix}$
- Desired trajectory 2: $q_{1d} = \begin{pmatrix} 0.4\sin(2ft) \\ 0.8\sin(ft) \end{pmatrix}$

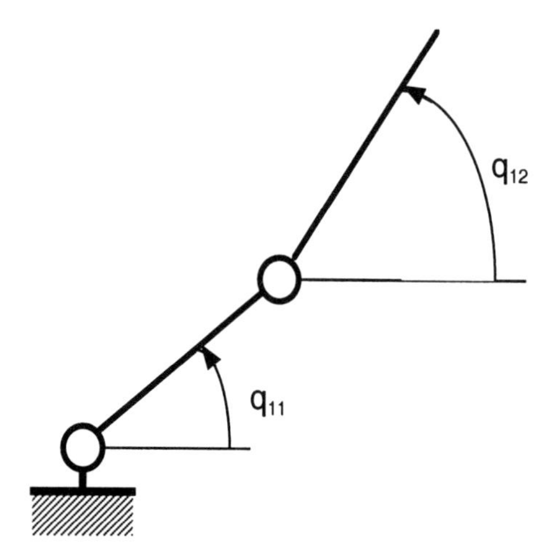


Figure 8.2: Joint angles on the Capri robot.

- Desired trajectory 3: $q_{1d} = \begin{pmatrix} \frac{b^5}{(s+b)^5}[g(t)] \\ -\frac{b^5}{(s+b)^5}[g(t)] \end{pmatrix}$

with $f = \omega(1 - \exp(-at))^4$, $a = 14$, $\omega = 9$ rad/s, $g(t)$ is a square function with magnitude 0.8 rad, period 5 s, and $b = 30$. The variable $s$ is the Laplace transform variable. The choice for $f$ allows one to smooth sufficiently the desired orbit to be tracked, as required by the theoretical developments. The other parameters values have been chosen so that the nonlinearities and the flexibilities effects are significant. Concerning the system in figure 8.1, two desired trajectories have been chosen: $q_{1d} = sin(\omega t)$ and $q_{1d} = \frac{b^5}{(s+b)^5}[g(t)]$. The parameters $\omega$ and $b$ have been varied as indicated in the figures captions. These time functions, which are sufficiently different to one another, have been chosen to permit to conclude about the capability of adaptation of the controllers to a modification of the desired motion. This is believed to constitute an important property in applications, since it dispenses the user from retuning the control gains between two different tasks. As a matter of fact, the following criteria have been retained to evaluate the performance of the controllers:

- The tracking error during the steady-state regime is an important parameter for performance evaluation. The quadratic errors sums $e_i = \int_{10}^{20} \tilde{q}_i^2(t)dt$ for each joint ($i = 1, 2$ for the Capri robot and $i = 3$ for the pulleys) and the maximum tracking error (pulleys) have been computed on-line.

- The shape and magnitude of the input signal.

- The capabilities of the various control schemes to provide an acceptable performance for any of the above desired motions, without having to retune the feedback gains.

The transient behaviour has not been included in this list. This will be explained from the observation of the experimental results. Let us emphasize that the presented results therefore concern two quite different plants (one nonlinear with high stiffness, the other one linear and with high flexibility), and with significantly different motions. They are consequently expected to provide an objective view of the capabilities of the various controllers.

**Remark 8.2 (Feedback gains tuning method)** Two methods have been employed to tune the gains. From a general point of view, one has to confess that one of the main drawbacks of nonlinear controllers such as backstepping and passivity-based ones, is that Lyapunov-like analysis does not provide the designer or the user with any acceptable way to tune the gains. The fact that increasing the gains accelerates the convergence of the Lyapunov function towards zero, is a nice theoretical result, that happens to be somewhat limited in practice.
Concerning the Capri robot, experiments were started with the first link fixed with respect to the base, i.e. with only the second link to be controlled. The gains of the PD input were chosen from the second-order approximation obtained by assuming an infinite joint stiffness. From the fact that the Slotine and Li scheme in (6.51) mainly consists of a PD action plus a nonlinear part, these values have been used as a basis for the tuning of the gains $\lambda$ and $\lambda_1$ in (6.51). The full-order system is linear of order 4 (a one degree-of-freedom flexible joint manipulator). The gains were tuned by essentially placing the closed-loop poles according to simple criteria like an optimal response time, nonoscillatory modes. In all cases, the desired trajectory 1 was used to determine a first set of gains. This provided a basis to choose the gains for the complete robot. Experiments were started with trajectory 1, and the gains were modified in real-time (essentially by increasing them in a heuristic manner) until the performance observed through the TRACE30W could no more be improved. Then trajectories 2 and 3 were tested, and the gains modified again if needed.
It has to be stressed that even in the linear case (like for the pulley-system), tuning the gains of such nonlinear controls is not evident. Indeed the gains appear quite nonlinearly in general in the state feedback, and their influence on the closed-loop dynamics is not obvious. For instance it is difficult to find a region in the gain space of the passivity-based controller in (8.3), such that the gains can be modified and at the same time the poles remain real.
In view of these limitations and of the lack of a systematic manner to calculate optimal feedback gains, advantage has been taken in [35] of the pulley-system linearity. Since this system is linear, the controllers in (8.1), (8.2) and (8.3)

reduce to linear feedbacks of the form $u = Gx + h(t)$, where $h(t)$ accounts for the tracking terms. De Larminat [97] has proposed a systematic (and more or less heuristic) method to calculate the matrix $G$ for LQ controllers. Actually one should notice that despite the fact that the nonlinear backstepping and passivity-based controllers have a linear structure when applied to a linear system, their gains appear in a very nonlinear way in the state feedback matrix $G$. As an example, the term multiplying $q_1$ for the scheme in (8.3) is equal to $-(\lambda\lambda_2 + k)\frac{\lambda_1\lambda}{k} + (\lambda_2 + I_2\lambda)\frac{\lambda I_1+\lambda_1}{I_1} + I_2\frac{\lambda_1\lambda}{I_1}$ (the gains $\lambda_1$ and $\lambda_2$ can be introduced in (6.103) and (6.104) respectively instead of using only one gain in both expressions, so that the passivity-based controller has three gains). The tuning method proposed in [97] that applies to LQ controllers allows one to choose the weighting matrices of the quadratic form to be minimized, in accordance with the desired closed-loop bandwidth (or cut-off frequency $\omega_c(CL)$). The advantages of this method are that the user focuses on one closed-loop parameter only to tune the gains, which is quite appreciable in practice. Therefore one gets an "optimal" state feedback matrix $G_{\mathrm{LQ}}$, with a controller $u = G_{\mathrm{LQ}}x$ in the case of regulation. Since the various controllers used in the experiments yield some state feedback matrices $G_{\mathrm{PD}}$, $G_{\mathrm{BACK1}}$, $G_{\mathrm{BACK2}}$ and $G_{\mathrm{MES}}$ respectively, which are (highly) nonlinear functions of the gains as shown above, we choose to calculate their gains so that the norms $||G_{\mathrm{LQ}} - G_{\mathrm{CONT}}||$ are minimum. This amounts to solving a nonlinear set of equations $f(Z) = 0$ where $Z$ is the vector of gains. This is in general a hard task, since we do not *a priori* know any root (otherwise the job would be done!). This has been done numerically by constructing a grid in the gain space of each scheme and minimizing the above norm with a standard optimization routine. The experimental results prove that the method may work well, despite possible improvements (especially in the numerical way to solve $f(Z) = 0$). Its extension towards the nonlinear case remains an open problem. ♠♠

The quadratic error sums $e_1$, $e_2$ are reported in tables 8.1 and 8.2. The error $e_3$ is in table 8.3. The maximum tracking errors $|q_1 - q_d|_{\max}$ for the pulley-system are reported in table 8.4. All the results for the pulley-system in tables 8.3 and 8.4 concern the desired motion $q_{1d} = sin(\omega t)$. In each case the presented figures represent an average of several experiments. Concerning trajectories 2 and 3 in tables 8.1 and 8.2, the results outside brackets have been obtained after having retuned the feedback gains. The ones between brackets have been obtained using the same gains as for trajectory 1. When they are not modified, it means that we have not been able to improve the results. A cross x indicates that no feedback gains have been found to stabilize the system.
The next results that concern the Capri robot are reported in figures 8.3 to 8.21. The tracking errors $\tilde{q}_{11}$, $\tilde{q}_{12}$ and the inputs (currents) $I_{c1}$ and $I_{c2}$ at each motor, are depicted in figures 8.3 to 8.13. Figures 8.14 to 8.21 contain results concerning the transient behaviour when the second link position tracking

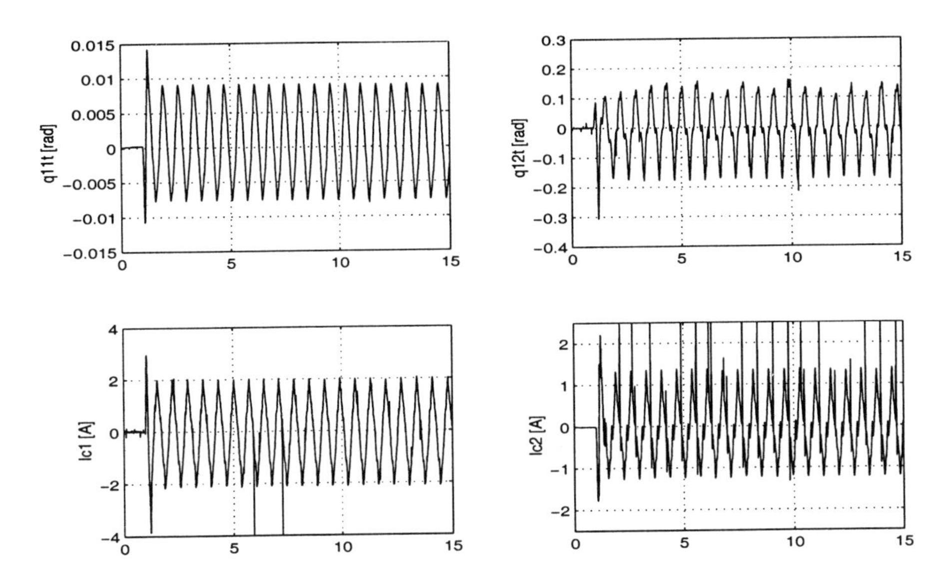


Figure 8.3: PD controller, desired trajectory 1

errors are initially of 0.4 rad. The inputs $I_{c1}$ and $I_{c2}$ are the calculated ones, not the true input of the actuators (they coincide as long as there is no saturation, i.e. $I_{c1} \leq 2$ A and $I_{c2} \leq 2$ A). The results concerning the pulley-system are in figures 8.29 to 8.28. The signals $q_d(t)$ and $q_1(t)$ are shown in the upper boxes, and the torque input $u$ is depicted in the lower boxes.
The following comments can be made:

### Adaptation to the desired motion

The gains of the PD controller that correspond to the tests on the Capri robot, reported in tables 8.1 and 8.2, are given in table 8.5. They show that significant changes have been necessary from one desired motion to the next. One sees that the PD gains have had to be modified drastically to maintain a reasonable performance level. On the contrary it is observable from tables 8.1 and 8.2 that even without any gain modification, the other controllers still perform well in general. In any case the modifications have seldom exceeded 50 % and concerned very few gains [36]. Since this is also true for the Slotine and Li controller, we conclude that the insensitivity of the performance with respect to desired motion changes is essentially due to the compensation of the nonlinearities.
The Slotine and Li controller seems to provide the most invariant performance with respect to the desired motion. This is especially apparent for trajectory 2 on the Capri experiments. In this case it provides the best error $e_2$, even after

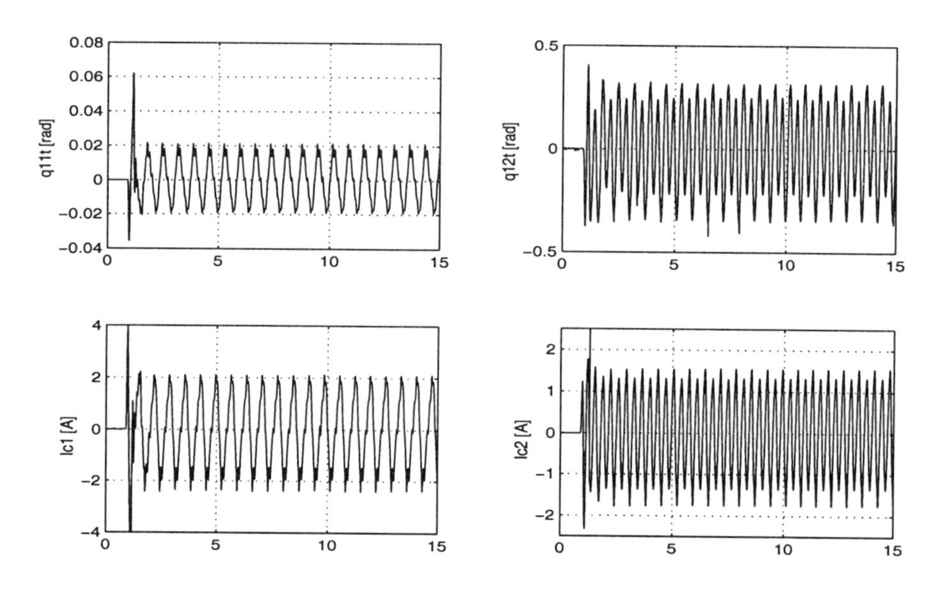


Figure 8.4: PD controller, desired trajectory 2

having retuned the gains for controllers 2 and 3. This may be explained by the fact that the input in (6.51) is much smoother than the others (see figure 8.7). This in turn may be a consequence of its simplicity, and from the fact that it does not use the noisy potentiometer signal.

### Backstepping controllers

For the Capri experiments, it has not been possible to find feedback gains that stabilize controller 1. On the contrary this has been possible for the pulley-system, see figures 8.30, 8.23, 8.26. This confirms the fact that the modification of the intermediate Lyapunov function (see (6.115) and (6.116)) may play a significant role in practice, and that the term $K(s_1 + \dot{s}_1)$ is a high-gain in the loop if $K$ is large.

### Compensation of nonlinearities

Although the PD algorithm provides a stable closed-loop behaviour in all cases (for the Capri experiments and at the price of very large gain modifications as we pointed out above), its performance is poor for trajectories 1 and 2. The behaviour is much better for trajectory 3. This can be explained since this is almost a regulation task. The improvements obtained with the Slotine and Li scheme show that the Coriolis and centrifugal terms may play an important role depending on the desired motion.

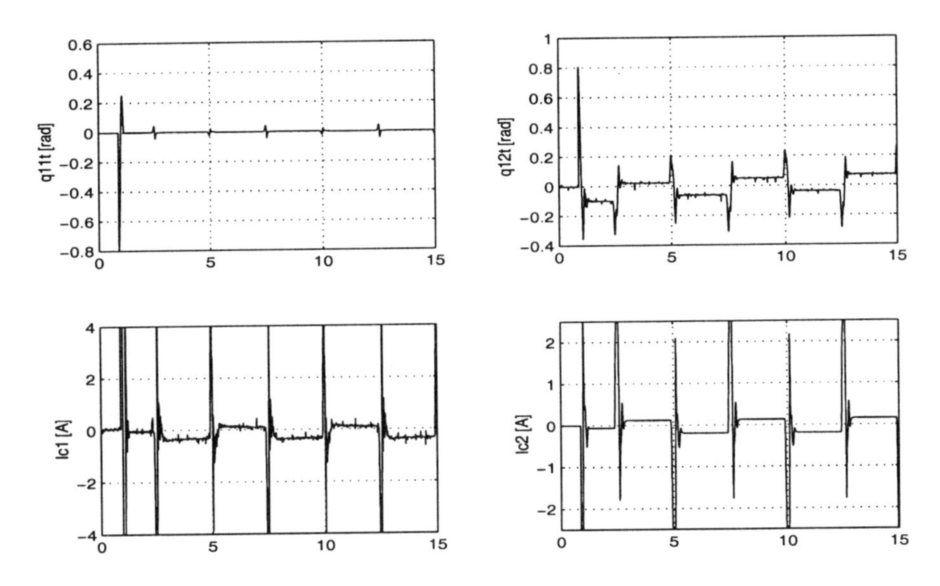


Figure 8.5: PD controller, desired trajectory 3

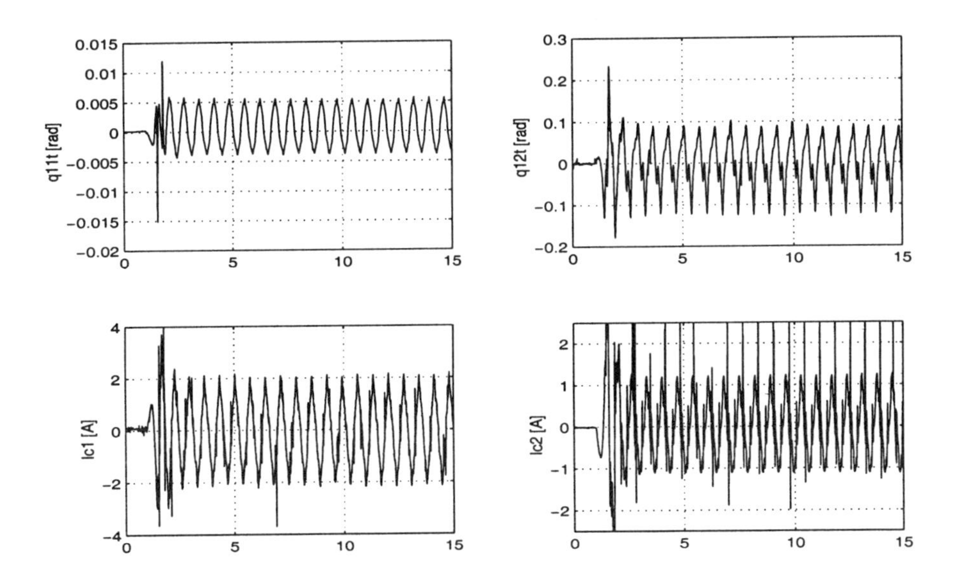


Figure 8.6: SLI controller, desired trajectory 1

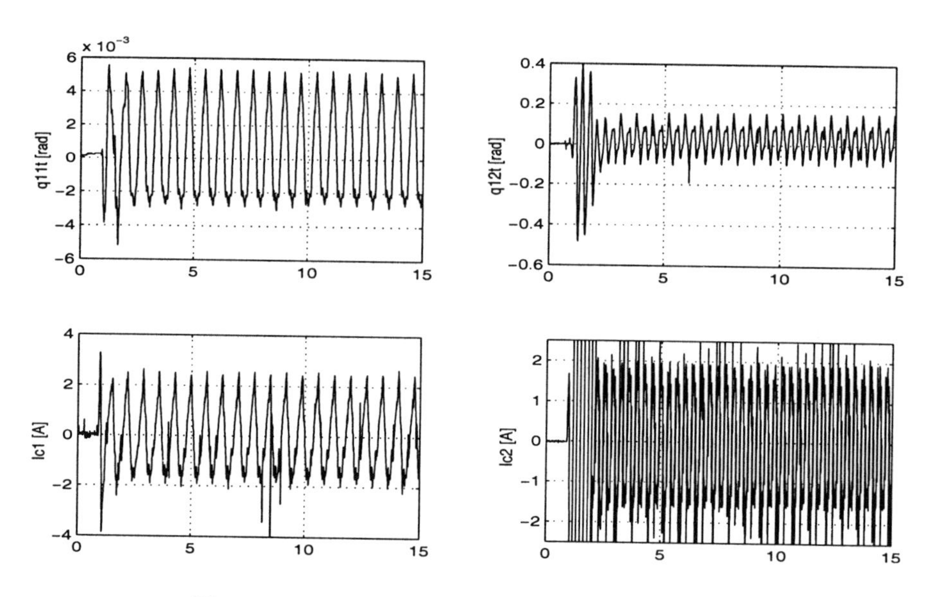


Figure 8.7: SLI controller, desired trajectory 2.

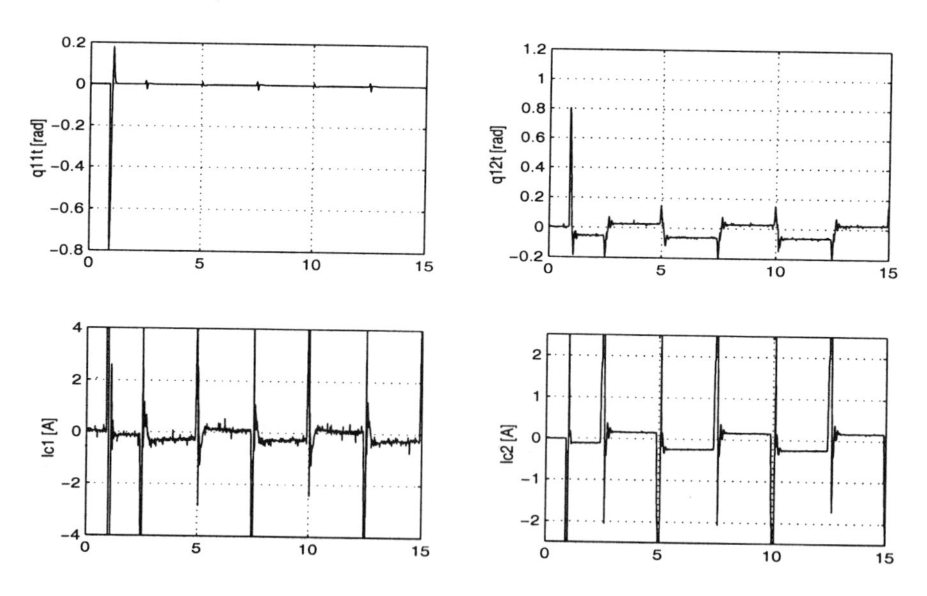


Figure 8.8: SLI controller, desired trajectory 3

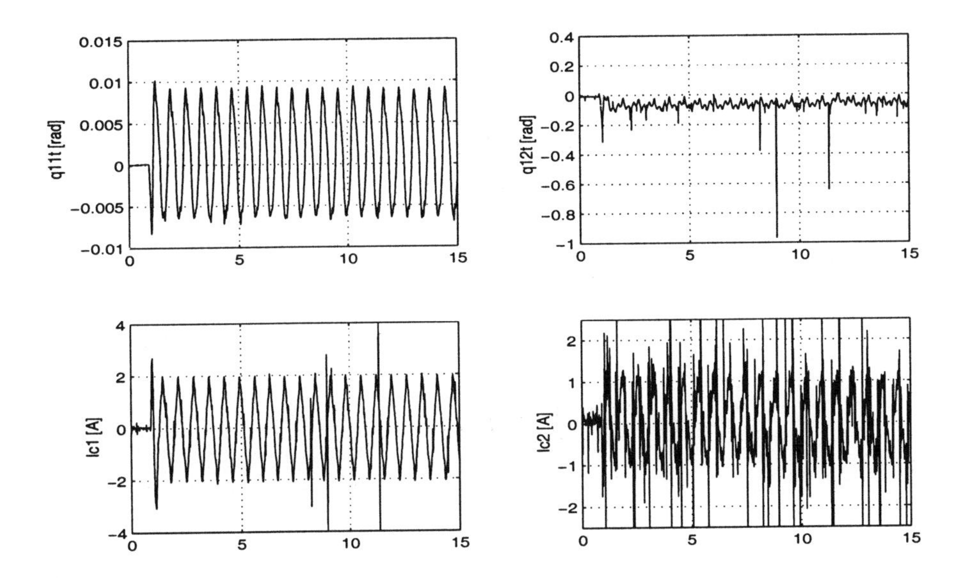


Figure 8.9: Controller 2, desired trajectory 1.

### Compensation of flexibilities

The PD and the Slotine and Li controls behave well for the Capri robot because the joint stiffness is large. The results obtained for the pulley-system show that the behaviour deteriorates a lot if $K$ is small, see tables 8.3 and 8.4.

### Controller complexity

The rather complex structure of the nonlinear controllers 1, 2 and 3 is not an obstacle to their implementation with the available real-time computer described above. In particular recall that the acceleraton and jerk are estimated by inverting the dynamics (see section (6.4)). Such terms have a complicated structure and depend on the system's physical parameters in a nonlinear way. Some experiments have shown that the sampling period (1 ms) could have been decreased to 0.5 ms.

### Torque input

The major problem that prevents certain controllers from behaving correctly is the input magnitude and shape. This has been noted above. The performance of controllers 2 and 3 may be less good than that of the Slotine and Li algorithm, mainly because of the chattering in the input, inducing vibrations in the mechanical structure. Chattering is particularly present during the regulation phases in $I_{c2}$ for trajectory 3 and controllers 2 and 3, see figures

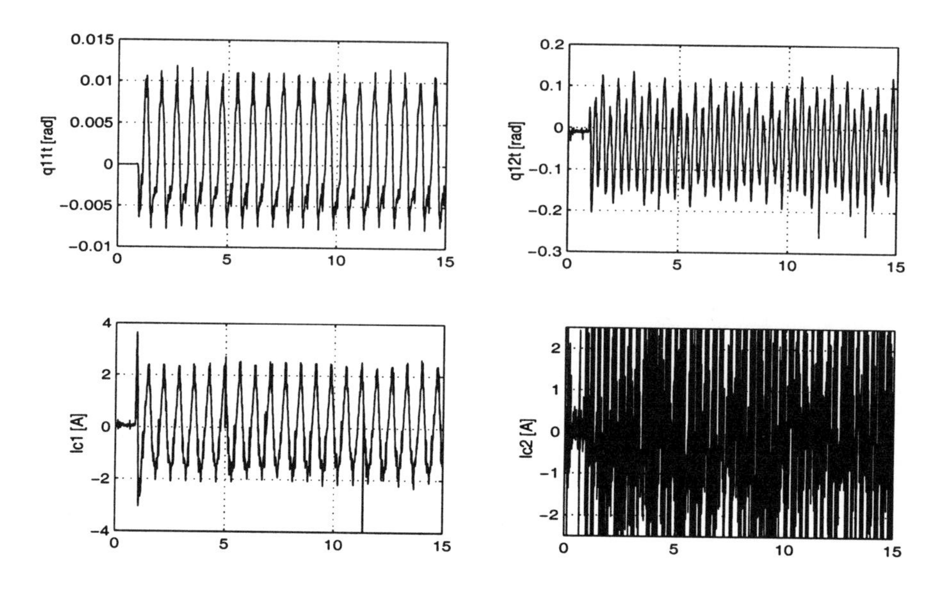


Figure 8.10: Controller 2, desired trajectory 2

8.11, 8.13. On the contrary figures 8.5 and 8.8 show smooth inputs. It may be expected from figures 8.18 to 8.21 that a less noisy velocity $\dot{q}_1$ obtained from a better position measurement would bring the shape of $I_{c2}$ close to the input in figures 8.16, 8.17. Indeed they differ only in terms of chatter. One concludes that an optical encoder to measure $q_1$ would be a better solution.

### Backstepping versus passivity-based controls

It is noteworthy that controllers 2 and 3 possess quite similar closed-loop behaviours, see figures 8.31 and 8.32, 8.24 and 8.25, 8.27 and 8.28 for the pulley-system, 8.9 and 8.33, 8.10 and 8.12, 8.11 and 8.13 for the Capri robot (although $I_{c2}$ chatters slightly less for controller 3, see figures 8.9 and 8.33, and 8.11 and 8.13). The advantage of passivity-based methods is that the controllers are obtained in one shot, whereas the backstepping approach a priori leads to various algorithms. This can be an advantage (more degrees of freedom), but also a drawback as controller 1 behaviour proves. Notice on figures 8.29, 8.30, 8.31 and 8.32 that controllers 2 and 3 allow one to damp the oscillations much better than controller 1 and the PD (it is possible that the PD gains could have been tuned in a better way for these experiments; see however the paragraph below on gain tuning for the pulley-system).

### Transient behaviour

The transient behaviour for the tracking error $\tilde{q}_{12}$ can be improved slightly when the flexibilities are taken into account in the controller design. This can be seen by comparing figures 8.6 and 8.7 with figures 8.9 and 8.10, 8.33 and 8.12. The tracking error tends to oscillate more for the Slotine and Li scheme than for the others. Notice that these results have been obtained with initial tracking errors close to zero. However the results in figures 8.14 to 8.21 prove that the controllers respond quite well to initial state deviation. The transient duration is around 0.5 s for all the controllers. The tracking errors have a similar shape once the transient has vanished. The only significant difference is in the initial input $I_{c2}$. The torque is initially much higher for nonzero initial conditions.

### Feedback gains tuning

The method described in remark 8.2 for tuning the gains in the case of the pulley-system provides nice preliminary results. The gains that have been used in all the experiments for the pulley-system have not been modified during the tests on the real device to tentatively improve the results. They have been kept constant. This tends to prove that such a method is quite promising since it relies on the choice of a single parameter (the closed loop bandwidth, chosen as $\omega_c(CL) = 11$ rad/s in the experiments) and is therefore quite attractive for potential users.

**Remark 8.3** .

- The actuators and

  current drivers neglected dynamics may have a significant influence on the closed-loop behaviour. A close look at tables 8.3 and 8.4 shows the existence of a resonance phenomenon in the closed-loop. This can be confirmed numerically by replacing $u$ with $u_f = \frac{u}{1+\tau s}$ which allows one to suspect that this actuator neglected dynamics may play a crucial role in the loop. It might be then argued that developing velocity observers for such systems may not be so important, whereas some neglected dynamics, whose influence has received less attention in the literature, have a significant effect.

- The peaks in the input $I_{c2}$ for trajectory 1 are due to the saturation of the DC tachometers when the trajectory is at its maximum speed. When the saturation stops, the velocity signal delivered by the tachometers has a short noisy transient that results in such peaks in the input. However this has not had any significant influence on the performance, since such peaks are naturally filtered by the actuators (let us recall that the *calculated* inputs are depicted).

### 8.1.5 Conclusions

In this section we have presented experimental results that concern the application of passivity-based (PD, Slotine and Li, the controller in subsection 6.4.1) and backstepping controllers, to two quite different laboratory plants which serve as flexible joint-rigid link manipulators. The major conclusion is that passivity-based controllers provide generally very good results. In particular the PD and Slotine and Li algorithms show quite good robustness and provide a high level of performance when the flexibility remains small enough. Tracking with high flexibility implies the choice of controllers which are designed from a model that incorporates the joint compliance. These experimental results illustrate nicely the developments of the foregoing chapter: one goes from the PD scheme to the one in subsection 6.4.1 by adding more complexity, but always through the addition of new dissipative modules to the controller, and consequently to the closed-loop system. These three schemes can really be considered to belong to the same "family", namely passivity-based controllers. It is therefore not surprizing that their closed-loop behaviour when applied to real plants reproduces this "dissipative modularity": the PD works well when nonlinearities and flexibilities remain small enough, the Slotine and Li algorithm improves the robustness with respect to nonlinearities, and the scheme in subsection 6.4.1 provides a significant advantage over the other two only if these two dynamical effects are large enough. Finally it is noteworthy that all controllers present a good robustness with respect to the uncertainties listed in subsection 8.1.3.

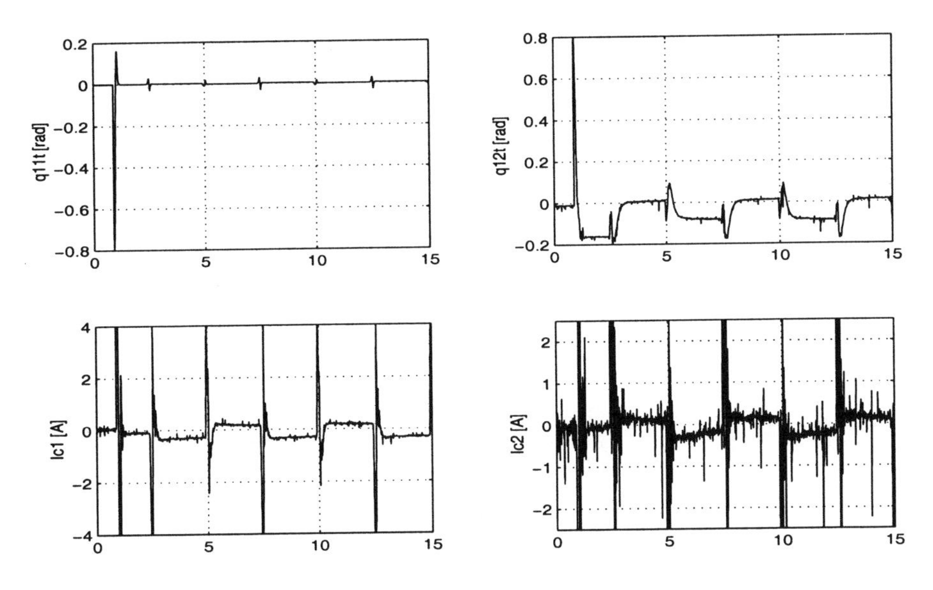


Figure 8.11: Controller 2, desired trajectory 3.

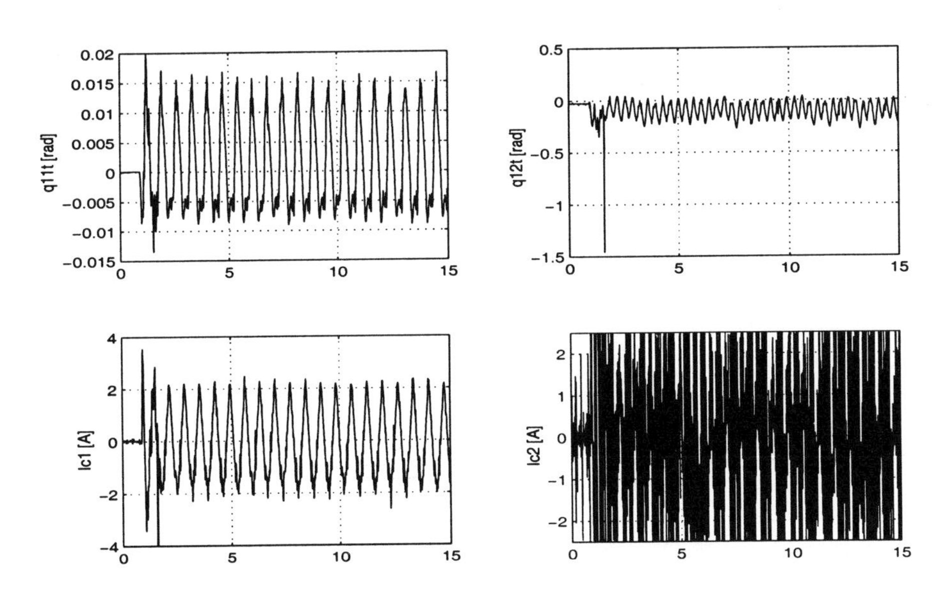


Figure 8.12: Controller 3, desired trajectory 2.

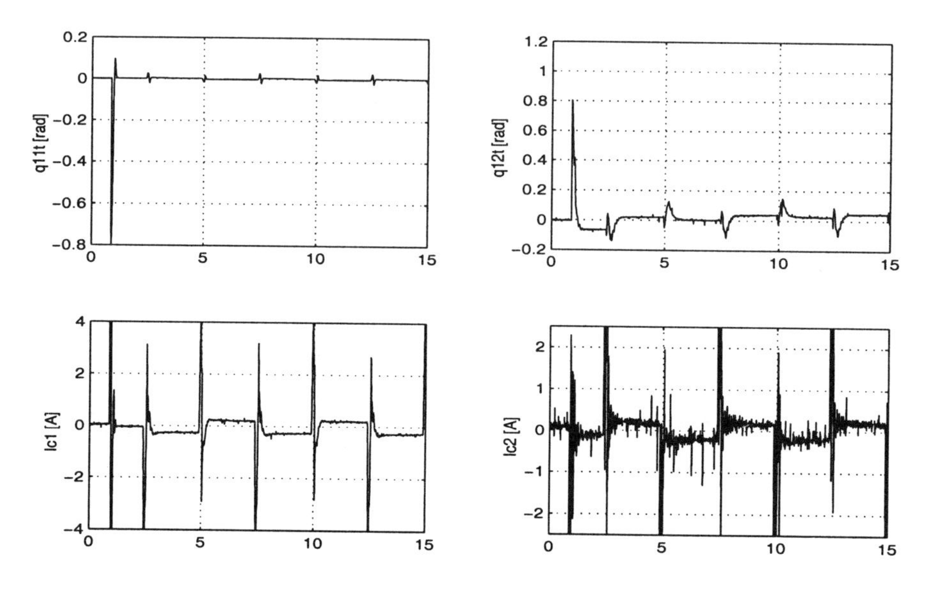


Figure 8.13: Controller 3, desired trajectory 3.

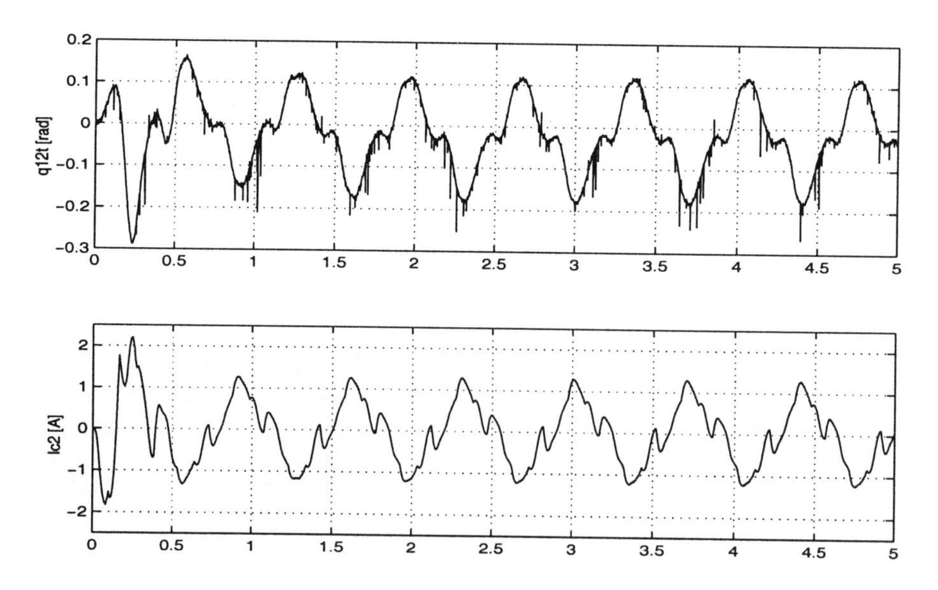


Figure 8.14: PD controller, desired trajectory 1, zero initial conditions.

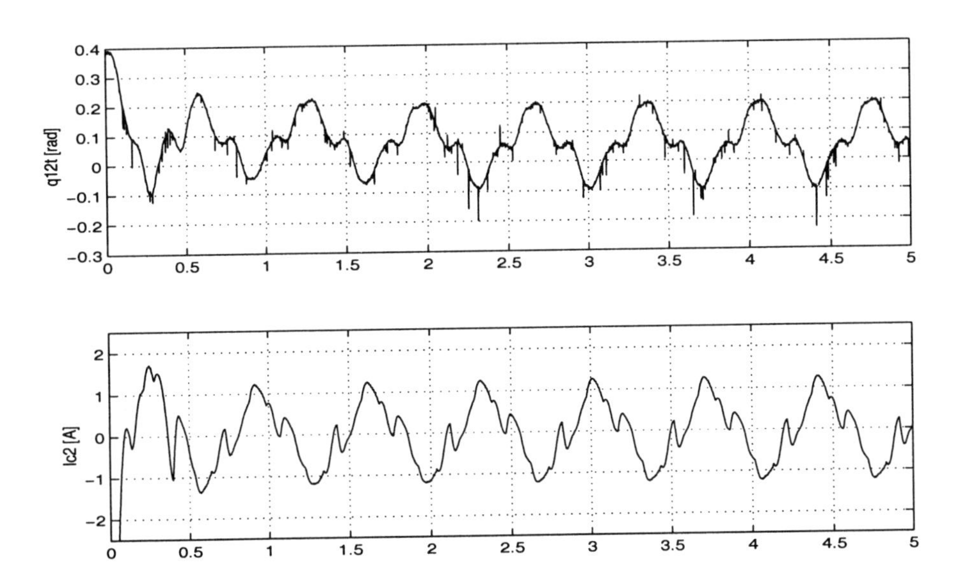


Figure 8.15: PD controller, desired trajectory 1, nonzero initial conditions.

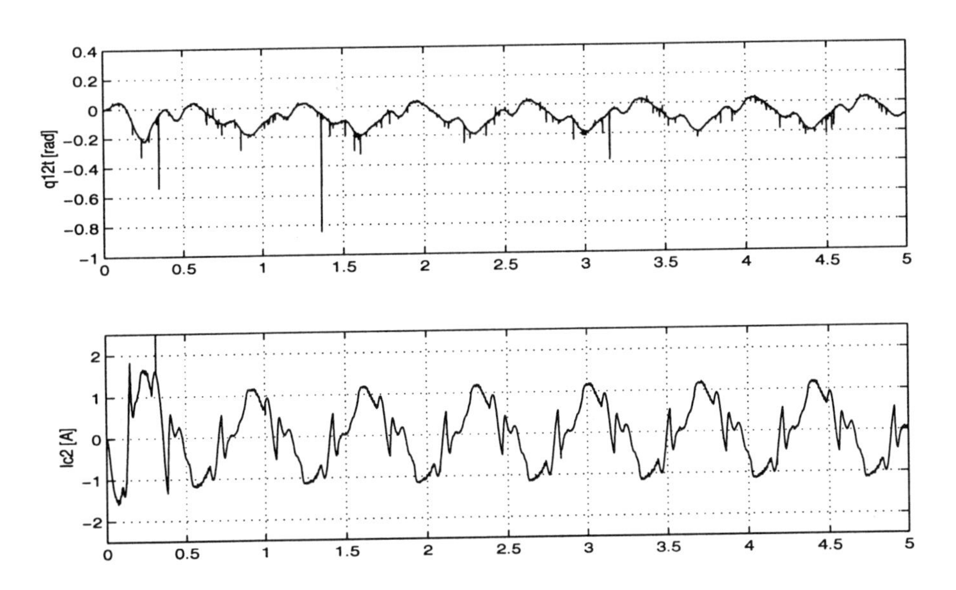


Figure 8.16: SLI controller, desired trajectory 1, zero initial conditions.

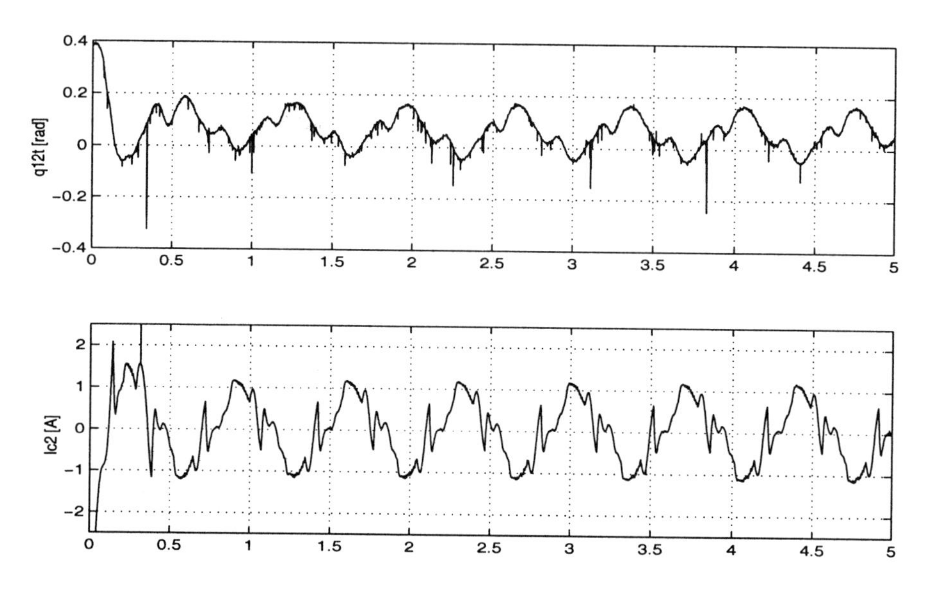


Figure 8.17: SLI controller, desired trajectory 1, nonzero initial conditions.

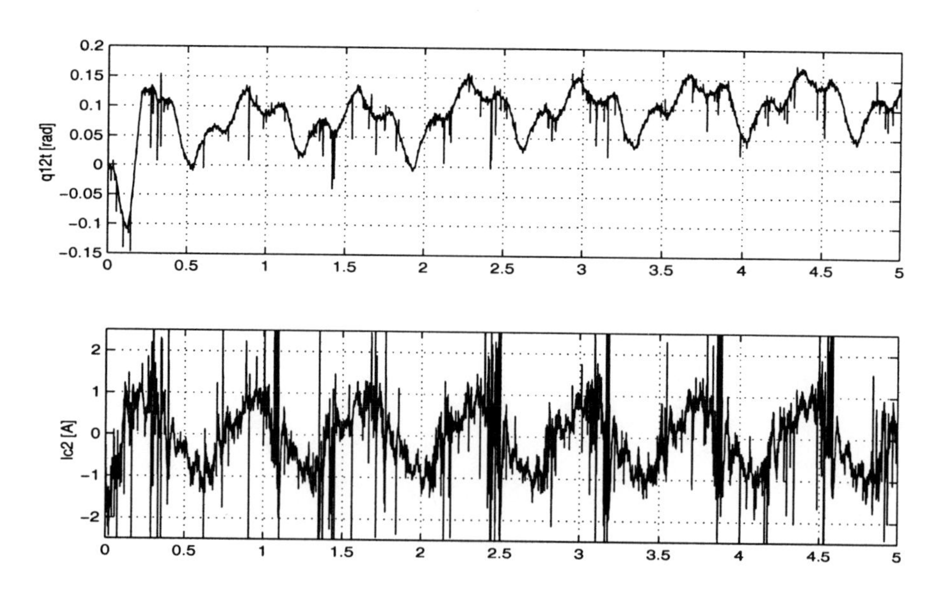


Figure 8.18: Controller 2, desired trajectory 1, zero initial conditions.

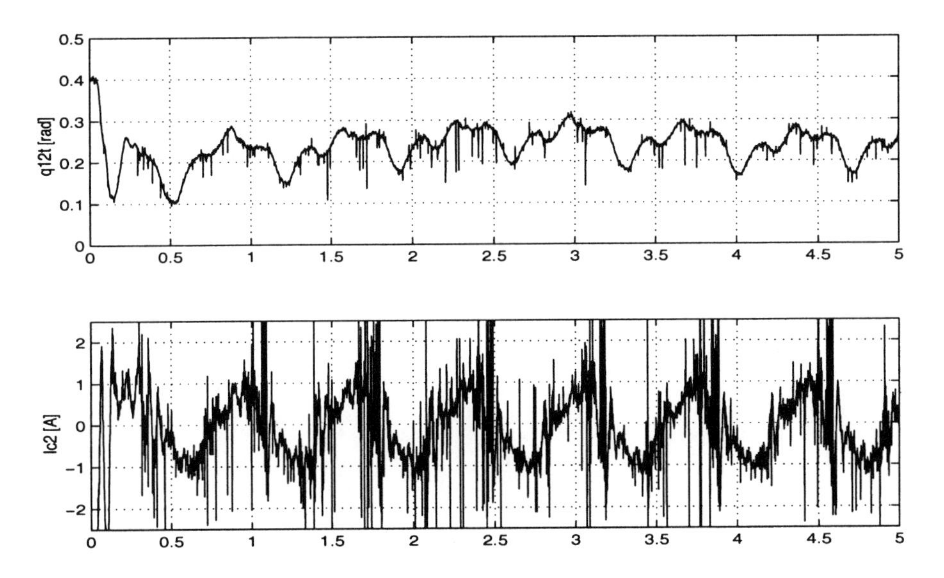


Figure 8.19: Controller 2, desired trajectory 1, nonzero initial conditions.

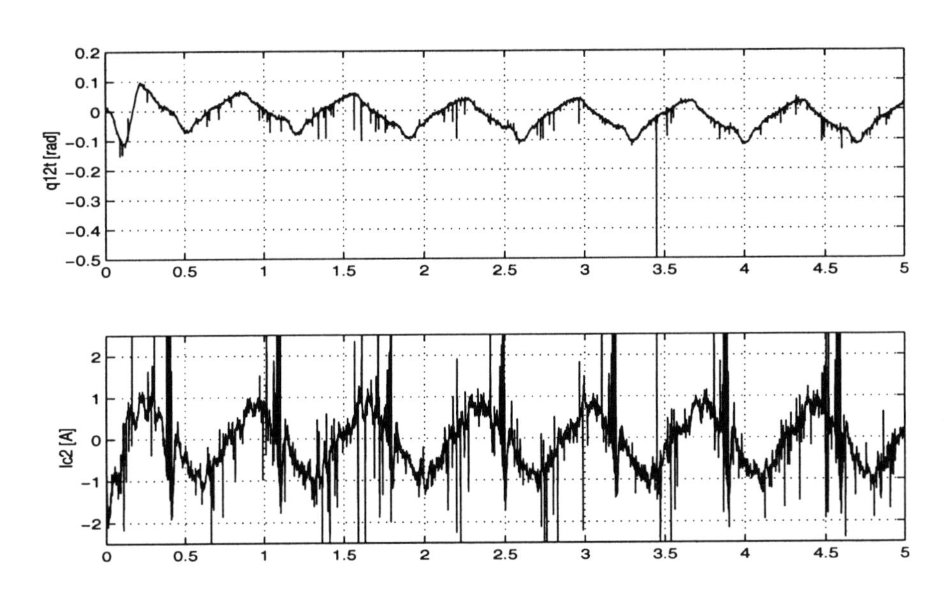


Figure 8.20: Controller 3, desired trajectory 1, zero initial conditions.

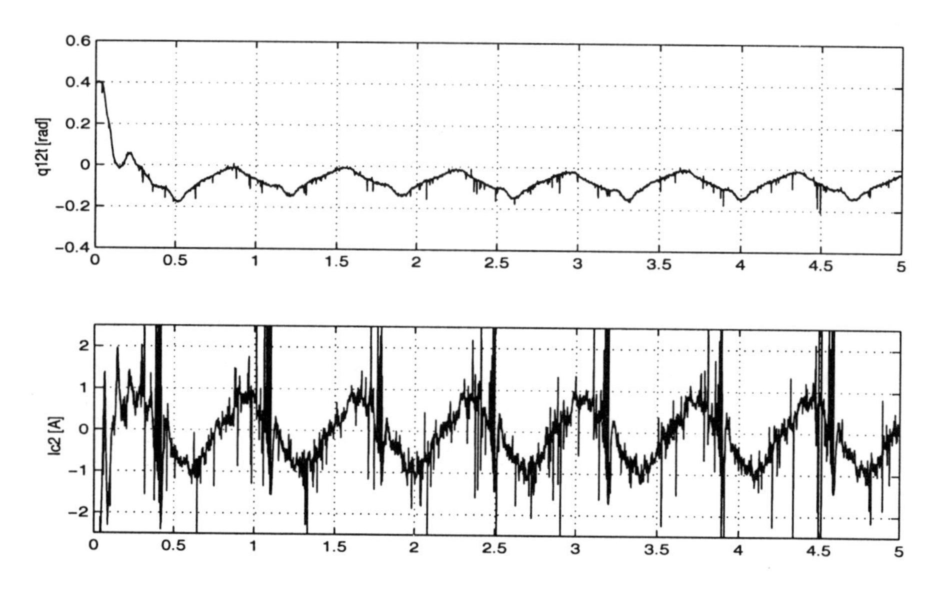


Figure 8.21: Controller 3, desired trajectory 1, nonzero initial conditions.

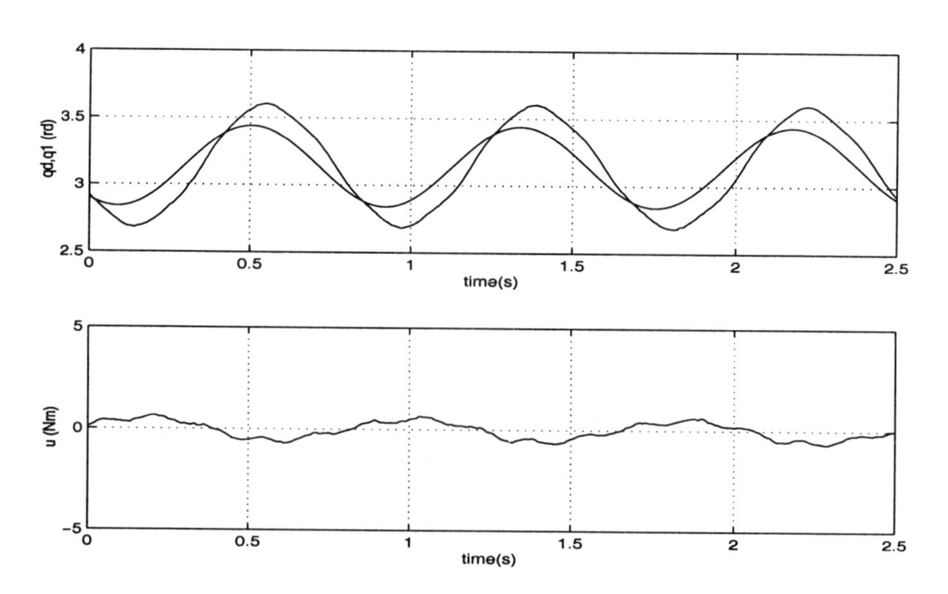


Figure 8.22: PD controller, $\omega = 7.5$ rad/s.

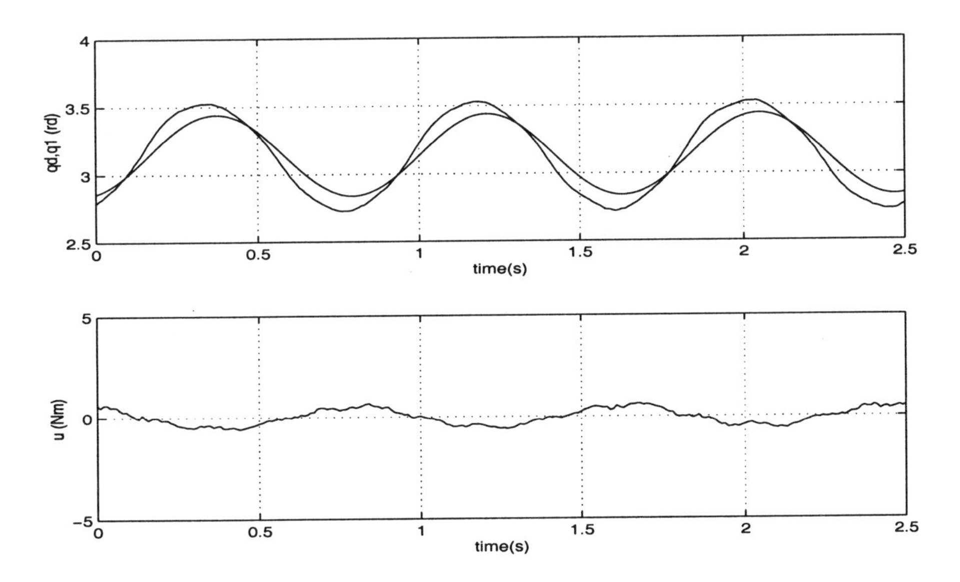


Figure 8.23: Controller 1, $\omega = 7.5$ rad/s.

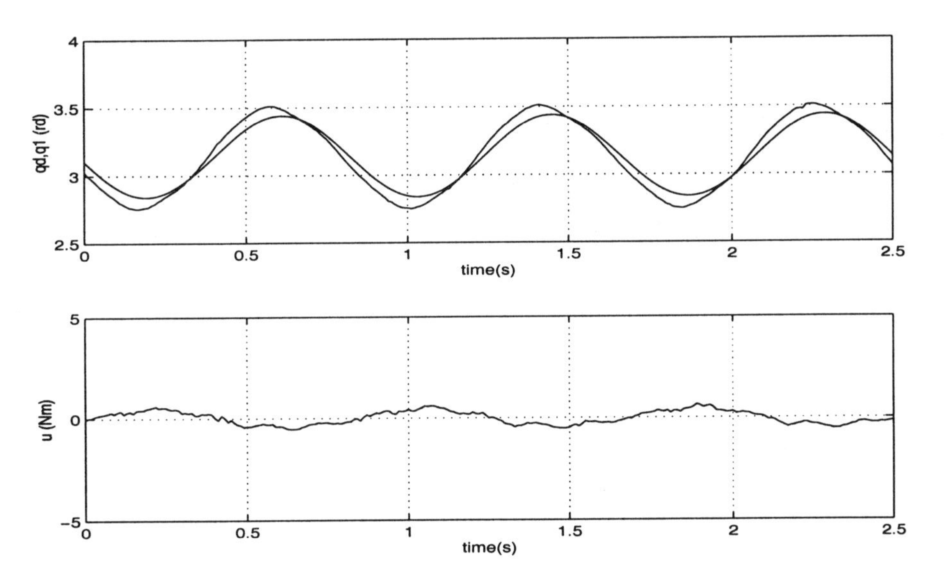


Figure 8.24: Controller 2, $\omega = 7.5$ rad/s.

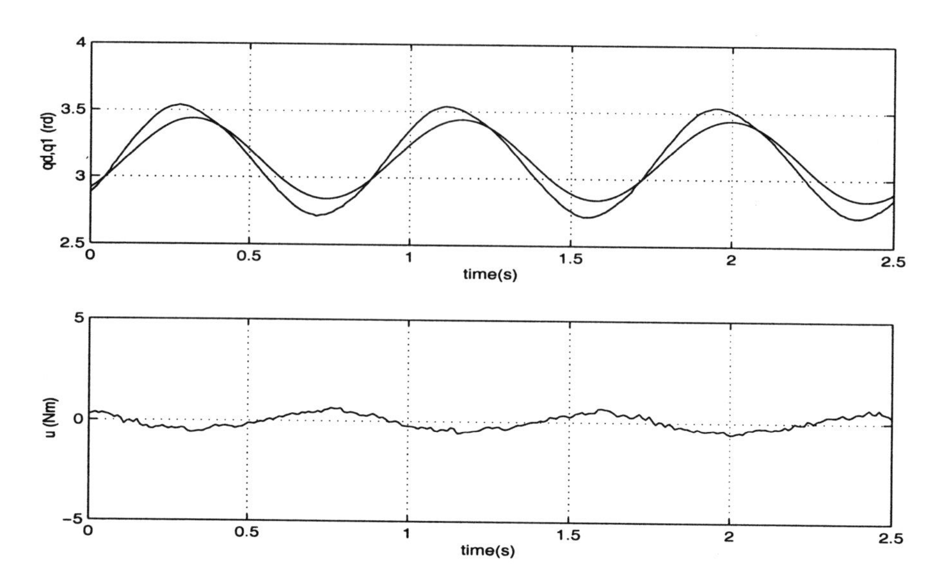


Figure 8.25: Controller 3, $\omega = 7.5$ rad/s.

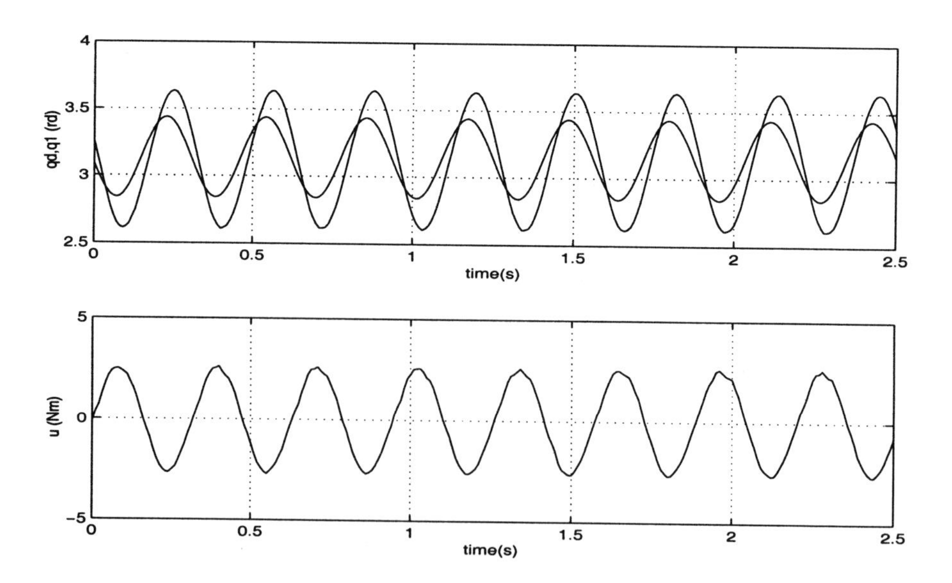


Figure 8.26: Controller 1, $\omega = 20$ rad/s.

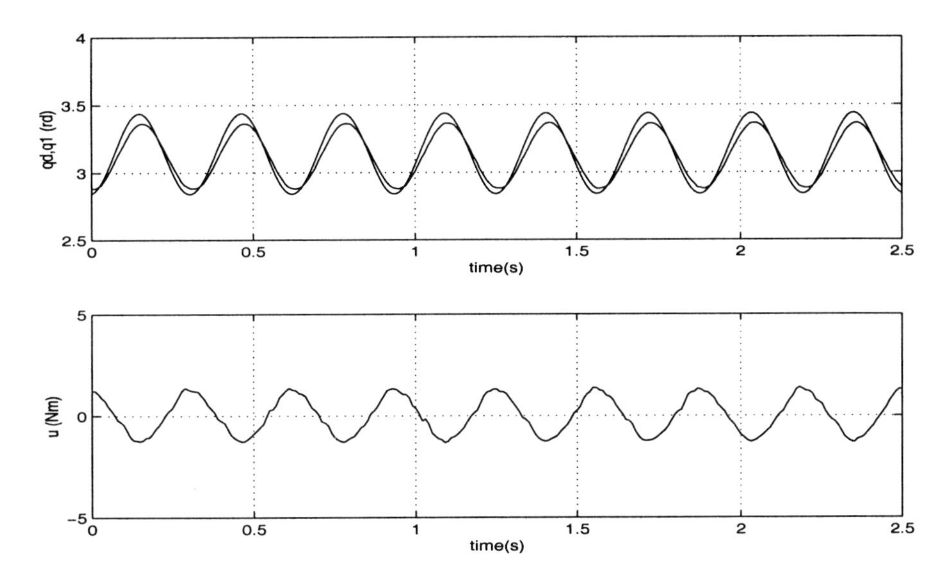


Figure 8.27: Controller 2, $\omega = 20$ rad/s.

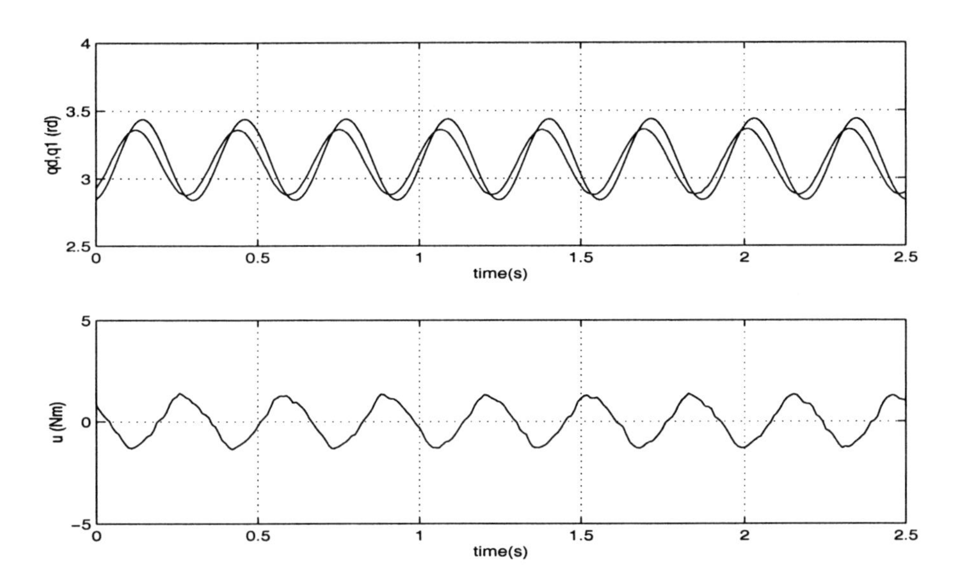


Figure 8.28: Controller 3, $\omega = 20$ rad/s.

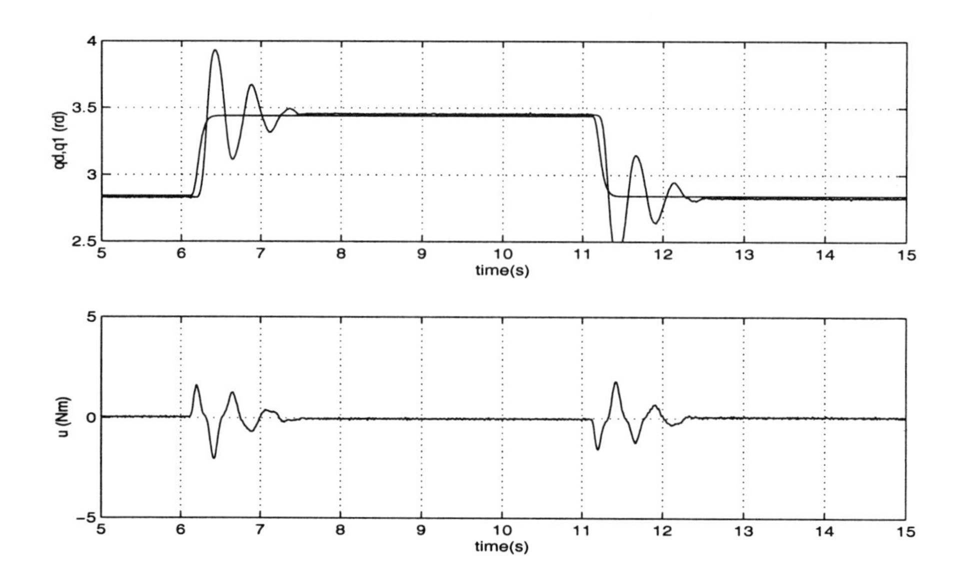


Figure 8.29: PD controller, $b = 40$.

| Controller | $e_1$ (traj. 1) | $e_2$ (traj. 1) | $e_1$ (traj. 2) | $e_2$ (traj. 2) |
|---|---|---|---|---|
| PD | 0.346 | 84.5 | 1.4 (1.6) | 360 (1000) |
| SLI | 0.11 | 37.9 | 0.02 (0.034) | 40 (51) |
| Controller 1 | x | x | x | x |
| Controller 2 | 0.34 | 12 | 0.3 | 75 (173) |
| Controller 3 | 0.64 | 9 | 0.224 (0.6) | 70 (150) |

Table 8.1:

# 8.2 Stabilization of the inverted pendulum

## 8.2.1 Introduction

The inverted pendulum is a very popular experiment used for educational purposes in modern control theory. It is basically a pole which has a pivot on a cart that can be moved horizontally. The pole moves freely around the cart and the control objective is to bring the pole to the upper unstable equilibrium position by moving the cart on the horizontal plane. Since the angular acceleration of the pole can not be controlled directly, the inverted pendulum is an underactuated mechanical system. Therefore, the techniques developed for fully-actuated mechanical robot manipulators can not be used to control the inverted pendulum.

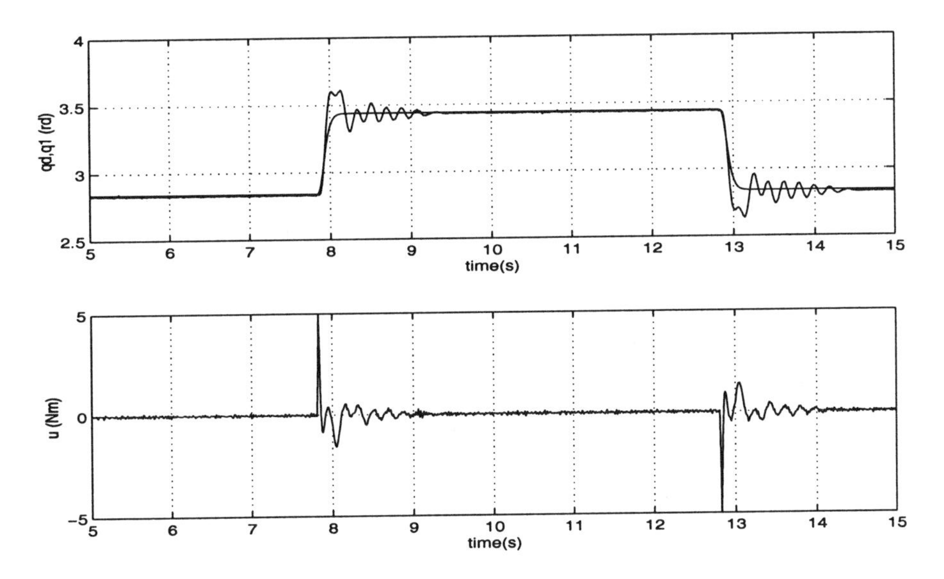


Figure 8.30: Controller 1, $b = 40$.

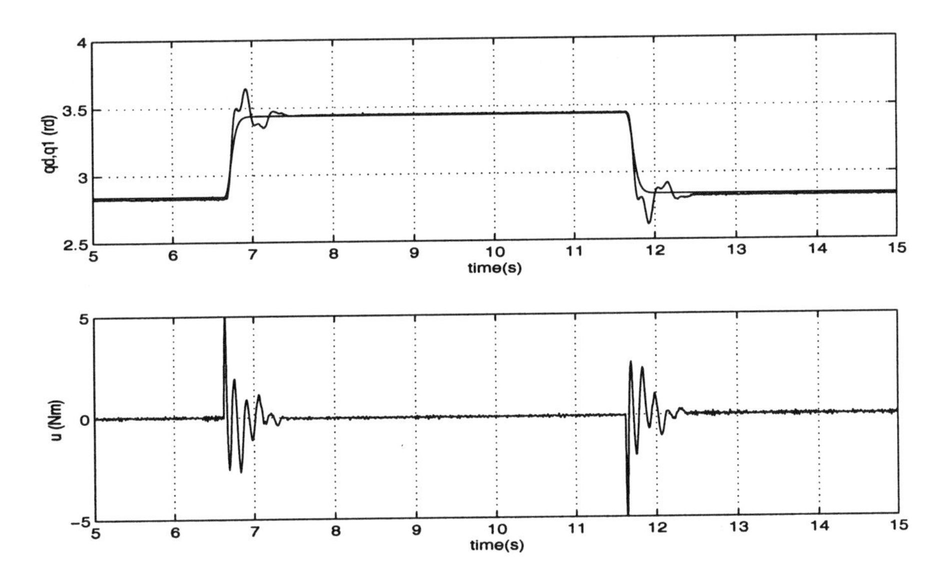


Figure 8.31: Controller 2, $b = 40$.

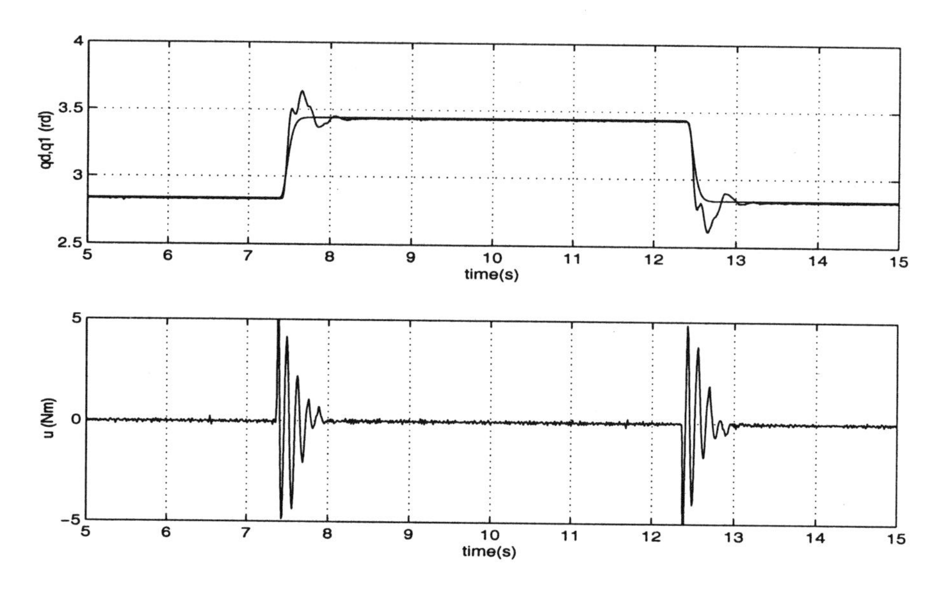


Figure 8.32: Controller 3, $b = 40$.

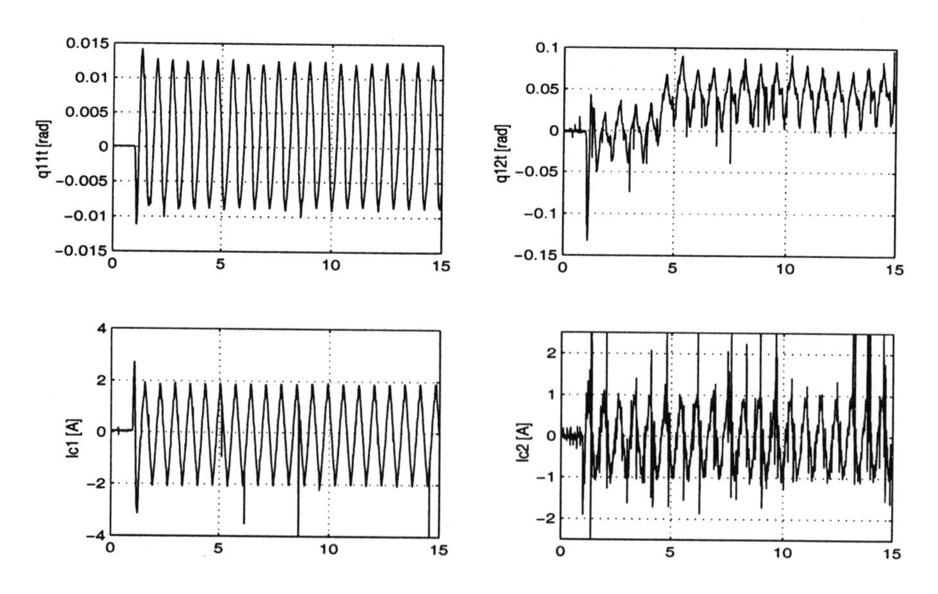


Figure 8.33: Controller 3, desired trajectory 1.

| Controller | $e_1$ (traj. 3) | $e_2$ (traj. 3) |
|---|---|---|
| PD | 0.3 (0.3) | 50 (50) |
| SLI | 0.055 (0.055) | 30 (30) |
| Controller 1 | x | x |
| Controller 2 | 0.135 (0.135) | 30 (30 |
| Controller 3 | 0.19 (0.19) | 15 (15) |

Table 8.2:

| $\omega$ (rad/s) | PD | Control. 1 | Control. 2 | Control. 3 |
|---|---|---|---|---|
| 2.5 | 0.70 | 0.21 | 0.25 | 0.33 |
| 5 | 3.54 | 2.57 | 1.54 | 2.78 |
| 7.5 | 20.86 | 8.53 | 4.17 | 7.92 |
| 10 | x | 20.60 | 13.00 | 19.03 |
| 12.5 | x | 48.07 | 35.15 | 36.05 |
| 15 | x | 63.44 | 53.33 | 31.03 |
| 20 | x | 37.70 | 2.97 | 8.58 |

Table 8.3:

| $\omega$ (rad/s) | PD | Controller 1 | Controller 2 | Controller 3 |
|---|---|---|---|---|
| 2.5 | 0.0630 | 0.0293 | 0.0374 | 0.0386 |
| 5 | 0.0943 | 0.1138 | 0.0840 | 0.0983 |
| 7.5 | 0.1946 | 0.1501 | 0.1040 | 0.1472 |
| 10 | x | 0.2428 | 0.1823 | 0.2150 |
| 12.5 | x | 0.4138 | 0.2965 | 0.2910 |
| 15 | x | 0.4494 | 0.3418 | 0.2581 |
| 20 | x | 0.2842 | 0.0842 | 0.1364 |

Table 8.4:

| PD Controller | traj. 1 | traj. 2 | traj. 3 |
|---|---|---|---|
| $\lambda_{21}$ | 1500 | 650 | 1500 |
| $\lambda_{22}$ | 250 | 10 | 250 |
| $\lambda_{11}$ | 30 | 4 | 30 |
| $\lambda_{12}$ | 5 | 3.5 | 5 |

Table 8.5:

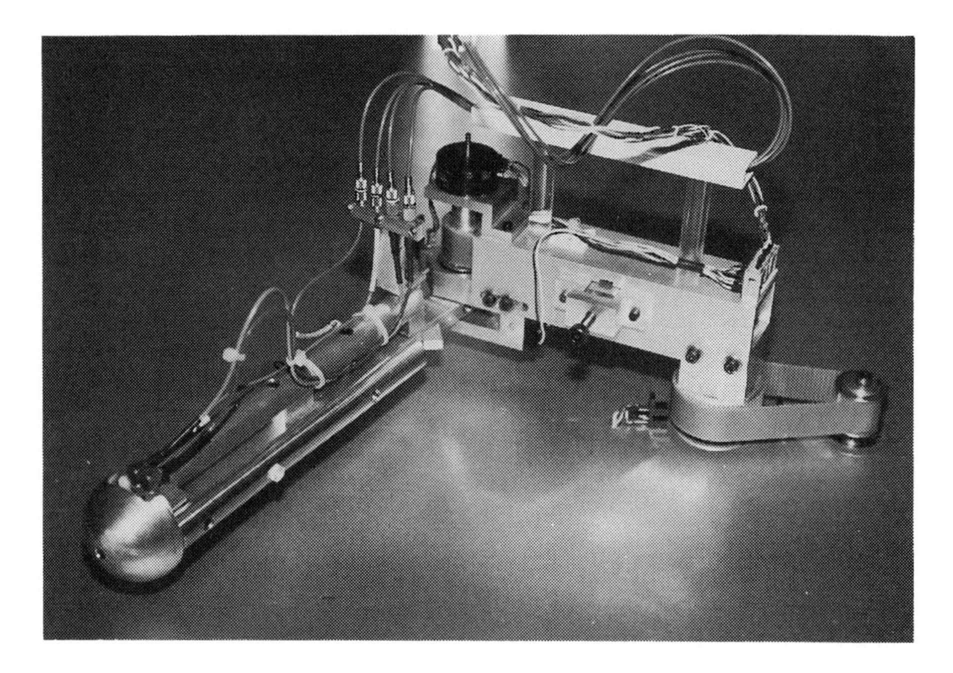

Figure 8.34: The Capri robot of the Laboratoire d'Automatique de Grenoble.

The cart and pole system is also known because the standard nonlinear control techniques are ineffective to control it. Indeed the relative degree [72] of the system is not constant (when the output is chosen to be the swinging energy of the pendulum), the system is not input-output linearizable. Jakubczyk and Respondek [73] have shown that the inverted pendulum is not feedback linearizable. An additional difficulty comes from the fact that when the pendulum swings past the horizontal the controllability distribution does not have a constant rank.

### 8.2.2 System's dynamics

Consider the cart and pendulum system as shown in figure (8.36). We will consider the standard assumptions, i.e. massless rod, point masses, no friction, etc.. $M$ is the mass of the cart, $m$ the mass of the pendulum, concentrated in the bob, $\theta$ the angle that the pendulum makes with the vertical and $l$ the length of the rod. The equations of motion can be obtained either by applying Newton's second law or by the Euler-Lagrange formulation.
The system can be written as:

$$M(q)\ddot{q} + C(q,\dot{q})\dot{q} + g(q) = \tau \tag{8.5}$$

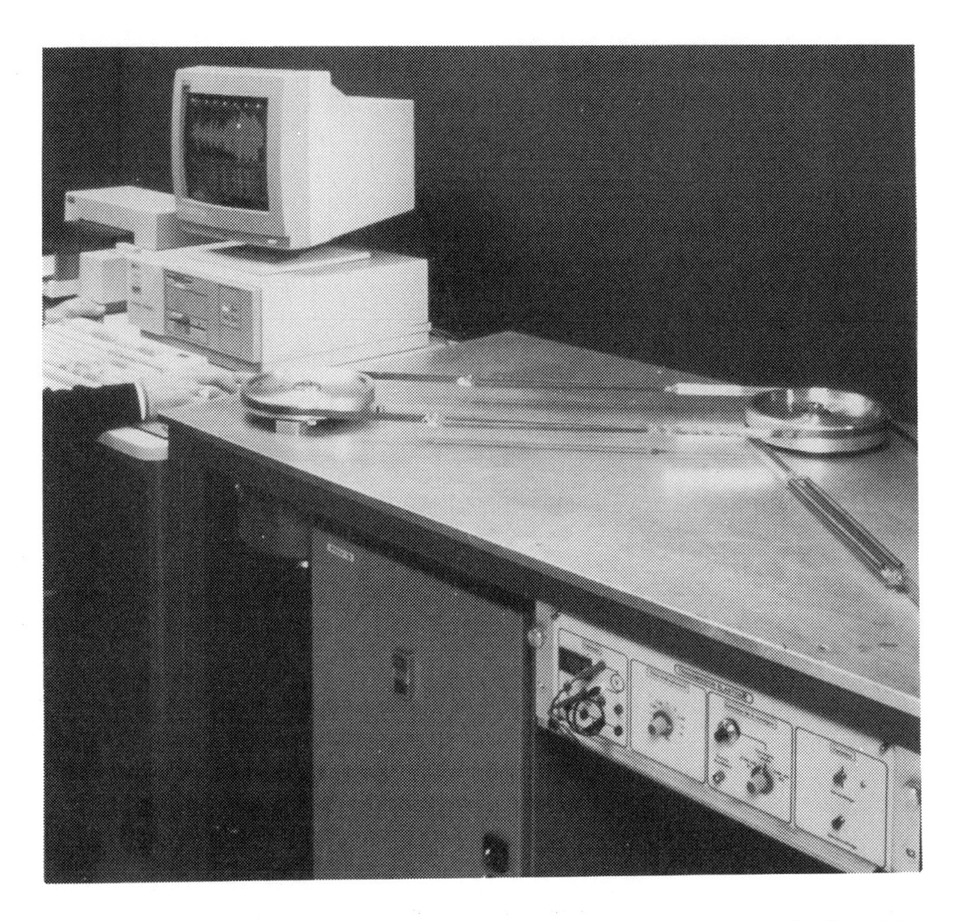

Figure 8.35: The pulley system of the Laboratoire d'Automatique de Grenoble.

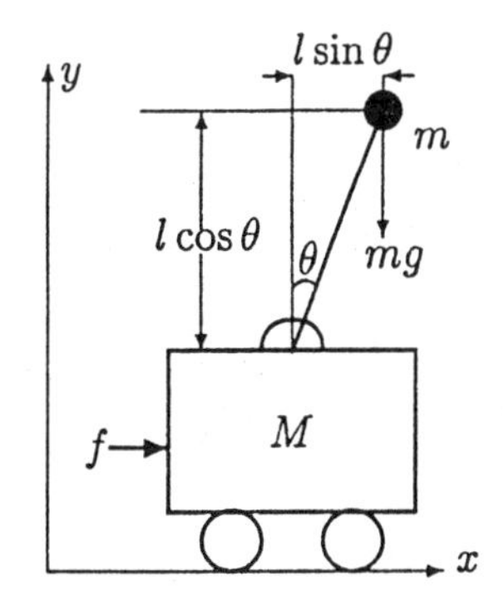


Figure 8.36: The cart pendulum system

where:

$$q = \begin{bmatrix} x \\ \theta \end{bmatrix} \quad M(q) = \begin{bmatrix} M+m & ml\cos\theta \\ ml\cos\theta & ml^2 \end{bmatrix} \tag{8.6}$$

$$C(q,\dot{q}) = \begin{bmatrix} 0 & -ml\sin\theta\dot{\theta} \\ 0 & 0 \end{bmatrix} \tag{8.7}$$

$$g(q) = \begin{bmatrix} 0 \\ -mgl\sin\theta \end{bmatrix} \quad \text{and} \quad \tau = \begin{bmatrix} f \\ 0 \end{bmatrix} \tag{8.8}$$

Note that $M(q)$ is symmetric and

$$\begin{aligned} \det(M(q)) &= (M+m)ml^2 - m^2l^2\cos^2\theta \\ &= Mml^2 + m^2l^2\sin^2\theta > 0 \end{aligned} \tag{8.9}$$

Therefore, $M(q)$ is positive definite for all $q$. From (8.6) and (8.7) it follows that

$$\dot{M} - 2C = \begin{bmatrix} 0 & ml\sin\theta\dot{\theta} \\ -ml\sin\theta\dot{\theta} & 0 \end{bmatrix} \tag{8.10}$$

which is a skew-symmetric matrix, see lemma 5.4. The potential energy of the pendulum can be defined as $U = mgl(\cos\theta - 1)$. Note that $U$ is related to $g(q)$ as follows:

$$g(q) = \frac{\partial U}{\partial q} = \begin{bmatrix} 0 \\ -mgl\sin\theta \end{bmatrix} \tag{8.11}$$

**Passivity of the inverted pendulum**

The total energy of the cart and pole system is given by

$$\begin{aligned} E &= K(q,\dot{q}) + U(q) \\ &= \tfrac{1}{2}\dot{q}^T M(q)\dot{q} + mgl(\cos\theta - 1) \end{aligned} \tag{8.12}$$

Therefore from (8.5), (8.6), (8.7), (8.8), (8.10) and (8.11) we obtain:

$$\begin{aligned} \dot{E} &= \dot{q}^T M(q)\ddot{q} + \tfrac{1}{2}\dot{q}^T \dot{M}(q)\dot{q} + \dot{q}^T g(q) \\ &= \dot{q}^T(-C(q,\dot{q})\dot{q} - g(q) + \tau + \tfrac{1}{2}\dot{M}\dot{q}) + \dot{q}^T g(q) \\ &= \dot{q}^T\tau \\ &= \dot{x}f \end{aligned} \tag{8.13}$$

Integrating both sides of the above equation we obtain

$$\begin{aligned} \int_0^t \dot{x} f dt &= E(t) - E(0) \\ &\geq -2mgl - E(0) \end{aligned} \tag{8.14}$$

Therefore, the system having $f$ as input and $\dot{x}$ as output is passive. Note that for $f = 0$ and $\theta \in [0, 2\pi[$ the system (8.5) has a subset of two equilibrium points. $(x, \dot{x}, \theta, \dot{\theta}) = (*, 0, 0, 0)$ is an unstable equilibrium point and $(x, \dot{x}, \theta, \dot{\theta}) = (*, 0, \pi, 0)$ is a stable equilibrium point. The total energy $E(q, \dot{q})$ is equal to 0 for the unstable equilibrium point and to $-2mgl$ for the stable equilibrium point. The control objective is to stabilize the system around its unstable equilibrium point, i.e. to bring the pendulum to its upper position and the cart displacement to zero simultaneously.

### 8.2.3 Stabilizing control law

Let us first note that in view of (8.12) and (8.6), if $\dot{x} = 0$ and $E(q, \dot{q}) = 0$ then

$$\frac{1}{2} ml^2 \dot{\theta}^2 = mgl(1 - \cos\theta) \tag{8.15}$$

The above equation defines a very particular trajectory which corresponds to a homoclinic orbit. Note that $\dot{\theta} = 0$ only when $\theta = 0$. This means that the pendulum angular position moves clockwise or counter-clockwise until it reaches the equilibrium point $(\theta, \dot{\theta}) = (0, 0)$. Thus our objective can be reached if the system can be brought to the orbit (8.15) for $\dot{x} = 0$, $x = 0$ and $E = 0$. Bringing the system to this homoclinic orbit solves the problem of "swinging up" the pendulum. In order to balance the pendulum at the upper equilibrium position the control must eventually be switched to a controller which guarantees (local) asymptotic stability of this equilibrium [193]. By guaranteeing convergence to the above homoclinic orbit, we guarantee that the trajectory will enter the basin of attraction of any (local) balancing controller. We do not consider in this book the design of the balancing controller.

The passivity property of the system suggests us to use the total energy E in (8.12) in the controller design. Since we wish to bring to zero $x, \dot{x}$ and $E$ we propose the following Lyapunov function candidate:

$$V(q, \dot{q}) = \frac{k_E}{2} E(q, \dot{q})^2 + \frac{k_v}{2} \dot{x}^2 + \frac{k_x}{2} x^2 \tag{8.16}$$

where $k_E$, $k_v$ and $k_x$ are strictly positive constants. Note that $V(q, \dot{q})$ is a positive semi-definite function. Differentiating $V$ and using (8.13) we obtain

$$\begin{aligned}\dot{V} &= k_E E\dot{E} + k_v\dot{x}\ddot{x} + k_x x\dot{x} \\ &= k_E E\dot{x}f + k_v\dot{x}\ddot{x} + k_x x\dot{x} \\ &= \dot{x}(k_E Ef + k_v\ddot{x} + k_x x)\end{aligned} \tag{8.17}$$

Let us now compute $\ddot{x}$ from (8.5). The inverse of $M(q)$ can be obtained from (8.6), (8.7) and (8.9) and is given by:

$$M^{-1} = \frac{1}{\det(M)}\begin{bmatrix} ml^2 & -ml\cos\theta \\ -ml\cos\theta & M+m \end{bmatrix} \tag{8.18}$$

with $\det(M) = ml^2(M + m\sin^2\theta)$.
Therefore we have

$$\begin{bmatrix} \ddot{x} \\ \ddot{\theta} \end{bmatrix} = [det(M(q))]^{-1}\left(\begin{bmatrix} 0 & m^2l^3\dot{\theta}\sin\theta \\ 0 & -m^2l^2\dot{\theta}\sin\theta\cos\theta \end{bmatrix}\begin{bmatrix} \dot{x} \\ \dot{\theta} \end{bmatrix} + \begin{bmatrix} -m^2l^2g\sin\theta\cos\theta \\ (M+m)mgl\sin\theta \end{bmatrix} + \begin{bmatrix} ml^2f \\ -mlf\cos\theta \end{bmatrix}\right)$$

Thus $\ddot{x}$ can be written as

$$\ddot{x} = \frac{1}{M + m\sin^2\theta}\left[m\sin\theta(l\dot{\theta}^2 - g\cos\theta) + f\right] \tag{8.19}$$

Introducing the above in (8.17) one has

$$\dot{V} = \dot{x}\left[f\left(k_E E + \frac{k_v}{M+m\sin^2\theta}\right) + \frac{k_v m\sin\theta(l\dot{\theta}^2 - g\cos\theta)}{M+m\sin^2\theta} + k_x x\right] \tag{8.20}$$

For simplicity and without loss of generality we will consider $M = m = l = 1$, thus

$$\dot{V} = \dot{x}\left[f\left(k_E E + \frac{k_v}{1+\sin^2\theta}\right) + \frac{k_v\sin\theta(\dot{\theta}^2 - g\cos\theta)}{1+\sin^2\theta} + k_x x\right] \tag{8.21}$$

We propose a control law such that

$$f\left(k_E E + \frac{k_v}{1+\sin^2\theta}\right) + \frac{k_v\sin\theta(\dot{\theta}^2 - g\cos\theta)}{1+\sin^2\theta} + k_x x = -k_{dx}\dot{x} \tag{8.22}$$

which will lead to

$$\dot{V} = -k_{dx}\dot{x}^2 \tag{8.23}$$

Note that other functions $f(\dot{x})$ such that $\dot{x}f(\dot{x}) > 0$ are also possible.
The control law in (8.22) will have no singularities provided that

$$\left(k_E E + \frac{k_v}{1+\sin^2\theta}\right) \neq 0 \tag{8.24}$$

The above condition will be satisfied if for some $\epsilon > 0$

$$|E| \leq \frac{\frac{k_v}{k_E} - \epsilon}{2} < \frac{\frac{k_v}{k_E}}{\left(1 + \sin^2\theta\right)} \tag{8.25}$$

Note that when using the control law (8.22), the pendulum can get stuck at the (lower) stable equilibrium point, $(x, \dot{x}, \theta, \dot{\theta}) = (0, 0, \pi, 0)$. In order to avoid this singular point, which occurs when $E = -2mgl$ (see (8.12)), we require $|E| < 2mgl$ i.e. $|E| < 2g$ (for $m = 1$, $l = 1$). Taking also (8.25) into account, we require

$$|E| < c = \min(2g, \frac{\frac{k_v}{k_E} - \epsilon}{2}) \tag{8.26}$$

Since V is a non-increasing function (see (8.23)), (8.26) will hold if the initial conditions are such that

$$V(0) < \frac{c^2}{2} \tag{8.27}$$

The above defines the region of attraction as will be shown in the next section.

## Domain of attraction

Condition (8.27) imposes bounds on the initial energy of the system. Note that the potential energy $U = mgl(\cos\theta - 1)$ lies between $-2g$ and 0, for $m = l = 1$. This means that the initial kinetic energy should belong to $[0, c + 2g)$. Note also that the initial position of the cart $x(0)$ is arbitrary since we can always choose an appropiate value for $k_x$ in $V$ (8.16). If $x(0)$ is large we should choose $k_x$ small. The convergence rate of the algorithm may however decrease when $k_x$ is small. Note that when the initial kinetic energy $K(q(0), \dot{q}(0))$ is zero, the initial angular position $\theta(0)$ should belong to $(-\pi, \pi)$. This means that the only forbidden point is $\theta(0) = \pi$.
When the initial kinetic energy $K(q(0), \dot{q}(0))$ is different from zero, i.e. $K(q(0), \dot{q}(0))$ belongs to $(0, c + 2g)$ (see (8.26) and (8.27)), then there are less restrictions on the initial angular position $\theta(0)$. In particular, $\theta(0)$ can even be pointing downwards, i.e. $\theta = \pi$ provided that $K(q(0), \dot{q}(0))$ is not zero. Despite the fact that our controller is local, its basin of attraction is far from being small. The simulation example and the real-time experiments will show this feature.
For future use we will rewrite the control law $f$ from (8.22) as:

$$f = \frac{k_v \sin\theta \left(g\cos\theta - \dot{\theta}^2\right) - \left(1 + \sin^2\theta\right)\left(k_x x + k_{dx}\dot{x}\right)}{k_v + \left(1 + \sin^2\theta\right) k_E E} \tag{8.28}$$

The stability analysis can be obtained by using the Krasovskii-LaSalle's invariance theorem. The stability properties are summarized in the following lemma.

**Lemma 8.1** Consider the inverted pendulum system (8.5) and the controller in (8.28) with strictly positive constants $k_E$, $k_v$, $k_x$ and $k_{dx}$. Provided that the state initial conditions satisfy inequalities (8.26) and (8.27), then the solution of the closed-loop system converges to the invariant set $M$ given by the homoclinic orbit (8.15) with $(x, \dot{x}) = (0, 0)$. Note that $f$ does not necessarily converge to zero.

**Proof** The proof can be found in [109].

### 8.2.4 Simulation results

In order to observe the performance of the proposed control law based on an energy approach of the system, we have performed simulations on MATLAB using SIMULINK.
We have considered the real system parameters $\bar{M} = M + m = 1.2$, $ml^2 = 0.0097$ and $ml = 0.04$, and $g = 9.804\ ms^{-2}$ of the inverted pendulum at the University of Illinois at Urbana-Champaign. Recall that the control law requires initial conditions such that (8.27) is satisfied. We have chosen the gains $k_E = 1$, $k_v = 1$, $k_x = 10^{-2}$ and $k_{dx} = 1$. These gains have been chosen to increase the convergence rate in order to switch to a linear stabilizing controller in a reasonable time.
The algorithm brings the inverted pendulum close to the homoclinic orbit but the inverted pendulum will remain swinging while getting closer and closer to the origin. Once the system is close enough to the origin, i.e. $(|x| \leq 0.1, |\dot{x}| \leq 0.2, |\theta| \leq 0.3, |\dot{\theta}| \leq 0.3)$, we switch to the linear LQR controller $f = -K[x \quad \dot{x} \quad \theta \quad \dot{\theta}]^T$ where $K = [44 \quad 23 \quad 74 \quad 11]$. Figure 2 shows the results for an initial position:

$$x = 0.1 \qquad \dot{x} = 0$$

$$\theta = \frac{2\pi}{3} \qquad \dot{\theta} = 0$$

Simulations showed that the nonlinear control law brings the system to the homoclinic orbit (see the phase plot in figure 8.42). Switching to the linear controller occurs at time $t = 120\ s$. Note that before the switching the energy $E$ goes to zero and that the Lyapunov function $V$ is decreasing and converges to zero.

### 8.2.5 Experimental results

We have performed experiments on the inverted pendulum setting at the University of Illinois at Urbana-Champaign.

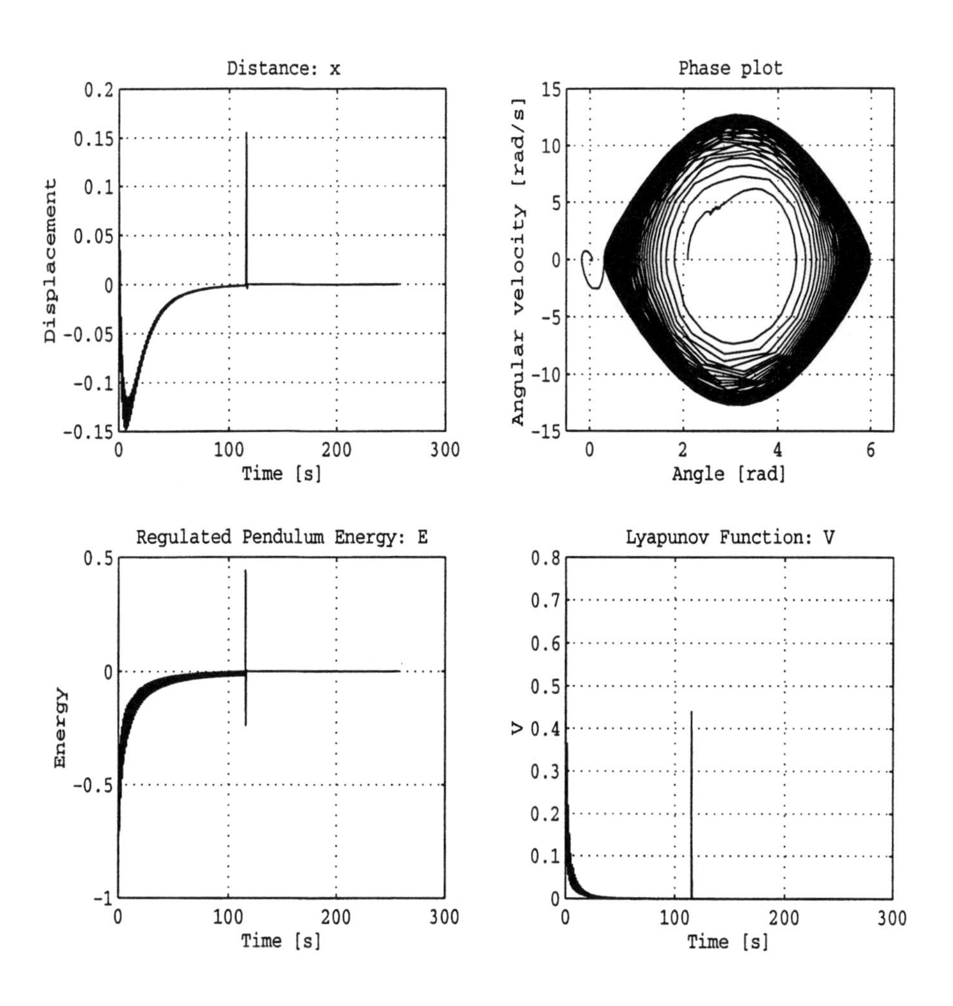


Figure 8.37: Simulation results for the inverted pendulum.

The parameters of the model used for the controller design and the linear controller gains $K$ are the same as in the previous section.
For this experiment we have chosen the gains $k_E = 1$, $k_v = 1.15$, $k_x = 20$ and $k_{dx} = 0.001$.
Figure 8.43 shows the results for an initial position:

$$x = 0 \qquad \dot{x} = 0$$

$$\theta = \pi + 0.1 \qquad \dot{\theta} = 0.1$$

Real-time experiments showed that the nonlinear control law brings the system to the homoclinic orbit (see the phase plot in figure 8.43). Switching to the linear controller occurs at time $t = 27\ s$. Note that the control input lies in an acceptable range.
Note that in both simulation and experimental results, the initial conditions lie slightly outside the domain of attraction. This proves that the domain of attraction in (8.26) and (8.27) is conservative.

### 8.2.6 Conclusions

We have presented a control strategy for the inverted pendulum that brings the pendulum to a homoclinic orbit, while the cart displacement converges to zero. Therefore the state will enter the basin of attraction of any locally convergent controller.
The control strategy is based on the total energy of the system, using its passivity properties. A Lyapunov function is obtained using the total energy of the system. The convergence analysis is carried out using the Krasovskii-LaSalle's invariance principle. The system nonlinearities have not been compensated before the control design which has enabled us to exploit the physical properties of the system in the stability analysis. The proposed control strategy is proved to be applicable to a wider class of underactuated mechanical systems (see [51] and [52]).

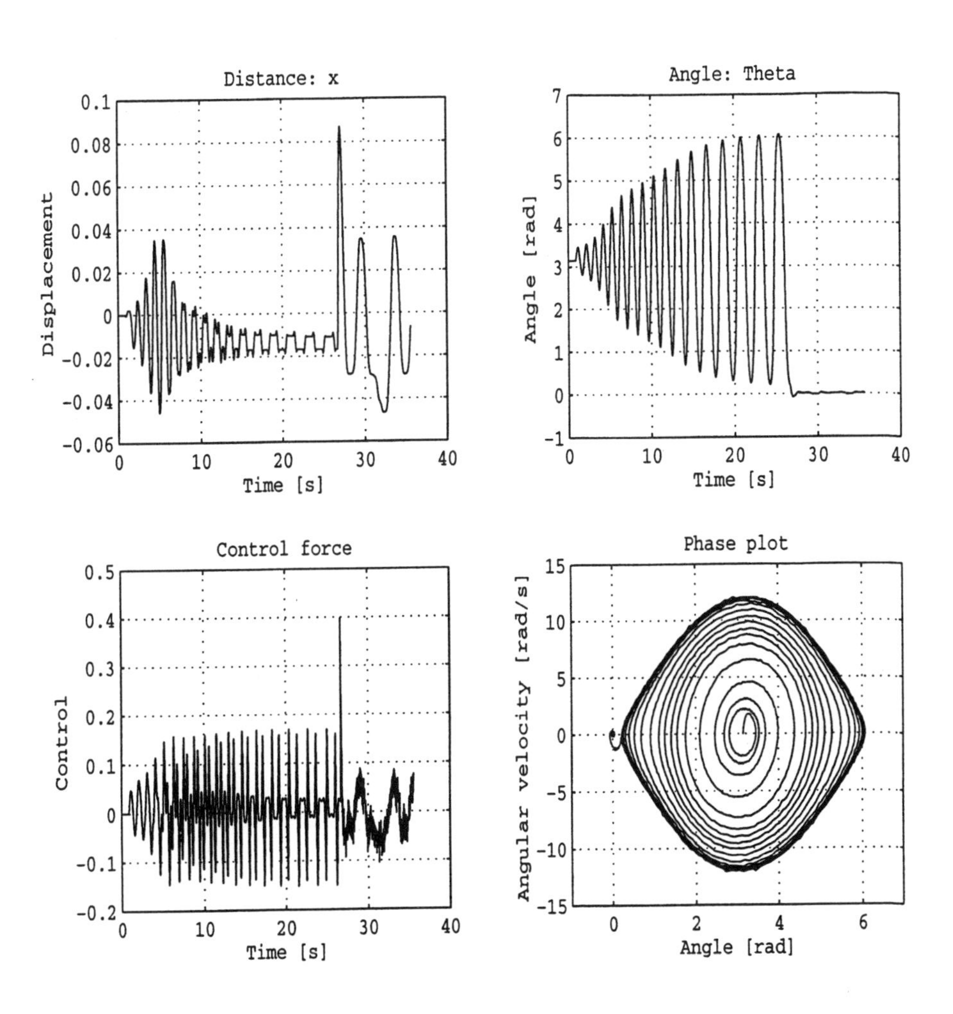


Figure 8.38: Experimental results for the inverted pendulum.

# Chapter 9

# Appendix

In this Appendix we present the background for the main tools used throughout the book; namely, Lyapunov theorem, differential geometry, and input–output methods. No proofs of the various theorems and lemmas are given, and the reader is referred to the cited literature.

## 9.1 Lyapunov stability

We will use throughout the appendix a rather standard notation and terminology. $\mathbb{R}^+$ will denote the set of nonnegative real numbers, and $\mathbb{R}^n$ will denote the usual $n$-dimensional vector space over $\mathbb{R}$ endowed with the Euclidean norm

$$\|x\| = \left(\sum_{j=1}^{n} |x_j|^2\right)^{1/2}.$$

Let us consider a nonlinear dynamic system represented as

$$\dot{x} = f(x,t) \tag{9.1}$$

where $f$ is a nonlinear vector function, and $x \in \mathbb{R}^n$ is the state vector.

### 9.1.1 Autonomous systems

The nonlinear system (9.1) is said to be *autonomous* (or *time-invariant*) if $f$ does not depend explicitly on time, i.e.,

$$\dot{x} = f(x); \tag{9.2}$$

otherwise the system is called *non-autonomous* (or *time-varying*). In this section, we briefly review the *Lyapunov theory* results for autonomous systems while non-autonomous systems will be reviewed in the next section. Lyapunov theory is the fundamental tool for stability analysis of dynamic systems. The basic stability concepts are summarized in the following definitions.

**Definition 9.1 (Equilibrium)** A state $x^\star$ is an *equilibrium* point of (9.2) if $f(x^\star) = 0$.

**Definition 9.2 (Stability)** The equilibrium point $x = 0$ is said to be *stable* if, for any $\rho > 0$, there exists $r > 0$ such that if $\|x(0)\| < r$, then $\|x(t)\| < \rho$ $\forall t \geq 0$. Otherwise the equilibrium point is unstable.

**Definition 9.3 (Asymptotic stability)** An equilibrium point $x = 0$ is *asymptotically stable* if it is stable, and if in addition there exists some $r > 0$ such that $\|x(0)\| < r$ implies that $x(t) \to 0$ as $t \to \infty$.

**Definition 9.4 (Marginal stability)**
An equilibrium point that is Lyapunov stable but not asymptotically stable is called *marginally stable.*

**Definition 9.5 (Exponential stability)**
An equilibrium point is *exponentially stable* if there exist two strictly positive numbers $\alpha$ and $\lambda$ independent of time and initial conditions such that

$$\|x(t)\| \leq \alpha \|x(0)\| \exp(-\lambda t) \qquad \forall t > 0 \tag{9.3}$$

in some ball around the origin.

♠♠

The above definitions correspond to *local* properties of the system around the equilibrium point. The above stability concepts become *global* when their corresponding conditions are satisfied for *any initial state.*

### Lyapunov linearization method

Assume that $f(x)$ in (9.2) is continuously differentiable and that $x = 0$ is an equilibrium point. Then, using Taylor expansion, the system dynamics can be written as

$$\dot{x} = \left.\frac{\partial f}{\partial x}\right|_{x=0} + o(x) \tag{9.4}$$

where $o$ stands for higher-order terms in $x$. Linearization of the original non-linear system at the equilibrium point is given by

$$\dot{x} = Ax \tag{9.5}$$

where $A$ denotes the Jacobian matrix of $f$ with respect to $x$ at $x = 0$, i.e.,

$$A = \left.\frac{\partial f}{\partial x}\right|_{x=0}.$$

A linear time-invariant system of the form (9.5) is (asymptotically) stable if $A$ is a (strictly) stable matrix, i.e., if all the eigenvalues of $A$ have (negative) nonpositive real parts. The stability of linear time-invariant systems can be determined according to the following theorem.

**Theorem 9.1** The equilibrium state $x = 0$ of system (9.5) is asymptotically stable if and only if, given any matrix $Q > 0$, the solution $P$ to the Lyapunov equation

$$A^T P + PA = -Q \tag{9.6}$$

is positive definite. If $Q$ is only positive semi-definite ($Q \geq 0$), then only stability is concluded. ♠♠

In the above theorem, notice that if $Q = L^T L$ with $(P, L)$ being an observable pair, then asymptotic stability is obtained again.
Local stability of the original nonlinear system can be inferred from stability of the linearized system as stated in the following theorem.

**Theorem 9.2** If the linearized system is strictly stable (unstable), then the equilibrium point of the nonlinear system is locally asymptotically stable (unstable).

The above theorem does not allow us to conclude anything when the linearized system is marginally stable. Then one has to rely on more sophisticated tools like the invariant manifold theory [88].

### Lyapunov direct method

Let us consider the following definitions.

**Definition 9.6 ((Semi-)definiteness)**
A scalar continuous function $V(x)$ is said to be locally *positive (semi-)definite* if $V(0) = 0$ and $V(x) > 0$ ($V(x) \geq 0$) for $x \neq 0$. Similarly, $V(x)$ is said to be *negative (semi-)definite* if $-V(x)$ is positive (semi-)definite.

**Definition 9.7 (Lyapunov function)** $V(x)$ is called a *Lyapunov function* for the system (9.2) if, in a ball $B$ containing the origin, $V(x)$ is positive definite and has continuous partial derivatives, and if its time derivative along the solutions of (9.2) is negative semi-definite, i.e., $\dot{V}(x) = (\partial V / \partial x) f(x) \leq 0$.

The following theorems can be used for local and global analysis of stability, respectively.

**Theorem 9.3 (Local stability)** The equilibrium point 0 of system (9.2) is (asymptotically) stable in a ball $B$ if there exists a scalar function $V(x)$ with continuous derivatives such that $V(x)$ is positive definite and $\dot{V}(x)$ is negative semi-definite (negative definite) in the ball $B$.

**Theorem 9.4 (Global stability)** The equilibrium point of system (9.2) is globally asymptotically stable if there exists a scalar function $V(x)$ with continuous first order derivatives such that $V(x)$ is positive definite, $\dot{V}(x)$ is negative definite and $V(x)$ is radially unbounded, i.e., $V(x) \to \infty$ as $||x|| \to \infty$.

#### Krasovskii-LaSalle's invariant set theorem

Krasovskii-LaSalle's results extend the stability analysis of the previous theorems when $\dot{V}$ is only negative semi-definite. They are stated as follows.

**Definition 9.8 (Invariant set)** A set $S$ is an *invariant set* for a dynamic system if every trajectory starting in $S$ remains in $S$.

Invariant sets include equilibrium points, limit cycles, as well as any trajectory of an autonomous system.

**Theorem 9.5 (Krasovskii-LaSalle)** Consider the system (9.2) with $f$ continuous, and let $V(x)$ be a scalar function with continuous first partial derivatives. Consider a region $\Gamma$ defined by $V(x) < \gamma$ for some $\gamma > 0$. Assume that the region $\Gamma$ is bounded and $\dot{V}(x) \leq 0\ \forall x \in \Gamma$. Let $\Omega$ be the set of all points in $\Gamma$ where $\dot{V}(x) = 0$, and $M$ be the largest invariant set in $\Omega$. Then, every solution $x(t)$ originating in $\Gamma$ tends to $M$ as $t \to \infty$. On the other hand, if $\dot{V}(x) \leq 0\ \forall x$ and $V(x) \to \infty$ as $||x|| \to \infty$, then all solutions globally asymptotically converge to $M$ as $t \to \infty$.

Another formulation of this result is as follows [135].

**Theorem 9.6** Under the same assumptions of theorem 9.5, let $K$ be the set of points not containing whole trajectories of the system for $\leq t \leq \infty$. Then if $\dot{V}(x) \leq 0$ outside of $K$ and $\dot{V}(x) = 0$ inside $K$, the system is asymptotically stable. ♠♠

Notice in particular that $\{x = 0\} \notin K$. $K$ can be a surface, a line, etc.

### 9.1.2 Non-autonomous systems

In this section we consider non-autonomous nonlinear systems represented by (9.1). The stability concepts are characterized by the following definitions.

**Definition 9.9 (Equilibrium)** A state $x^\star$ is an *equilibrium* point of (9.1) if $f(x^\star, t) = 0\ \forall t \geq t_0$.

**Definition 9.10 (Stability)** The equilibrium point $x = 0$ is *stable* at $t = t_0$ if for any $\rho > 0$ there exists an $r(\rho, t_0) > 0$ such that $||x(t_0)|| < \rho\ \forall t \geq t_0$. Otherwise the equilibrium point $x = 0$ is unstable.

**Definition 9.11 (Asymptotic stability)** The equilibrium point $x = 0$ is *asymptotically stable* at $t = t_0$ if it is stable and if it exists $r(t_0) > 0$ such that $||x(t_0)|| < r(t_0) \Rightarrow x(t) \to 0$ as $t \to \infty$.

**Definition 9.12 (Exponential stability)** The equilibrium point $x = 0$ is *exponentially stable* if there exist two positive numbers $\alpha$ and $\lambda$ such that $||x(t)|| \leq \alpha ||x(t_0)|| \exp(-\lambda(t - t_0))$ $\forall t \geq t_0$, for $x(t_0)$ sufficiently small.

**Definition 9.13 (Global asymptotic stability)** The equilibrium point $x = 0$ is *globally asymptotically stable* if it is stable and $x(t) \to 0$ as $t \to \infty$ $\forall x(t_0)$.

The stability properties are called *uniform* when they hold independently of the initial time $t_0$ as in the following definitions.

**Definition 9.14 (Uniform stability)** The equilibrium point $x = 0$ is *uniformly stable* if it is stable with $r = r(\rho)$ that can be chosen independently of $t_0$.

**Definition 9.15 (Uniform asymptotic stability)** The equilibrium point $x = 0$ is *uniformly asymptotically stable* if it is uniformly stable and there exists a ball of attraction $B$, independent of $t_0$, such that $x(t_0) \in B \Rightarrow x(t) \to 0$ as $t \to \infty$.

### Lyapunov linearization method

Using Taylor expansion, the system (9.1) can be rewritten as

$$\dot{x} = A(t)x + o(x,t) \tag{9.7}$$

where

$$A(t) = \left.\frac{\partial f}{\partial x}\right|_{x=0}.$$

A linear approximation of (9.1) is given by

$$\dot{x} = A(t)x. \tag{9.8}$$

The result of theorem 9.1 can be extended to linear time-varying systems of the form (9.8) as follows.

**Theorem 9.7** A necessary and sufficient condition for the uniform asymptotic stability of the origin of system (9.8) is that a matrix $P(t)$ exists such that

$$V = x^T P(t) x > 0$$

and

$$\dot{V} = x^T (A^T P + PA + \dot{P}) x \leq k(t) V,$$

where $\lim_{t\to\infty} \int_{t_0}^{t} k(\tau) d\tau = -\infty$ uniformly with respect to $t_0$. ♠♠

We can now state the following result.

**Theorem 9.8** If the linearized system (9.8) is uniformly asymptotically stable, then the equilibrium point $x = 0$ of the original system (9.1) is also uniformly asymptotically stable.

### Lyapunov direct method

We present now the Lyapunov stability theorems for non-autonomous systems. The following definitions are required.

**Definition 9.16 (Function of class $\mathcal{K}$)** A continuous *function* $\kappa : [0, k) \to \mathbb{R}^+$ is said to be *of class* $\mathcal{K}$ if

(i) $\kappa(0) = 0$,

(ii) $\kappa(\chi) > 0 \quad \forall \chi > 0$,

(iii) $\kappa$ is nondecreasing.

Statements (ii) and (iii) can also be replaced with

(ii') $\kappa$ is strictly increasing,

so that the inverse function $\kappa^{-1}$ is defined. The function is said to be of class $\mathcal{K}_\infty$ if $k = \infty$ and $\kappa(\chi) \to \infty$ as $\chi \to \infty$. ♠♠

Based on the definition of function of class $\mathcal{K}$, a modified definition of exponential stability can be given.

**Definition 9.17 ($\mathcal{K}$-exponential stability)** The equilibrium point $x = 0$ is *K-exponentially stable* if there exist a function $\kappa(\cdot)$ of class $\mathcal{K}$ and a positive number $\lambda$ such that $\|x(t)\| \leq \kappa(\|x(t_0)\|) \exp(-\lambda(t - t_0)) \ \forall t \geq t_0$, for $x(t_0)$ sufficiently small.

**Definition 9.18 (Positive definite function)** A *function* $V(x, t)$ is said to be locally (globally) *positive definite* if and only if there exists a function $\alpha$ of class $\mathcal{K}$ such that $V(0, t) = 0$ and $V(x, t) \geq \alpha(\|x\|) \ \forall t \geq 0$ and $\forall x$ in a ball $B$.

**Definition 9.19 (Decrescent function)** A *function* $V(x, t)$ is locally (globally) *decrescent* if and only if there exists a function $\beta$ of class $\mathcal{K}$ such that $V(0, t) = 0$ and $V(x, t) \leq \beta(\|x\|) \ \forall t > 0$ and $\forall x$ in a ball $B$.

♠♠

The main Lyapunov stability theorem can now be stated as follows.

**Theorem 9.9** Assume that $V(x,t)$ has continuous first derivatives around the equilibrium point $x = 0$. Consider the following conditions on $V$ and $\dot{V}$ where $\alpha$, $\beta$ and $\gamma$ denote functions of class $\mathcal{K}$:

$$\begin{array}{rl} (i) & V(x,t) \geq \alpha(||x||) > 0 \\ (ii) & \dot{V}(x,t) \leq 0 \\ (iii) & V(x,t)\ le\beta(||x||) \\ (iv) & \dot{V} \leq -\gamma(||x||) < 0 \\ (v) & \lim_{x\to\infty} \alpha(||x||) = \infty. \end{array} \tag{9.9}$$

Then the equilibrium point $x = 0$ is:

- stable if conditions (i) and (ii) hold;
- uniformly stable if conditions (i)–(iii) hold;
- uniformly asymptotically stable if conditions (i)–(iv) hold;
- globally uniformly asymptotically stable if conditions (i)–(v) hold.

♠♠

**Barbalat's lemma**

Krasovskii-LaSalle's results are only applicable to autonomous systems. On the other hand, Barbalat's lemma can be used to obtain stability results when the Lyapunov function derivative is negative semi-definite.

**Lemma 9.1 (Barbalat)** If the differentiable function $f$ has a finite limit as $t \to \infty$, and if $\dot{f}$ is uniformly continuous, then $\dot{f} \to 0$ as $t \to \infty$. ♠♠

This lemma can be applied for studying stability of non-autonomous systems with Lyapunov theorem, as stated by the following result.

**Lemma 9.2** If a scalar function $V(x,t)$ is lower bounded and $\dot{V}(x,t)$ is negative semi-definite, then $\dot{V}(x,t) \to 0$ as $t \to \infty$ if $\dot{V}(x,t)$ is uniformly continuous in time.

## 9.2 Differential geometry theory

In this section we briefly review the fundamental results of *differential geometry theorem*. Consider a nonlinear affine single-input/single-output system of the form

$$\dot{x} = f(x) + g(x)u \tag{9.10}$$

$$y = h(x) \tag{9.11}$$

where $h(x) : \mathbb{R}^n \to \mathbb{R}$ and $f(x), g(x) : \mathbb{R}^n \to \mathbb{R}^n$ are smooth functions. For ease of presentation we assume that system (9.10) and (9.11) has an equilibrium at $x = 0$.

**Definition 9.20 (Lie derivative)** The *Lie derivative* of $h$ with respect to $f$ is the scalar

$$L_f h = \frac{\partial h}{\partial x} f,$$

and the higher derivatives satisfy the recursion

$$L_f^i h = L_f(L_f^{i-1} h)$$

with $L_f^0 h = h$.

**Definition 9.21 (Lie bracket)** The *Lie bracket* of $f$ and $g$ is the vector

$$[f, g] = \frac{\partial g}{\partial x} f - \frac{\partial f}{\partial x} g,$$

and the recursive operation is established by

$$ad_f^i g = [f, ad_f^{i-1} g].$$

♠♠

Some properties of Lie brackets are:

$$[\alpha_1 f_1 + \alpha_2 f_2, g] = \alpha_1 [f_1, g] + \alpha_2 [f_2, g]$$

$$[f, g] = -[g, f],$$

and the Jacobi identity

$$L_{ad_g} h = L_f(L_g h) - L_g(L_f h).$$

To define nonlinear changes of coordinates we need the following concept.

**Definition 9.22 (Diffeomorphism)** A function $\phi(x) : \mathbb{R}^n \to \mathbb{R}^n$ is said to be a *diffeomorphism* in a region $\Omega \in \mathbb{R}^n$ if it is smooth, and $\phi^{-1}(x)$ exists and is also smooth. ♠♠

A sufficient condition for a smooth function $\phi(x)$ to be a diffeomorphism in a neighbourhood of the origin is that the Jacobian $\partial\phi/\partial x$ be nonsingular at zero.

The conditions for feedback linearizability of a nonlinear system are strongly related with the following theorem.

**Theorem 9.10 (Frobenius)** Consider a set of linearly independent vectors $\{f_1(x), \ldots, f_m(x)\}$ with $f_i(x) : \mathbb{R}^n \to \mathbb{R}^n$. Then, the following statements are equivalent:

(i) *(complete integrability)* there exist $n - m$ scalar functions $h_i(x) : \mathbb{R}^n \to \mathbb{R}$ such that

$$L_{f_j} h_i = 0 \qquad 1 \leq i \quad j \leq n - m$$

where $\partial h_i / \partial x$ are linearly independent;

(ii) *(involutivity)* there exist scalar functions $\alpha_{ijk}(x) : \mathbb{R}^n \to \mathbb{R}$ such that

$$[f_i, f_j] = \sum_{k=1}^{m} \alpha_{ijk}(x) f_k(x).$$

♠♠

### 9.2.1 Normal form

In this section we present the normal form of a nonlinear system which has been instrumental for the development of the feedback linearizing technique. For this, it is convenient to define the notion of relative degree of a nonlinear system.

**Definition 9.23 (Relative degree)** The system (9.10) and (9.11) has *relative degree* $n^*$ at $x = 0$ if

(i) $L_g L_f^k h(x) = 0$, $\forall x$ in a neighbourhood of the origin and $\forall k < n^* - 1$;

(ii) $L_g L_f^{n^*-1} h(x) \neq 0$. ♠♠

It is worth noticing that in the case of linear systems, e.g., $f(x) = Ax$, $g(x) = Bx$, $h(x) = Cx$, the integer $n^*$ is characterized by the conditions $CA^k B = 0$ $\forall k < n^* - 1$ and $CA^{n^*-1} B \neq 0$. It is well known that these are exactly the conditions that define the relative degree of a linear system.
Another interesting interpretation of the relative degree is that $n^*$ is exactly the number of times we have to differentiate the output to obtain the input explicitly appearing.
The functions $L_f^i h$ for $i = 0, 1, \ldots, n^* - 1$ have a special significance as demonstrated in the following theorem.

**Theorem 9.11 (Normal form)** If system (9.10) and (9.11) has relative degree $n^* \leq n$, then it is possible to find $n - n^*$ functions $\phi_{n^*+1}(x), \ldots, \phi_n(x)$

so that

$$\phi(x) = \begin{pmatrix} h(x) \\ L_f h(x) \\ \vdots \\ L_f^{n^*-1} h(x) \\ \phi_{n^*+1}(x) \\ \vdots \\ \phi_n(x) \end{pmatrix}$$

is a diffeomorphism $z = \phi(x)$ that transforms the system into the following normal form

$$\begin{aligned} \dot{z}_1 &= z_2 \\ \dot{z}_2 &= z_3 \\ &\vdots \\ \dot{z}_{r-1} &= z_r \\ \dot{z}_r &= b(z) + a(z)u \\ \dot{z}_{r+1} &= q_{r+1}(z) \\ &\vdots \\ \dot{z}_n &= q_n(z). \end{aligned}$$

Moreover, $a(z) \neq 0$ in a neighbourhood of $z_0 = \phi(0)$. ♠♠

### 9.2.2 Feedback linearization

From the above theorem we see that the state feedback control law

$$u = \frac{1}{a(z)}(-b(z) + v) \tag{9.12}$$

yields a closed-loop system consisting of a chain of $r$ integrators and an $(n-r)$-dimensional autonomous system. In the particular case of $n^* = n$ we *fully linearize* the system. The first set of conditions for the triple $\{f(x), g(x), h(x)\}$ to have relative degree $n$ is given by the partial differential equation

$$\frac{\partial h}{\partial x}\left( g(x) ad_f g(x) \dots ad_f^{n-2} g(x) \right) = 0.$$

Frobenius theorem shows that the existence of solutions to this equation is equivalent to the involutivity of $\{g(x), ad_f g(x), \dots, ad_f^{n-2} g(x)\}$. It can be shown that the second condition, i.e., $L_g L_f^{n-1} h(x) \neq 0$ is ensured by the linear independence of $\{g(x), ad_f g(x), \dots, ad_f^{n-1} g(x)\}$.
The preceding discussion is summarized by the following key theorem.

**Theorem 9.12** For the system (9.10) there exists an output function $h(x)$ such that the triple $\{f(x), g(x), h(x)\}$ has relative degree $n$ at $x = 0$ if and only if:

(i) the matrix $\{g(0), ad_f g(0), \ldots, ad_f^{n-1} g(0)\}$ is full rank

(ii) the set $\{g(x), ad_f g(x), \ldots, ad_f^{n-2} g(x)\}$ is involutive around the origin.

The importance of the preceding theorem can hardly be overestimated. It gives (a priori verifiable) necessary and sufficient conditions for full linearization of a nonlinear affine system. However, it should be pointed out that this control design approach requires on one hand the solution of a set of partial differential equations. On the other hand, it is intrinsically nonrobust since it relies on exact cancellation of nonlinearities. In the linear case this is tantamount to pole-zero cancellation.

### 9.2.3 Stabilization of feedback linearizable systems

If the relative degree of the system $n^* < n$ then, under the action of the feedback linearizing controller (9.12), there remains an $(n - n^*)$-dimensional subsystem. The importance of this subsystem is underscored in the proposition below.

**Theorem 9.13** Consider the system (9.10) and (9.11) assumed to have relative degree $n^*$. Further, assume that the trivial equilibrium of the following $(n - n^*)$-dimensional dynamical system is *locally* asymptotically stable:

$$\begin{aligned} \dot{z}_{n^*+1} &= q_{n^*+1}(0, \ldots, 0, z_{n^*+1}, \ldots, z_n) \\ &\vdots \\ \dot{z}_n &= q_n(0, \ldots, 0, z_{n^*+1}, \ldots, z_n) \end{aligned} \tag{9.13}$$

where $q_{n^*+1}, \ldots, q_n$ are given by the normal form. Under these conditions, the control law (9.12) yields a *locally* asymptotically stable closed-loop system. ♠♠

The $(n - n^*)$-dimensional system (9.13) is known as the *zero dynamics*. It represents the dynamics of the unobservable part of the system when the input is set equal to zero and the output is constrained to be identically zero. It is worth highlighting the qualifier *local* in the above theorem; in other words, it can be shown that the conditions above are not enough to ensure *global* asymptotic stability. This is illustrated by the following example

### 9.2.4 Further reading

The original Lyapunov theorem is contained in [116], while stability of nonlinear dynamic systems is widely covered in [98] [100]. The proofs of the theorems concerning Lyapunov stability theorem can be found in [206] [88] [64]. An extensive presentation of differential geometry methods can be found in [72] and the references therein. For the extension to the multivariable case and further details we refer the reader again to [72] as well as to [148].

# Bibliography


[1] A. Ailon, R. Lozano and M. Gil, 1997,"Point-to-point regulation of a robot with flexible joints including effects of actuator dynamics". IEEE Transactions on Automatic Control, Vol 42, No 4, pp 559-564.

[2] R. Abraham and J.E. Marsden 1978 *Foundations of Mechanics*, Second edition, Benjamin Cummings, Reading, MA USA.

[3] B.D.O. Anderson, R.B. Bitmead, C.R. Johnson, P.V. Kokotovic, R.L. Kosut, I.M.Y. Mareels, L. Praly, and B.D. Riedle, 1986 *Stability Analysis of Adaptive Systems: Passivity and Average Analysis*, The MIT Press Cambridge, MA.

[4] B.D.O. Anderson and S. Vongpanitlerd, 1973 *Network Analysis and Synthesis: A Modern Systems Theory Approach*, Englewood Cliffs, New Jersey; Prentice Hall.

[5] B.D.O. Anderson and J.B. Moore, 1971 *Linear Optimal Control. Prentice-Hall*, Englewood Cliff, N.Y..

[6] B.D.O. Anderson, 1967"A system theory criterion for positive real matrices", SIAM J. Control, vol.5, No 2, pp 171-182.

[7] D. Angeli E. Sontag, Y. Wang, 1998 "A remark on integral input to state stability", in Proc. IEEE Conf. Decision and Control, Tampa, December, pp. 2491-2496.

[8] H. Arai, S. Tachi, 1991 "Position Control of a Manipulator with Passive Joints using Dynamic Coupling", IEEE Trans. Robotics and Automation, vol.7, No 4, pp.528-534, August.

[9] S. Arimoto, 1996, *Control Theory of Nonlinear Mechanical Systems: A Passivity-Based and Circuit-Theoretic Approach*, Oxford University Press, Oxford, UK.

[10] K.J. Astrom, K. Furuta, 1996 "Swinging Up a pendulum by Energy Control", In IFAC'96, Preprints 13th World Congress of IFAC, vol.E, pp.37-42, San Francisco, California.

[11] B. d'Andréa Novel, G. Bastin, B. Brogliato, G. Campion, C. Canudas, H. Khalil, A. de Luca, R. Lozano, R. Ortega, P. Tomei, B. Siciliano, 1996 *Theory of robot control*, C. Canudas de Wit, G. Bastin, B. Siciliano (Eds.), Springer Verlag, CCES, London.

[12] V.I. Babitsky, 1998 *Theory of vibro-impact systems and applications*, Springer Berlin, Foundations of Engineering Mechanics.

[13] N.E. Barabanov, 1988 "On the Kalman problem", Sibirskii Matematischeskii Zhurnal, vol.29, pp.3-11, May-June. Translated in Siberian Mathematical Journal, pp.333-341.

[14] N.E. Barabanov, A.K. Gelig, G.A. Leonov, A.L. Likhtarnikov, A.S. Matveev, V.B. Smirnova, A.L. Fradkov, 1996 "The frequency theorem (Kalman-Yakubovich lemma) in control theory", Automation and Remote Control, vol.57, no 10, pp.1377-1407.

[15] S. Battilotti, L. Lanari, 1995 "Global set point control via link position measurement for flexible joint robots", Systems and Control Letters, vol.25, pp.21-29.

[16] S. Battilotti, L. Lanari, R. Ortega, 1997 "On the role of passivity and output injection in the ouput feedback stabilisation problem: Application to robot control", European Journal of Control, vol.3, pp.92-103.

[17] H. Berghuis, H. Nijmeijer, 1993 "A passivity approach to controller-observer design for robots", IEEE Transactions on Robotics and Automation, vol.9, no 6, pp.741-754.

[18] H. Berghuis, R. Ortega, H. Nijmeijer, 1992 "A robust adaptive controller for robot manipulators", Proc. IEEE Int. Conference on Robotics and Automation, Nice, France, April, pp.1876-1881.

[19] J. Bernat, J. Llibre, 1996 "Counterexample to Kalman and Markus-Yamabe conjectures in dimension larger than 3", Dynamics of Continuous, Discrete and Impulsive Systems, vol.2, pp.337-379.

[20] G.M. Bernstein and M.A. Lieberman. *A method for obtaining a canonical Hamiltonian for nonlinear LC circuits* IEEE Trans. on Circuits and Systems, CAS-35, 3, pp.411-420, 1989

[21] A.M. Block, 1996 *Mechanical Design and Control of the Pendubot*, MSc Thesis, University of Illinois at Urbana-Champaign.

[22] S. A. Bortoff, 1992 *Pseudolinearization using Spline Functions with Application to the Acrobot*, PhD Thesis, Dept. of Electrical and Computer Engineering, University of Illinois at Urbana-Champaign.

[23] F. H. Branin, 1977 "The network concept as a unifying principle in engineering and the physical sciences", in Problem Analysis in Science and Engineering, edited by F. H. Branin and K. Huseyin, 41-111, Academic Press, New York.

[24] P.C. Breedveld, 1984 *Physical systems theory in terms of bond graphs*, Ph.D. thesis, University of Twente , The Netherlands.

[25] M. Bridges, D.M. Dawson, 1995 "Backstepping control of flexible joint manipulators: a survey", Journal of Robotic Systems, vol.12, no 3, pp.199-216.

[26] R.W. Brockett, 1977 "Control theory and analytical mechanics", in *Geometric Control Theory*, C. Martin and R. Herman (Eds.), pp.1-46, Math.Sci.Press, Brookline, 1977.

[27] B. Brogliato, 1999 *Nonsmooth Mechanics*, Springer Verlag, London, Communications and Control Engineering Series, 2nd edition.

[28] B. Brogliato, I.D. Landau, R. Lozano, 1991 "Adaptive motion control of robot manipulators: a unified approach based on passivity", Int. J. of Robust and Nonlinear Control, vol.1, no 3, July-September, pp.187-202.

[29] B. Brogliato, R. Lozano, 1996 "Correction to "Adaptive control of robot manipulators with flexible joints", IEEE Transactions on Automatic Control, vol.41, no 6, pp.920-922.

[30] B. Brogliato, R. Lozano, 1992 "Passive least-squares-type estimation algorithm for direct adaptive control", Int. J. of Adaptive Control and Signal Processing, January, vol.6, no 1, pp.35-44.

[31] B. Brogliato, R. Lozano, I.D. Landau, 1993 "New relationships between Lyapunov functions and the passivity theorem", Int. J. Adaptive Control and Signal Processing, vol.7, pp.353-365.

[32] B. Brogliato, S.I. Niculescu, M.D.P. Monteiro Marques, 2000 "On the tracking control of a class of complementary-slackness mechanical systems", Systems and Control Letters, vol.39.

[33] B. Brogliato, S.I. Niculescu, P. Orhant, 1997 "On the control of finite dimensional mechanical systems with unilateral constraints", IEEE Transactions on Automatic Control, vol.42, no 2, pp.200-215.

[34] B. Brogliato, R. Ortega, R. Lozano, 1995 "Global tracking controllers for flexible-joint manipulators: a comparative study", Automatica, vol.31, no 7, pp.941-956.

[35] B. Brogliato, D. Rey, 1998 "Further experimental results on nonlinear control of flexible joint manipulators", American Control Conference, June 24-26.

[36] B. Brogliato, D. Rey, A. Pastore, J. Barnier, 1998 "Experimental comparison of nonlinear controllers for flexible joint manipulators", Int. J. of Robotics Research, vol.17, no 3, March, pp.260-281.

[37] V.A. Brusin, 1976 "The Lurie equation in the Hilbert space and its solvability'" (in Russian), Prikl. Mat. Mekh., vol.40, no 5, pp.947-955.

[38] C.I. Byrnes, A. Isidori, J.C. Willems, 1991 "Passivity, Feedback equivalence, and the global stabilization of minimum phase nonlinear systems", IEEE Transactions on Automatic Control, vol.36, no 11, pp. 1228-1240.

[39] G. Campion, B. d'Andréa-Novel and G. Bastin, 1990 "Controllability and stae-feedback stabilizability of non-holonmic mechanical systems", in *Advanced Robot Control*, C. de Witt (Ed.), LNCIS 162, pp.106-124, 1990

[40] L.O. Chua, J.D. McPherson, 1974 "Explicit topological formulation of lagrangian and Hamiltonian equations for nonlinear networks", IEEE trans. on Circuits and Systems, vol.21, no 2, pp.277-285.

[41] C.C. Chung and J. Hauser, 1995 "Nonlinear control of a swinging pendulum", Automatica, vol.31, no 6, 851-862.

[42] A. Cima, A. Gasull, E. Hubbers, F. Manosas, 1997 "A polynomial counterexample to the Markus-Yamabe conjecture", Advances in Mathematics, vol.131, pp.453-457, article no AI971673.

[43] J. Collado, R. Lozano, R. Johansson, 1999 "On Kalman-Yakubovich-Popov lemma for stabilizable systems". *Proceedings of the IEEE Conference on Decision and Control*, December 1999, Phoenix, USA.

[44] M. Dalsmo, A.J. van der Schaft, 1999 "On representations and integrability of mathematical structures in energy-conserving physical systems", SIAM J. Control and Optimization, vol.37, no 1, pp.54-91.

[45] C.A. Desoer, M. Vidyasagar, 1975 *Feedback systems: Input-Output properties*, Academic Press, New-York.

[46] C.A. Desoer, E.S. Kuh, 1969 *Basic circuit theory* McGraw Hill Int.

[47] A. Dzul, R. Lozano, 1999 "Passivity of a helicopter model", Technical Report, University of Technology of Compiègne, Heudiasyc. November.

[48] O. Egeland, 1993 *Servoteknikk*, Tapir Forlag, ISBN 82-519-1142-7.

[49] O. Egeland, J.-M. Godhavn, 1994 "Passivity-based adaptive attitude control of a rigid spacecraft", IEEE Transactions on Automatic Control, Vol 39, pp 842-846, April.

[50] F.J. Evans, J.J. van Dixhoorn, 1974 "Towards more physical structure in systems' theory", in *Physical Structure in Systems' Theory*, J.J. van Dixhoorn and F.J. Evans (Eds.), pp.1-15, Academic press, London.

[51] I. Fantoni, R. Lozano, M. Spong, 1998 "Passivity based control of the pendubot", IEEE Transactions on Automatic Control, to appear.

[52] I. Fantoni, R. Lozano, F. Mazenc, K. Y. Pettersen, 2000, "Stabilization of an undereactuated hovercraft". Int. J. of Robust and Nonlinear Control, to appear.

[53] E.D. Fasse, P.C. Breedveld, 1998 "Modelling of Elastically Coupled Bodies: Part I: General Theory and Geometric Potential Function Method", ASME J. of Dynamic Systems, Measurement and Control, vol.120, pp.496-500, December.

[54] E.D. Fasse, P.C. Breedveld, 1998 "Modelling of Elastically Coupled Bodies: Part II: Exponential- and Generalized-Coordinate Methods", ASME J. of Dynamic Systems, Measurement and Control, Vol.120, pp.501-506, December.

[55] P. Faurre, M. Clerget, F. Germain, 1979 *Opérateurs Rationnels Positifs. Application à l'hyperstabilité et aux processus aléatoires*, Méthodes Mathématiques de l'Informatique, Dunod, Paris.

[56] R. Garrido-Moctezuma, D. Suarez, R. Lozano, 1998, "Adaptive LQG control of PR plants". Int. J. of Adaptive Control, Vol 12, pp 437-449.

[57] A. Kh. Gelig, G.A. Leonov, V.A. Yakubovich, 1978 *The stability of nonlinear systems with a nonunique equilibrium state*, (in Russian), Nauka, Moscow.

[58] F. Génot, B. Brogliato, 1999 "New results on Painlevé paradoxes", European Journal of Mechanics A/Solids, vol.18, no 4, pp.653-677.

[59] H. Goldstein, 1980 *Classical Mechanics*, Second Edition, Addison Wesley.

[60] C. Gutierrez, 1995 "A solution to the bidimensional global asymptotic stability conjecture", Ann. Inst. Henri Poincaré, vol.12, no 6, pp.627-671.

[61] J.-T Gravdahl and O. Egeland, 1999 "Compressor Surge and Rotating Stall: Modeling and Control". *Advances in Industrial Control*, Springer-Verlag London.

[62] W.M. Haddad, D.S. Bernstein, Y.W. Wang, 1994 "Dissipative $H_2/H_\infty$ Controller synthesis", IEEE Transactions on Automatic Control, vol.39, pp.827-831.

[63] W. Haddad and D. Bernstein, 1991 "Robust stabilization with positive real uncertainty: Beyond the small gain theorem", Systems and Control Letters, vol.17, pp.191-208.

[64] W. Hahn, 1967 *Stability of Motion*, Springer-Verlag, New York, NY.

[65] D.J. Hill, P.J. Moylan, 1980 "Connections between Finite-Gain and asymptotic stability" IEEE-Transactions on Automatic Control, no 5 pp.931-936, October.

[66] D.J. Hill, P.J. Moylan, 1976 "The stability of nonlinear dissipative systems", IEEE Transactions on Automatic Control, pp.708-711, October.

[67] D.J. Hill, P.J. Moylan, 1980 "Dissipative dynamical systems: basic input-output and state properties", J. Franklin Institute, vol.309, pp.327-357.

[68] D.J. Hill, P.J. Moylan, 1975 "Cyclo-dissipativeness, dissipativeness and losslessness for nonlinear dynamical systems", Technical Report EE7526, November, The university of Newcastle, Dept. of Electrical Engng., New South Wales, Australia.

[69] N. Hogan, 1985 "Impedance Control : an Approach to Manipulation: Part 2 - Implementation", Trans. ASME, J. Dynamic Systems and Measurements, vol.107, no 1, pp.8-16.

[70] C.H. Huang, P.A. Ioannou, J. Maroulas, and M.G. Safonov, 1999 "Design of strictly positive real systems using constant output feedback", IEEE Transactions on Automatic Control, vol.44, no 3, pp.569-573, March.

[71] P. Ioannou, G. Tao, 1987 "Frequency domain conditions for SPR functions", IEEE Transactions on Automatic Control, vol.32, pp.53-54, January.

[72] A. Isidori, 1989 *Nonlinear Control Systems*, 2nd edition,, Springer.

[73] B. Jakubczyk, W. Respondek, 1980 "On the linearization of control systems", Bull. Acad. Polon. Sci. Math., vol.28, pp.517-522.

[74] E.A. Johannessen, 1997 *Synthesis of dissipative output feedback controllers*, Ph.D. Dissertation, NTNU Trondheim.

[75] E. Johannessen, O. Egeland, 1995 "Synthesis of positive real $H_\infty$ controller", Proceedings of the American Control Conference, Washington, pp. 2437-2438.

[76] R. Johansson, 1990 "Adaptive control of robot manipulator motion", IEEE Transactions on Robotics and Automation, vol.6, pp.483-490.

[77] R. Johansson, A. Robertsson, R. Lozano, 1999 "Stability analysis of adaptive output feedback control", IEEE Conference on Decision and Control, Phoenix, Arizona, December.

[78] D.L. Jones and F.J. Evans, 1973 "A Classification of physical variables and its application in variational methods", J. of the Franklin Institute, vol.295, no 6, pp.449-467.

[79] T. Kailath, 1980 *Linear systems*, Prentice-Hall.

[80] R.E. Kalman, 1963 "Lyapunov Functions for the Problem of Lurie in Automatic Control", Proc. Nat. Acad. Sci. U.S.A., vol.49, no 2, pp.201-205.

[81] R.E. Kalman, 1964 "When is a linear control system optimal?", ASME Ser. D, J. of Basic Engng., vol.86, pp.1-10.

[82] I. Kanellakopoulos, P. Kokotovic, A.S. Morse, 1991 "Systematic design of adaptive controllers for feedback linearizable systems", IEEE Transactions on Automatic Control, vol.36, no 11, pp.1241-1253.

[83] L.V. Kantorovich, G.P. Akilov, 1982 *Functional Analysis*, Second Edition, Pergamon Press.

[84] D.C. Karnopp, D.L. Margolis and R.C. Rosenberg, 1990 "Systems Dynamics : A Unified Approach", Second Edition, John Wiley & Sons, New-York, USA.

[85] A. Kelkar, S. Joshi, 1996 *Control of nonlinear multibody flexible space structures*, LNCIS 221, Springer Verlag, London.

[86] R. Kelly, (1998), "Global positioning of robot manipulators via PD control plus a nonlinear control action". IEEE Transactions on Automatic Control, vol.43, pp.934-938.

[87] R. Kelly, and V. Santibañez, (1998) "Global regulation of elastic joints robots based on energy shaping", IEEE Transactions on Automatic Control, vol.43, pp.1451-1456.

[88] H.K. Khalil, 1992 *Nonlinear systems*, MacMillan, NY.

[89] D.E. Koditschek, 1988 "Application of a new Lyapunov function to global adaptive attitude tracking", 27th IEEE Conference on Decision and Control, Austin, December.

[90] W.S. Koon, J.E. Marsden, 1997 "Poisson reduction for nonholonomic systems with symmetry", Proc. of the *Workshop on Nonholonomic Constraints in Dynamics*, Calgary, August 26-29.

[91] M. Krstic, I. Kanellakopoulos, P. Kokotovic, 1994 "Nonlinear design of adaptive controllers for linear systems", IEEE Transactions on Automatic Control, vol.39, pp.752-783.

[92] M. Krstic, P. Kokotovic, I. Kanellakopoulos, 1993 "Transient performance improvement with a new class of adaptive controllers", Systems and Control Letters, vol.21, pp.451-461.

[93] C. Lanczos, 1970 *The Variational Principles of Mechanics*, Dover, NY, 4th Edition.

[94] P. Lancaster and M. Tismenetsky, 1985 *The Theory of Matrices*, New York, Academic Press.

[95] I.D. Landau, 1979 *Adaptive Control. The Model Reference Approach*, Marcel Dekker, New York.

[96] I.D. Landau, R. Lozano, M. M'Saad, 1997 *Adaptive Control*, Springer-Verlag, CCES, London.

[97] P. de Larminat, 1993 *Automatique. Commande des systèmes linéaires*, Hermès, Paris.

[98] J. La Salle, S. Lefschetz, 1961 *Stability by Liapunov's Direct Method*, Academic Press, New York, NY.

[99] L. Lefèvre, 1998 *De l'introduction d'éléments fonctionnels au sein de la théorie des bond graphs*, Ph.D. Ecole Centrale de Lille, France.

[100] S. Lefschetz, 1962 *Stability of Nonlinear Control Systems*, Academic Press, New York, NY.

[101] P. Libermann, C.M. Marle, 1987 *Symplectic geometry and Analytical Mechanics*, Reidel, Dordrecht.

[102] A. L. Likhtarnikov, V.A.Yakubovich, 1977 "The frequency theorem for one-parameter semigroups", (in Russian), Izv. Akad. Nauk SSSR, Ser. Mat., no 5, pp.1064-1083.

[103] Z. Lin, A. Saberi, M. Gutmann, Y. A. Shamash,1996 "Linear controller for an inverted pendulum having restricted travel: A high-and-low approach", Automatica, vol.32, no 6, 933-937.

[104] J. Loncaric, 1987 "Normal form of stiffness and compliant matrices", IEEE J. of Robotics and Automation, vol.3, no 6, pp.567-572.

[105] R. Lozano, S.M. Joshi, 1990 "Strictly positive real functions revisited", IEEE Transactions on Automatic Control, vol.35, pp.1243-1245, November.

[106] R. Lozano, B. Brogliato, 1992 "Adaptive control of first order nonlinear system without a priori information on the parameters", IEEE Transactions on Automatic Control, vol.37, no 1, January.

[107] R. Lozano, S. Joshi, 1988 "On the design of dissipative LQG type controllers" Proceedings of the 27th IEEE Conference on Decision and Control, Austin, Texas, pp.1645-1646, December.

[108] R. Lozano, I. Fantoni, 1998 "Passivity based control of the inverted pendulum", IFAC NOLCOS, The Netherlands, July.

[109] R. Lozano, I. Fantoni, 2000, "Stabilization of the inverted pendulum around its homoclinic orbit", To appear in Systems and Control Letters.

[110] R. Lozano, A. Valera, P. Albertos, S. Arimoto, T. Nakayama, 1999 "PD Control of robot manipulators with joint flexibility, actuators dynamics and friction", Automatica, vol.35, pp.1697-1700.

[111] R. Lozano, B. Brogliato, 1991 "Adaptive motion control of flexible joint manipulators", American Control Conference, Boston, USA, June.

[112] R. Lozano, B. Brogliato, I.D. Landau, 1992 "Passivity and global stabilization of cascaded nonlinear systems", IEEE Transactions on Automatic Control, vol.37, no 9, pp.1386-1388.

[113] R. Lozano, B. Brogliato, 1992 "Adaptive control of robot manipulators with flexible joints", IEEE Transactions on Automatic Control, vol.37, no 2, pp.174-181.

[114] R. Lozano, C. Canudas de Wit, 1990 "Passivity-based adaptive control for mechanical manipulators using LS type estimation", IEEE Transactions on Automatic Control, vol.35, pp.1363-1365.

[115] A. de Luca, 1988 "Dynamic control of robots with joint elasticity", Proc. IEEE Int. Conference on Robotics and Automation, Philadelphia, USA, pp.152-158.

[116] A.M. Lyapunov, 1907 *The General Problem of Motion Stability*, in Russian, 1892; translated in French, Ann. Faculté des Sciences de Toulouse, pp.203-474.

[117] N.H. McClamroch, 1989 "A singular perturbation approach to modeling and ocntrol of manipulators constrained by a stiff environment", 28$^{th}$ IEEE Conference on Decision and Control, December.

[118] N.H. McClamroch, D. Wang, 1988 "Feedback stabilization and tracking of constrained robots", IEEE Trans. on Automatic Control, vol.33, no 5, pp.419-426, May.

[119] R. Marino, P. Tomei, 1995 *Nonlinear Control Design. Geometric, Adaptive, Robust*, Prentice Hall.

[120] C.C.H. Ma, M. Vidyasagar, 1986 "Nonpassivity of linear discrete-time systems", Systems and Control Letters, vol.7, pp.51-53.

[121] R. Mahony, T. Hamel, A. Dzul, 1999 "Hover control via approximate lyapunov control for a model helicopter", Conference on Decision and Control, Phoenix, December.

[122] R. Mahony, R. Lozano, 1999 "An energy based approach to the regulation of a model helicopter near to hover", Proceedings of the European Control Conference, ECC'99, Karlsruhe, Germany.

[123] C.M. Marle, 1997 "Various approaches to conservative and nonconsevative nonholonomic systems", Proc. of the *Workshop on Nonholonomic Constraints in Dynamics*, Calgary, August 26-29.

[124] B.M. Maschke, A.J. van der Schaft, 1992 "Port controlled Hamiltonian systems: modeling origins and system theoretic properties", Proc. 2nd IFAC Symp. on Nonlinear Control Systems design, NOLCOS'92, pp.282-288, Bordeaux, June.

[125] B.M. Maschke, A. van der Schaft, P.C. Breedveld, 1992 "An Intrinsic Hamiltonian Formulation of Network Dynamics: Nonstandard Poisson Structures and Gyrators", Journal of the Franklin Institute, vol.329, no 5, pp.923-966.

[126] B.M. Maschke, A.J. van der Schaft, 1994 "A Hamiltonian approach to stabilization of non-holonomic mechanical systems", Proc. 33rd IEEE Conf. on Decision and Control, Lake Buena Vista, Florida, pp.14-16, February.

[127] B.M. Maschke, P. Chantre, 1994 "Bond graph modeling and parameter identification of a heat exchanger", Proc. ASME Int. Mechanical Engineering Congress and Exhibition, vol.55-2, pp.645-652, Chicago, USA, Nov. 6-11.

[128] B.M. Maschke, A.J. van der Schaft, P.C. Breedveld, 1995 "An intrinsic Hamiltonian formulation of the dynamics of LC-circuits", Trans. IEEE on Circuits and Systems, I: Fundamental Theory and Applications, vol.42, no 2, pp.73-82, February.

[129] B.M. Maschke, 1996 "Elements on the modelling of multibody systems", Modelling and Control of Mechanisms and Robots, pp.1-38, C.Melchiorri and A.Tornambè (Eds.), World Scientific Publishing Ltd.

[130] B.M. Maschke, A.J. van der Schaft, 1997 "Interconnected Mechanical Systems Part I: Geometry of interconnection and implicit Hamiltonian systems", in Modelling and Control of Mechanical Systems, A.Astolfi, C.Melchiorri and A.Tornambè (Eds.), pp.1-16, Imperial College Press.

[131] B.M. Maschke, A.J. van der Schaft, 1996 "Interconnection of systems: the network paradigm", Proc. 35th Conference on Decision and Control, Kobe, Japan, pp.207-212.

[132] B.M. Maschke, A.J. van der Schaft, 1997 "Interconnected Mechanical Systems Part II: The dynamics of spatial mechanical networks", *Modelling and Control of Mechanical Systems*, A.Astolfi, C.Melchiorri and A.Tornambè (Eds.), pp.17-30, Imperial College Press.

[133] F. Mazenc, L. Praly, 1996 "Adding integrators, saturated controls, and stabilization for feedforward systems", IEEE Transactions on Automatic Control, vol.41, no 11, pp.1559-1578, November.

[134] G. Meisters, 1996 "A biography of the Markus-Yamabe conjecture", available at http://www.math.unl.edu/ gmeister/Welcome.html , expanded form of a talk given at the conference Aspects of Mathematics – Algebra, Geometry and Several Complex Variables, June 10-13, the Universiy of Hong-Kong.

[135] Y. Merkin, 1997 *Introduction to the Theory of Stability*, Springer Verlag, TAM 24.

[136] J. De Miras, A. Charara, 1998 "A vector oriented control for a magnetically levitated shaft", IEEE Trans. on Magnetics, vol.34, no 4, pp.2039-2041.

[137] J. De Miras, A. Charara, 1999 "Vector desired trajectories for high rotor speed magnetic bearing stabilization", IFAC'99, 14th World Congress, July, China.

[138] J.J. Moreau, 1979 "Application of convex analysis to some problems of dry friction", in Trends in Applications of Pure Mathematics to Mechanics, vol. II (H. Zorski, Ed.), Pitman Publishing Ltd, London, pp.263-280.

[139] A.S. Morse, 1992 "High-order parameter tuners for the adaptive control of linear and nonlinear systems", Proc. of the US-Italy joint seminar "Systems, models and feedback: theory and application", Capri, Italy.

[140] P.J. Moylan, B.D.O. Anderson, 1973 "Nonlinear regulator theory and an inverse optimal control problem", IEEE Transactions on Automatic Control, vol.18, pp.460-465.

[141] P.J. Moylan, D.J. Hill, 1978 "Stability criteria for large-scale systems", IEEE Transactions on Automatic Control, vol.23, no 2, pp.143-149.

[142] R.M. Murray, Z. Li, S.S. Sastry, 1994 *A Mathematical Introduction to Robotic Manipulation*, CRC Press, Boca Raton, Florida.

[143] K.S. Narendra and J.H. Taylor, 1973 *Frequency domain criteria for absolute stability*, Academic Press.

[144] K.S. Narendra, A. Annaswamy, 1989 *Stable adaptive systems*, Prentice Hall.

[145] S.I. Niculescu, R. Lozano, 2000 "On the passivity of linear delay systems", To appear in IEEE Transactions on Automatic Control.

[146] S.I. Niculescu, 1997, "Systémes á retard: Aspects qualitatifs sur la stabilité et la stabilisation", Diderot Editeur, Arts et Sciences.

[147] S.I. Niculescu, E. I. Verriest, L. Dugard, J. M. Dion, 1997 "Stability and robust stability of time-delay systems: A guided tour", in *Stability and Control of Time-Delay Systems* (L. Dugard and E. I. Verriest, Eds.), LNCIS 228, Springer-Verlag, London, pp.1-71.

[148] H. Nijmeier, A.J. van der Schaft, 1990 *Nonlinear Dynamical Control Systems*, Springer Verlag, New-York.

[149] R. Ortega, 1993 "On Morse's new adaptive controller: parameter convergence and transient performance", IEEE Transactions on Automatic Control, vol.38, pp.1191-1202.

[150] R. Ortega, G. Espinosa, 1993 "Torque regulation of induction motors", Automatica, vol.29, pp.621-633.

[151] B. Paden, R. Panja, 1988 "Globally asymptotically stable PD+ controller for robot manipulators", Int. J. of Control, vol.47, pp.1697-1712.

[152] L. Paoli, M. Schatzman, 1993 "Mouvement à un nombre fini de degrés de liberté avec contraintes unilatérales: cas avec perte d'énergie", Mathematical Modelling and Numerical Analysis (Modèlisation Mathématique et Analyse Numérique), vol.27, no 6, pp.673-717.

[153] P.C. Parks, 1966 "Lyapunov redesigns of model reference adaptive control systems", IEEE Transactions on Automatic Control, vol.11, pp.362-367.

[154] H.M. Paynter, 1961 *Analysis and Design of Engineering Systems*, M.I.T. Press, Cambridge, MA, 1961.

[155] V.M. Popov, 1973 *Hyperstability of Control Systems*, Berlin, Springer-Verlag.

[156] V.M. Popov, 1959 "Critéres de stabilité pour les systèmes non linéaires de réglage automatique, basés sur l'utilisation de la transformée de Laplace", (in Romanian), St. Cerc. Energ., IX, no 1, pp.119-136.

[157] V.M. Popov, 1959 "Critères suffisants de stabilité asymptotique globale pour les systèmes automatiques non linéaires à plusieurs organes d'exécution", (in Romanian), St. Cerc. Energ., IX, no 4, pp.647-680.

[158] V.M. Popov, 1964 "Hyperstability and optimality of automatic systems with several control functions", Rev. Roum. Sci. Techn., Sér. Electrotechn. et Energ., vol.9, no 4, pp.629-690.

[159] L. Praly, 1995 "Stabilisation du système pendule-chariot: Approche par assignation d'energie", Personal communication.

[160] R.W. Prouty, 1995 *Helicopter Dynamics and Control*, Krieger Publishing Company. reprint with additions, original edition 1986.

[161] A. Rantzer, 1996, "On the Kalman-Yakubovich-Popov lemma", Systems and Control Letters, vol 28, pp7-10.

[162] S.S. Rao, 1990 *Mechanical vibrations*, Reading, Massachussetts, Addisson Wesley.

[163] V. Rasvan, 1978, "Some system theory ideas connected with the stability problem"", Journ. of Cybernetics vol 8, pp.203-215.

[164] V. Rasvan, S. Niculescu, R. Lozano, 2000,"Delay systems: passivity, dissipativity and hyperstability", Tech. Report Heudiasyc-UTC.

[165] H.R. Rota, P.J. Moylan, 1993 "Stability of locally dissipative interconnected systems", IEEE Transactions on Automatic Control, vol.38, no 2, pp.308-312.

[166] W. Rudin, 1976, *Principles of Mathematical Analysis*, McGraw Hill, 3rd Edition.

[167] W. Rudin, 1987 *Real and Complex Analysis*, McGraw Hill series in Higher Maths, 3rd edition.

[168] N. Sadegh, R. Horowitz, 1987 "Stability analysis of adaptive controller for robotic manipulators", IEEE Int. Conference on Robotics and Automation, Raleigh, USA, 1987.

[169] N. Sadegh, R. Horowitz, 1990 "Stability and robustness analysis of a class of adaptive controllers for robotic manipulators", Int. J. of Robotics research, vol.9, pp.74-92.

[170] M.G. Safonov, E.A. Jonckeere, M. Verma, and D.J.N. Limebeer, 1987 "Synthesis od positive real multivariable feedback systems", Int. J. Control, vol.45, pp.817-842.

[171] V. Santibañez, R. Kelly, 1997 "Analysis of energy shaping based controllers for elastic joint robots via passivity theory", Proceedings of the 36th IEEE Conference on Decision and Control, San Diego, December.

[172] S.S. Sastry, 1984 "Model reference adaptive control- stability, parameter convergence and robustness", IMA J. Math. Control Info., vol.1, pp.27-66.

[173] A.J. van der Schaft, 1984 *System Theoretical description of Physical Systems*, CWI Tracts 3, CWI Amsterdam, The Netherlands.

[174] A.J. van der Schaft, 1986 "On feedback control of Hamiltonian systems, in *Theory and Application of Nonlinear Control Systems*, C.I. Byrnes and A. Lyndquist (Eds.), Elsevier Sc. Publ,, B.V. North Holland, pp. 273-290, 1986.

[175] A.J. van der Schaft, 1987 "Equations of motion for Hamiltonian systems with constraints", J. Phys. A: Math. Gen., vol.20, pp.3271-3277.

[176] A.J. van der Schaft, 1989 "System theory and mechanics", in *Three Decades of Mathematical System Theory*, H. Nijmeier and J.M. Schumacher (Eds.), LNCIS 135, Springer, London.

[177] A.J. van der Schaft, B.M. Maschke, 1994 "On the Hamiltonian formulation of non-holonomic mechanical systems", Reports on Mathematical Physics, vol.34, no 2, pp.225-233.

[178] A.J. van der Schaft, B.M. Maschke, 1995 "The Hamiltonian formulation of energy conserving physical systems with ports", Archiv für Elektronik und Übertragungstechnik, Vol.49, 5/6, pp.362-371.

[179] A.J. van der Schaft, B.M. Maschke, 1995 "Mathematical modeling of constrained Hamiltonian systems", Proc. 3rd IFAC Symp. on Nonlinear Control Systems design, NOLCOS'95, Tahoe City, California, USA, June 26-28.

[180] A.J. van der Schaft, 2000 *L2-gain and passivity Techniques in Nonlinear Control*, 2nd edition, Springer, London, CCES.

[181] H. Sira-Ramirez, R. Castro-Linares, 1999 "On the regulation of a helicopter system: A trajectory planning approach for the Liouvillian model", Proceedings of the European Control Conference ECC'99, Karlsruhe, Germany, Session DM-14.

[182] H. Sira-Ramirez, 1998, "On the passivity-based regulation of a class of delay-differential systems", in Proceedings 37th IEEE Conf. on Decision and Control, Tampa FL, pp.297-298.

[183] J.J. Slotine, W. Li, 1988 "Adaptive manipulator control: A case study", IEEE Transactions on Automatic Control, vol.33, pp.995-1003.

[184] J.J.E. Slotine, W. Li, 1991 *Applied Nonlinear Control*, Prentice-Hall, Englewood Cliffs, NJ.

[185] E. Sontag, 1998 *Mathematical Control Theory: Deterministic Finite Dimensional Systems.* Springer-Verlag, New York 1990. Second Edition 1998.

[186] E. Sontag, 1995 "State-space and I/O stability for nonlinear systems", in *Feedback Control, Nonlinear Systems, and Complexity*, Lecture Notes in Control and Information Science, B. Francis and A. Tannenbaum (Eds.), Springer-Verlag, Berlin.

[187] E. Sontag, 1995, "An abstract approach to dissipation", Proc. IEEE Conf. Decision and Control, New Orleans, December, pp. 2702-2703.

[188] M.W. Spong, 1987 "Modeling and control of elastic joint robots", ASME J. of Dyn. Syst. Meas. and Control, vol.109, pp.310-319.

[189] M.W. Spong, R. Ortega, R. Kelly, 1990 "Comments on Adaptive manipulator control: A case study", IEEE Transactions on Automatic Control, vol.35, pp.761-762.

[190] M.W. Spong, M. Vidyasagar, 1989 *Robot dynamics and control*, Wiley, New-York.

[191] M.W. Spong, A.M. Block, 1995 "The Pendubot: A Mechatronic System for Control Research and Education", 34th IEEE Conf. on Decision and Control, New Orleans, December.

[192] M.W. Spong, L. Praly, 1996 "Control of underactuated mechanical systems using switching and saturation", Proceedings of the Block Island Workshop on Control Using Logic Based Switching, Springer-Verlag.

[193] M.W. Spong, 1994 "The Swing Up Control of the Acrobot", IEEE Int. Conf. on Robotics and Automation, San Diego, CA.

[194] M.W. Spong, M. Vidyasagar, 1989 *Robot Dynamics and Control*, John Wiley & Sons, Inc., New York.

[195] M.W. Spong, 1989 "Adaptive control of flexible joint manipulators", Systems and Control Letters, vol.13, pp.15-21.

[196] M. W. Spong, 1995 "Adaptive control of flexible joint manipulators: comments on two papers", Automatica Vol. 31, 4, pp. 585-590.

[197] J. Steigenberger, 1995 "Classical framework for nonholonomic mechanical control systems", Int. J. of Robust and Nonlinear Control, vol.5, pp.331-342.

[198] W. Sun, P.P. Khargonekar, D. Shim, 1984 "Solution to the positive real control problem for linear time-invariant systems", IEEE Transactions on Automatic Control, vol.39, pp.2034-2046, October.

[199] G. Szegö, R.E.Kalman, 1963 "Sur la stabilité absolue d'un système d'équations aux différences finies", C. R. Acad. Sci. Paris, vol.257, no 2, pp.388-390.

[200] M. Takegaki, S. Arimoto, 1981 "A new feedback method for dynamic control of manipulators", ASME J. Dyn. Syst. Meas. Control, vol.102, pp.119-125.

[201] G. Tao, P. Ioannou, 1988 "Strictly positive real matrices and the Lefshetz-Kalman-Yakubovich Lemma", IEEE Transactions on Automatic Control, vol.33, pp.1183-1185, December.

[202] J.H. Taylor, 1974 "Strictly positive real functions and Lefschetz-Kalman-Yakubovich (LKY) lemma", IEEE Transactions on Circuits Systems, pp.310-311, March.

[203] P. Tomei, 1991 "A simple PD controller for robots with elastic joints," IEEE Transactions on Automatic Control, vol.36, pp.1208-1213.

[204] A. Szatkowski, 1979 "Remark on 'Explicit topological formulation of Lagrangian and Hamiltonian equations for nonlinear networks", IEEE Trans. on Circuits and Systems, vol.26, no 5, pp.358-360.

[205] M. Vidyasagar, 1981 *Input-Output Analysis of Large-Scale Interconnected Systems*, LNCIS, Springer-Verlag, London.

[206] M. Vidyasagar, 1993 *Nonlinear Systems Analysis*, 2nd Edition, Prentice Hall.

[207] Q. Wang, H. Weiss, J.L. Speyer, 1994 "System characterization of positive real conditions", IEEE Transactions on Automatic Control, vol.39, pp.540-544, March.

[208] Q. Wei, W.P. Dayawansa, W.S. Levine, 1995 "Nonlinear controller for an inverted pendulum having restricted travel", Automatica, vol.31, no 6, 841-850.

[209] J.T. Wen, 1988 "Time domain and frequency domain conditions for strict positive realness", IEEE Transactions on Automatic Control, vol.33, pp.988-992 , November.

[210] J.T. Wen, 1989 "Finite dimensional controller design for infinite dimensional systems: the circle criterion approach", Systems and Control Letters, vol.13, pp.445-454.

[211] J.C. Willems, 1971 *The Analysis of Feedback Systems*, MIT Press, Cambridge, MA.

[212] J.C. Willems, 1972 "Dissipative dynamical systems, Part I: General Theory", Arch. Rat. Mech. An., vol.45, pp.321-351.

[213] J.C. Willems, 1972 "Dissipative dynamical systems, Part II: Linear Systems with quadratic supply rates", Arch. Rat. Mech. An., vol.45, pp.352-393.

[214] J.C. Willems, 1971 "The generation of Lyapunov functions for input-output stable systems", SIAM J. Control, vol.9, pp.105-133, February.

[215] V.A. Yakubovich, 1962 "La solution de quelques inégalités matricielles rencontrées dans la théorie du réglage automatique", (in Russian), Doklady A.N. SSSR, t.143, no 6, pp.1304-1307.

[216] V.A. Yakubovich, 1975 "The frequency theorem for the Hilbert space of states and controls and its application to some problems of optimal control synthesis I, II" (in Russian) , Siberian Math. J., vol.15, no 3, pp.639-668, 1974; vol.16, no 5, pp. 1081-1102.

[217] V.A. Yakubovich, 1962 "Frequency conditions for the absolute stability of nonlinear automatic control systems", in Proceedings of Intercollegiate Conference on the Applications of Stability Theory and Analytic Mechanics (Kazan', 1962), Kazan Aviats. Inst., pp.123-134.

[218] G. Zames, 1966 "On the input–output stability of nonlinear time-varying feedback systems, part I", IEEE Trans. on Automatic Control, vol.11, pp.228-238.

[219] G. Zames, 1966 "On the input–output stability of nonlinear time-varying feedback systems, part II", IEEE Trans. on Automatic Control, vol.11, pp.465-477.

# Index